W9-CRB-528

Agri-Food Quality II:
Quality Management of Fruits and Vegetables

Agri-Food Quality II

Quality Management of Fruits and Vegetables

Edited by

M. Hägg
Centre of Metrology and Accreditation, Finland

R. Ahvenainen
Technical Research Centre of Finland, Finland

A. M. Evers
University of Helsinki, Finland

K. Tiilikkala
Agricultural Research Centre of Finland, Finland

ROYAL SOCIETY OF CHEMISTRY

The proceedings of the Second International Conference 'Agri-Food Quality II' organised by the Agricultural Research Centre of Finland (MTT), the Technical Research Centre of Finland (VTT), the University of Turku, the University of Helsinki, the Ministry of Agriculture and Forestry, and Lännen Tehaat Oy held on 22–25 April 1998 in Turku, Finland.

Sponsored by Suomen Akatemia, Ministry of Agriculture and Forestry, Elintarvikkeiden tutkimussäätiö, Elintarviketieteiden seura ry, City of Turku, Finnair, Marli Oy, Saarioinen, Cultor Corp., Lännen Tehaat Oy, COST Actions 915 and 99.

Compiled by T. Miettinen, Agricultural Research Centre of Finland.

Agri-Food Quality II berries logo by J.M. Pihlava, Agricultural Research Centre of Finland.

Special Publication No. 229

ISBN 0-85404-788-3

A catalogue record for this book is available from the British Library

Published by The Royal Society of Chemistry,
Thomas Graham House, Science Park, Milton Road,
Cambridge CB4 0WF, UK

For further information see our web site at www.rsc.org

Printed and bound by Athenaeum Press Ltd, Gateshead, Tyne & Wear

Preface

The international conference "Agri-Food Quality II: Quality Management of Fruits and Vegetables – from Field to Table" was held in Turku, Finland, April 22–25, 1998.

This conference was a follow-up conference to the first Agri-Food Quality conference held in 1995 in Norwich, UK. In the preface of the preceding book the editors hoped that Agri-Food Quality would develop into a series of conferences. It looks as if that wish will come true. This conference was held in 1998 in Finland, and there are plans for the following Agri-Food Quality conference to be held in the year 2000.

The theme of this conference was quality management of plant based food materials throughout the production chain, from field to table. The papers discuss developments and information relating to the improvement of vegetable and fruit quality through plant breeding, modification of cultivation technology and optimisation of pre- and post-harvest practice. There is an increasing need for scientists to discuss improvements and investigations from the beginning of the food chain to the end. This conference provided a starting point for many scientists to develop new investigations, which deal with the whole or part of the food chain.

We sincerely hope that the papers from this proceedings will stimulate many new investigations about the whole food chain, with scientists from different part of the world dealing with different parts of the food chain.

We would like to thank the contributors of this volume for their excellent work.

Contents

Session IV **Effects of Post-harvest Practice on Quality**

Session VII **Quality Improvements and Functional Foods**

Session I Quality Challenges in Future

TECHNOLOGICAL INNOVATIONS IN A CHANGING ENVIRONMENT

Wim M.F. Jongen

Food Science Group
Wageningen Agricultural University
P.O. Box 8129, 6700 EV Wageningen, The Netherlands

1 INTRODUCTION

The many changes in the market for new food products call for a repositioning of existing food production systems and raise the question whether the concepts currently used can survive the challenges of the future. What we see is that apart from market saturation, a number of other developments have a large influence on the situation in the market. Generally the consumer is better educated and more demanding. Also consumers are less predictable in their purchase behaviour, eat more outside homes and are more conscious about health related aspects. As a result there is a continuous need for new products and a more differentiated food product assortment. Related to this development, product life cycles become shorter and efficiency and flexibility of food production systems become even more important. The complexity of the issue of product innovation and product acceptance requires an integrated view. In fact there is a need for new concepts in which the various disciplinary approaches are combined into one integrated, techno-managerial approach. Technological inventions, such as in the field of biotechnology, have to be translated into products, which are attractive to consumers. Vice versa, changing consumer values and habits will stimulate innovation in food production technologies in order to produce suitable new products. This interdependency between consumers' wants and needs on the one hand and technologies and research on the other hand has been recognised by many food companies, but is not implemented systematically yet.

Another message is that there is a need for a chain oriented approach to product innovation which considers the whole food supply chain from breeders, via processing up to the consumer in one integrating concept. Traditionally, food supply chains have been characterised by two distinct features:

1. The one-way communication through the chain from producers of raw materials to the users of end products (the consumers) and

2. The poor understanding of the concept of product quality. Quality was and in a number of cases still is, predominantly based on technical criteria and producers focus(ed) in particular on costs and productivity.

Actors in the food supply chain have this approach to quality in common. In addition, each actor in a food supply chain will use specific quality criteria such as homogeneity and storability of raw materials at industrial level and ease of handling at the retail level. Sometimes these specific criteria can conflict with an appropriate quality of the end product. As a result there is a need for a unified concept of product quality and acceptance throughout the production chain. It requires a chain reversal in which the consumer has become the focal point. Food production systems of the future cannot be any longer solely production driven but should be characterised as being primarily consumer driven.

2 CHAIN REVERSAL AS A STARTING POINT

The large number of changes and the costs associated with product innovation make it necessary to develop a new approach towards the production of food products namely chain reversal in which the consumer has become the starting point of thinking. This makes it necessary to have good knowledge about the market and consumer preferences and to have quality control throughout the chain (1). Definition of product quality is the first step in this new approach and requires translation of consumers' perception and preferences into product characteristics that are measurable and can be specified. With this knowledge the acceptable band width of variation for specific quality criteria can be determined and processing conditions can be optimized. Also the demands for raw material properties can be (re)formulated. Using this concept raw material composition and properties can be coupled directly to the quality of the end-product. Also a distinction can be made between the possibilities to achieve the desired end-product quality via modulation of processing conditions and optimization of raw material properties e.g. via biotechnology (2). In figure 1 a schematic picture is drawn of the traditional(A) and innovative (B) chain oriented approach.

Chain Approaches

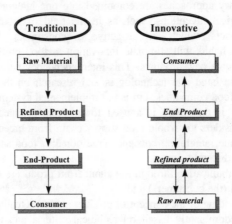

Figure 1

3 PRODUCT QUALITY

The understanding of the term Product Quality has been subject of a large number of studies and the contributions of Juran, Deming, Crosby and many others to improve our thinking about quality are invaluable. Juran (3) has included the consumer in his concept of product quality and prevention instead of control of quality. His definition is: 'Quality is fitness for use'. With respect to the situation for food products a better wording would be: 'Quality is to meet the expectations of the consumer'. Two important aspects should be underlined here. The first one is that the consumer is the starting point of thinking about quality and the second is that the consumer does not work with accurate specifications. *The* consumer doesn't exist. There is no average consumer. There is a specific consumer which, in a specific situation and at a certain moment, has a specific need to which the producer can respond.

The consumer buys and consumes a product for a number of reasons. Partly these reasons refer to product properties for another part these refer to the production system. I propose to use here the terms intrinsic and extrinsic factors. Intrinsic factors refer to physical product properties such as flavor, texture, appearance, keepability and nutritional value. Properties that are direct and/or indirect measurable and that can be objectivated. A food product as such has no quality, it has physical properties which are turned into quality attributes by the perception of the consumer. An example of this is the texture of an apple. From a physico-chemical point of view the texture of a product can be described in terms of cell wall composition and structure. The consumer perceives this during consumption and describes his experience in terms of crispiness or mealiness and toughness. The total of quality attributes (intrinsic factors) determines the quality of a given product. Extrinsic factors refer to the production system and include factors such as the amount of pesticides used during growing, the type of packaging material used, a specific processing technology or the use of biotechnology to modify product properties. They do not necessarily have a direct influence on physical product properties but influence the acceptance of the product for consumers. The total of intrinsic and extrinsic factors determines the intention to purchase (4). In figure 2 an attempt is made to visualize this concept.

Consumer Perception and Acceptance

Figure 2

Quality management can be a confusing concept, partly because people view quality relative to their role in the production chain and there is no unique definition of quality. The question is then which criteria and/or variables do give an adequate description of quality for a specific actor in the chain. In figure 3 a scheme is depicted representing the production-distribution cycle indicating the key steps in innovation and production (4).

Quality Dimensions in the Production-Distribution Cycle

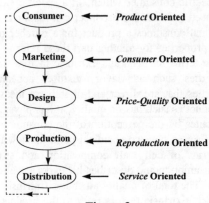

Figure 3

The consumer is the driving force for product innovation and the consumer generally views quality from the product-based perspective as has been described above. Indeed products should meet consumer needs and it is the task of the marketing department to identify the needs specified for the type of products that can be produced. A product that meets consumers' expectations can rightly be described as having good 'quality' and therefore marketing people work with a user-based definition of quality. The next step is to translate consumers' expectations into product characteristics and process specifications. Making this translation is the role of R&D, product design and engineering people. Product specifications address such things as size, from finish, taste, safety, keepability, nutritional value. Process specifications include the type of equipment needed and processing conditions required. Product designers must balance price and quality to meet marketing objectives. Therefore the focus here is primarily variation in product quality that can occur as a consequence of e.g. seasonal variations and /or cultivar differences. The implications are that also for processed food products large variations may occur if processing conditions are not appropriately adjusted. Nevertheless the producer must adhere to the design specifications so the quality focus is reproduction oriented. The production-distribution cycle is completed when the product has been moved via wholesale and retail outlets to the consumer. Distribution does not end the relationship between the consumer and the producer so service-oriented concepts of quality cannot be ignored by the producer.

4 QUALITY AS STEERING FACTOR FOR INNOVATION IN THE SUPPLY CHAIN

4.1 Production system characteristics

Product innovation in the food industry has to deal with a number of characteristics that are specific for food products and which make it complicated to achieve quality control throughout the chain.

1. Food products are perishable and depending on the conditions product properties can change very fast and result in spoilage. This means strict quality demands for production, storage and passage time throughout the production chain.
2. Production and harvest of plant foods is seasonal whereas consumers ask for constant availability. So there is a requirement for adequate storage and/or transport methods and facilities.
3. There is a clear awareness amongst consumers about the relationship between diet and health. This includes both the absence of unwanted components such as pathogens and toxicants, either naturally present or added, as well as assuring the levels of components that are wanted such as vitamins and minerals. Adequate risk prevention requires systematic quality control.
4. Small scale production is an obvious characteristic of plant foods and a complicating factor in assuring homogeneous products e.g. milk or raw materials for processing. It requires specialized organisational structures.
5. The number of retail outlets is very large which complicates an adequate control of distribution quality (service oriented). Beside that retail outlets also have to deal with a large number of wholesalers. This makes it virtually impossible to define accurate specifications with respect to quality demands for a given product.
6. Fresh food products are perishable and have only a very limited shelf-life. Besides, it is too expensive for retailers to maintain large amounts of produce in store. This has resulted in systems with very high delivery frequencies.
7. The primary aim of food industry is to provide the community with food products of good quality for a reasonable price. So generally food products have low added value. This is a major drawback in innovation.

4.2 Chain ruptures and product quality

One characteristic of the traditional supply chain is the use of differing concepts of product quality by actors in a specific production chain. In figure 4 an example is given of the large number of quality criteria a plant food product has to meet and how these differ going from the breeder to the consumer. From the figure it may become obvious that no single product can meet all the requirements indicated and that there is a clear need for one integrated concept of product quality throughout any specific supply chain.

Actors in the Chain and Quality Perception

Breeder	**Vitality Seeds, Yield**
Grower	**Productivity, Uniformity, Disease resistance**
Auction	**Uniformity, Reliable Supply, Constant Quality**
Distribution	**Keepability, Availability, Damage Sensitivity**
Retailer	**Good Shelf-life, Diversity, Appearance, Low waste**
Consumer	**Tasty, Healthy, Sustainable Convenience, Constant Quality**

Figure 4

Another characteristic that holds has to do with the notion that for all plant products during growth product quality is built up whereas storage and handling are primarily concerned with prevention of quality loss. This means that differing concepts have to be applied.

During growth knowledge about the relationships between agronomical conditions, cultivar properties and harvest moment is of great importance for a good shelf-life during storage and handling for fresh products and for adequate processing conditions for fabricated foods. Quality loss is inevitable because harvesting is in fact a situation of imposed stress and the post-harvest conditions are vital for maintenance of good end-product quality. We are actually talking about living organisms that respire and are metabolically active. The composition and structure of the raw material and the associated processing properties are strongly dependent upon these processes. For a chain oriented system of quality control we need to have methods that can measure external and internal quality and predict keepability. Also we need methods that can relate raw material composition and end-product quality to the processing conditions (2). One excellent example comes from recent work of van Kooten and his coworkers on cucumber. Using the efficiency of the photosynthesis system as a parameter they have developed a measurement system that can be used to measure temperature failure during storage and predict shelf-life in the chain.

4.3 Innovation cycles

As already mentioned in the introduction the market for plant food products has become more and more competitive and changing consumer demands have become a constant drive for companies to innovate. From a chain oriented perspective three major innovation cycles can be distinguished (Figure 5). The first one deals with developments in the market. Here we see a decreasing life cycle of products and fast changing preferences of consumers. The consumer generally behaves more impulsively and has become a moving target. The second innovation cycle refers to the technologies associated with processing and production systems. New technologies and approaches such as high pressure cooking, hurdle technologies, minimal processing etc. will increase the potential to meet new consumers' demands. Generally innovation in technologies is more slow than changes in the market situation. The third cycle deals with primary production and is actually the slowest cycle even with the use of modern biotechnology. Short-term changes in the market are impossible to follow. From a chain oriented perspective it is of utmost importance for breeders to have a strategic view on market developments and to identify market niches where they can be strong and ahead of their competitors. Consequently it is of great importance to establish strategic collaboration within the chain.

Innovation cycle
Plant Food Products

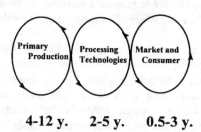

4-12 y.   2-5 y.   0.5-3 y.

Figure 5

5 LINKING CONSUMERS' WANTS TO TECHNOLOGIES AND RESEARCH

In the previous paragraphs we have seen that we are heading towards a situation in which continuous innovation has become a prerequisite for companies to stay in business. This raises the question how can these changes be turned into a challenge and whether the traditional innovation concepts are still valid. To answer these questions at a strategic level the *DFE* concept can be followed. The *DFE* concept is a three step approach which enables a link to be establish between changes in the market and the consequences for production systems. The first step is the question about the *Desirability* of new products (market socio-economics and consumer preferences). The second deals with the *Feasibility* of the production (technological possibilities and barriers). The third one follows with the question of *Efficacy* (organization of production chain).

In the context of this paper we will focus on the question how can consumers' wants be linked to necessary technological developments. Adequate prediction of the type of technological development needed to respond to market changes is crucial because choices have to be made with regard to the type of raw material and the (combinations of) processing technologies to be used.

Recently a series of overview studies on future consumer issues have been performed for the Dutch National Agricultural Research Council (NRLO). Meulenberg (5) has analysed the socio-economic developments in the food market and translated them into consumer categories. Subsequently Jongen et al. (6) have used these categories and developed a model for translation of consumer preferences and perceptions into desired technological developments. The model is based on a systems analysis which uses the consumer as the focal point and a stepwise approach is followed in which 7 successive steps are to be distinguished. These steps seem to be a useful framework for an integral model of product innovation in food production systems. These seven steps can be summarised as follows:

I.   *A thorough analysis of the socio-economic developments in specified markets*
II.   *Translation of consumer preferences and perceptions into consumer categories*
III.   *Translation of consumer categories into product assortments*
IV.   *Grouping of product assortments in product groups according to the stages of the production chain*
V.   *Identification of processing technologies relevant for specified product groups*
VI.   *Analysis of state of the art in relevant processing technologies*
VII.   *Matching state of the art of specified processing technologies with future needs*

Following this model the study showed that linking Research & Development programmes within companies successfully to market dynamics requires a number of new technological developments to be put into place. Also in connection with this model some conclusions and suggestions have been made. The study emphasises the need for 'dedicated' production systems which follow more closely market dynamics. A breakthrough must be realised in thinking from craft to "design for manufacture" making use of information technology and computer management systems. From a chain perspective de-coupling moments must be as late as technologically feasible. Another outcome of the study was the conclusion that product innovation has to become more effective with respect to the success ratio and must become better structured. Systems such as Quality Function Deployment and Effective Consumer Response (ECR) can be valuable tools in doing this. These systems have been developed for use in the computer and automobile industry and should be developed and evaluated in order to be better suitable for use in food product innovation. As mentioned earlier the complexity of the concept of product quality in the food sector requires a specific approach and these systems have to be adapted in order to become effective tools for innovation of food production systems.

6 SUMMARY

In this introductory paper a number of aspects relevant to innovation in food production systems are addressed. It is obvious that our knowledge about the relationships between market changes, consumer behaviour, food products and processing technologies is still by and large insufficient. Nevertheless there are promising developments and it seems that a turning point has been reached with respect to our thinking about innovation. Food production systems of the future cannot be any longer solely production driven but should be characterised as being primarily consumer driven. Another message is that there is a need for a chain oriented approach to product innovation which considers the whole food supply chain from breeders, via processing up to the consumer. One outcome should be a unified concept of product quality and acceptance throughout the production chain. Moreover, there is a need for the development of new concepts in which the various disciplinary approaches are combined into one integrated, techno-managerial approach. Technological inventions, such as in the field of biotechnology, have to be translated into products, which are attractive to consumers. Vice versa, changing consumer values and habits will stimulate innovation in food production technologies in order to produce suitable new products. This interdependency between consumers' wants and needs on the one hand and technologies and research on the other hand has been recognised by many food companies, but is not implemented systematically yet. This interrelationship between technology and consumer behaviour should receive more attention in the modelling of food product innovation.

The stepwise approach as proposed in this paper seems promising and deserves further attention to develop it into a useful approach to strategic investment in product innovation, in particular in future technologies and R&D programmes.

References

1. M.T.G. Meulenberg, Agricultural Marketing and Consumer Behaviour in a Changing World, Kluwer Academic Publishers, 1997, 95.
2. W.M.F. Jongen, Agri-Food Quality, The Royal Society of Chemistry, Cambridge, UK,1996, 263.
3. J.M. Juran, Juran on leadership for Quality- An executive handbook, The free press, New York, USA, 1989,
4. W.M.F. Jongen, Minimal processing and ready made foods, Proc. of SIK, 1996, 7.
5. M.T.G. Meulenberg, In Dutch: De levensmiddelenconsument van de toekomst. NRLO-Rapport, no. 96/4, Nationale Raad voor Landbouwkundig Onderzoek, Den Haag, The Netherlands,1996.
6. W.M.F. Jongen, A.R. Linnemann, G. Meerdink and R. Verkerk, In Dutch: Consumentgestuurde Technologie-ontwikkeling; Van wenselijkheid naar haalbaarheid en doeltreffendheid bij productie van levensmiddelen. NRLO-rapport no. 97/22, Nationale Raad voor Landbouwkundig Onderzoek, Den Haag, 1997.

THE NUTRITIONAL ENHANCEMENT OF PLANT-DERIVED FOODS IN EUROPE ("NEODIET")

D G Lindsay
Biochemistry Department
Institute of Food Research
NORWICH NR4 7

1. INTRODUCTION

In 1996 the EU financed a workshop on secondary metabolites in plant food which focussed particularly on compounds with potential health benefits. The workshop highlighted the potentially important public health and commercial needs that could be derived through enhancing the levels of these metabolites in plants, and which were only likely to be met through a concerted European research effort. This was felt to be essential if scientific and technological advances in Europe were to keep pace with developments which were occurring elsewhere in the world.

The workshop concluded that:

- Little was known about the natural range of levels of secondary plant metabolites which had potential health benefits;
- More investigation was needed into what impact enhancing these metabolites would have in terms of:
    - -health benefits
    - -agronomic benefits, and
    - -overall food quality.

In exploiting any such potential impact it was acknowledged that much more needed to be known about their biosynthesis, and how they might be effectively manipulated to enhance specific metabolites with health benefits. It was also recognised that it would be important to study how food processing technologies could be exploited to ensure that the levels in food were maximised and that their bioavailability subsequent to uptake and delivery to target organs and cells was investigated.

As a direct result of the workshop it was decided to propose a Concerted Action focussed on the enhancement of nutrients and protective factors in plant foods. The proposal was successful in receiving support from the 4th call in the EU's FAIR RTD programme and began in October 1997. It will run until March 2001.

2.OBJECTIVES

The overall objectives of the NEODIET concerted action programme are to set up a network of collaboration between research scientists in academia and industry, who are working both at the EU and national level, in the area of improving the nutritional value of plant foods through the use of genetic modification and food processing technologies.

It is to be expected that NEODIET will ensure that a much higher profile will be given to this topic throughout Europe and will bring together scientists to co-operate in a multi-disciplinary mode to address the challenges. It should enable data to be made easily available to research workers and encourage further research activities. The strong involvement of the agrochemical and food industries in the project will help to stimulate technological development.

The benefits of the application of biotechnology in agricultural production will be seen in a positive light providing direct health benefits to consumers and assisting in the improvement of public health without the need for a major change in food consumption habits.

The project will set up a user-friendly database which will enable all those European scientists with an interest in the field to determine what data are available and to update it as their own research outputs become available. The database will draw heavily on the work already undertaken in the NETTOX concerted action project, funded under the EU's previous Agriculture and Agro-industrial research programme (AAIR), which has set up a system for handling information on the levels and toxicity of individual chemicals in plant foods. The Co-ordinator of this database is also a member of the Steering Committee.

3. PRIORITY AREAS

The Steering Committee have decided that the priority areas for the work of NEODIET should be:
• Folates
• Vitamins C and E
• Carotenoids
• Glucosinolates
• Phytosterols
• Flavonoids
• Simple phenols
• Certain trace elements

This list was ranked in descending order based on the evidence that these substances provide:
• important health benefits;
• that there is a reasonable potential for genetic manipulation and enhancement of their levels in foods;
• that there is an industrial interest;
• there are wider food and agronomic quality benefits, and
• that there is currently a sufficient level of research activity underway in Europe.

Evidence for health benefits is constantly being updated. It is hoped that the priorities for NEODIET will closely reflect the evidence on health benefits which is being collated through the FAIR concerted action on "Functional Food Science in Europe" (FUFOSE). Although FUFOSE is concentrating on physiological function, rather than compounds, the involvement of a representative of ILSI Europe, who are co-ordinating this project, on the NEODIET Steering Committee will ensure that the latest position on health benefits is available to the Committee.

Without any interest on the part of industry there would be no potential for exploitation of this work. Their interests are represented through the involvement of ILSI Europe, and a representative of the European Plant Industry Forum (PIP), which is a group of scientists working in industry with a particular interest in the application of biotechnology in plant breeding. In addition the food sector is represented in the Steering Committee through the involvement of the chairperson of the Task Group on Processing.

4. CAROTENOIDS

4.1 Biosynthesis

The nature of the challenges faced in developing this field is well illustrated through the attempts which have been made to manipulate carotenoid levels in food plants.

A simplified version of the pathways leading to the synthesis of the carotenoids principally found in food plants is shown in figure 1.

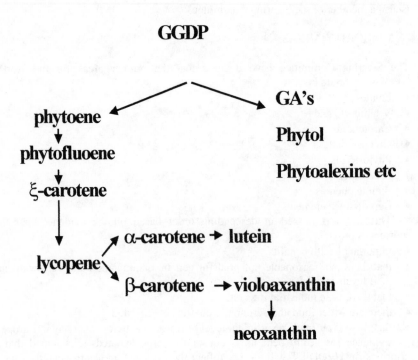

Figure 1: Biosynthesis of carotenoids from GDPP

4.2 Pathway manipulation

A crucial enzyme involved in their synthesis is phytoene synthase (PSY 1) which converts geranylgeranylphosphate(GGPP) into phytoene. Inhibition of PSY1 through the use of antisense gene-silencing techniques resulted in a 90% reduction in carotenoid levels compared with wild type in the tomato [1].

Overexpression of PSY1 results in dwarfed tomato plants[2]. This has been explained by the increased flow of metabolites into the carotenoid branch of the isoprenoid pathway leading to a reduction in the metabolites available for gibberellin synthesis. The general, isoprenoid pathway is utilised by plants for the synthesis of many important products including sterol, chlorophyll, quinone, and phytoalexin production, as well as for the gibberellins. It illustrates the complex interactions which are occurring and the potential disbenefits arising from the manipulation of the level of any one enzyme.

Improvements in the process can be made by ensuring that genes are expressed in a tissue-specific manner through the insertion of specific promoters. This has been achieved for rice where carotenoid levels have been increased significantly through the use of the daffodil *psy* gene[3]. In the case of tomato a two-fold increase in total carotenoid levels occurs using this psy transformant which is significantly lower than for rice. In rice the daffodil *psy* cDNA insertion is under the control of an endosperm-specific promoter. The choice of promoter will very much affect the timing and tissue-specific expression of a gene.

It has been possible to induce xanthophyll accumulation in potato tubers that do not normally contain ß-carotene[4]. The use of bacterial psy genes instead of the daffodil *psy* gene in tomatoes produces a seven-fold increase in ß-carotene levels but this is at the expense of the ripening pigment lycopene which is the predominate carotenoid in tomato.

This indicates the potential, given the right set of carotenoid biosynthesis genes, for the correct target sequences and promoters to accumulate carotenoids in other plant tissues/organs/crops which currently contain little or none. This remains a major challenge. Much more needs to be known about the exact regulatory factors which affect carotenoid biosynthesis

4.3 Health effects

Apart from the limited information that is available at present on the complex inter-relationships between the metabolic products in the isoprenoid pathway, there is still confusion over the health benefits from increasing carotenoid levels in the diet. In theory given the ability of the carotenoids to break radical chain reactions and act as powerful antioxidants against free radical damage *in vivo* arising from normal oxidative metabolism, they should show protective effects both against cardiovascular disease and cancer. ß–carotene has been shown to be integrated within the plasma lipoproteins, especially the LDL fraction[5], and should be well positioned to quench any free radical damage caused by oxidised cholesterol. But it has still to be effectively demonstrated that carotenoids are protective against cardiovascular effects. So far there is no evidence to support this view[6].

Similarly if it is assumed that carotenoids are incorporated within the lipoprotein micelles of cell membranes they are well place to protect against free radical damage affecting either mitochondrial or nuclear DNA and show protective effects against cancer. However when a study was undertaken in smokers to establish whether or not ß-carotene protected against free radical damage to lung tissue from smoking, it was found that those people who received supplements of ß-carotene were found to have an increased risk of developing lung cancer[7,8]. Conversely there is evidence that humans on low level supplementation with carotenoids show less DNA damage when isolated lymphocytes are subjected to oxidative stress [9,10].

Numerous explanations have been offered to explain this somewhat surprising observation which is inconsistent with many studies in animals which have shown ß-carotene to have protective effects. Most antioxidants have the capacity to act as prooxidants under certain conditions, especially if the redox potential of the cell is altered and transition metal ions are present in a free state. Since doses administered in the study were higher than those which would be usual in foods, it will be of interest to summarise the data on the carotenoids health effects at doses which are realistic in terms of delivery as a part of the plant foods opposed to supplementation.

This experience demonstrates the potential hazards which might arise when vitamins or other protective factors are administered as supplements rather than being ingested at augmented levels within the plant food. In such situations the metabolites may be limited naturally, either through plant metabolic controls or in subsequent bioavailability and their rate of uptake after consumption

References

1. C.R. Bird, J.A. Ray, J.D.Fletcher, J.M.Boniwell, A.S. Bird, C. Teulieres, T. Blain, P.M. Bradley and W. Schuch. Biotech., 1991, **9**, 635.
2. R. Frey and D.Grierson, Trends Genet., 1993, **9**, 438.
3. P.K.Burkhardt, P.Bayer, J.Wuenn, A. Kloetei, *et al.* Plant J., 1997, **11**, 1071.
4. C.R.Brown, C.G.Edwards, C-P.Yang, and B.B. Dean. J.Am. Soc. Hort. Sci., 1993, **118**, 145.
5. P.P.Reddy, B.A. Clevidence, E. Berlin, P.R.Taylor, J.C.Bieri and J.C.Smith. FASEB J. 1989, **3**, A955.
6. EU AIR Programme: Final report on "Increased Fruit and Vegetable Consumption within EU. Potential Health Benefits". AIR2-CT93-0888.
7. The Alpha Tocopherol, Beta-Carotene Cancer Prevention Study group. N. Engl. J. Med., 1994, **330**, 1150.
8. G.S.Omenn, G.E.Goodman, M.Thornquist, J.Grizzle, L.Rosenstock, S. Barnhardt *et al.* N. Engl. J. Med., 1996, **334**, 1029.
9. S. Astley, D. Hughes, A. Wright, A. Peerless and S. Southon. Biochem. Soc. Trans., 1996, **24**, 526S.
10. B.L.Pool-Zobel. A. Hub, H. Müller, I. Wollowski, and G. Rechkemmer. Carcinogenesis. 1997, **18**, 1847.

EFFECTS OF PROCESSING ON THE LEVEL AND BIOACTIVITY OF ANTIOXIDANTS IN FOOD PRODUCTS

Matthijs Dekker, Addie A. van der Sluis, Ruud Verkerk and Wim M.F. Jongen

Food Science Group
Wageningen Agricultural University
PO Box 8129, 6700 EV Wageningen, The Netherlands

1 INTRODUCTION

The intake of antioxidants has been associated with lower incidence in various ageing diseases. An important group of antioxidants are the flavonoids. Important sources of flavonols and flavones in the Dutch diet are tea (48%), onions (29%) and apples (7%)[1]. The intake of these compounds with the diet has been associated with a lower incidence of cardiovascular diseases in epidemiological studies. Compositional data used in these studies is often obtained from unprocessed materials, while most foods are consumed after some sort of processing, either by the producer or by the consumer. The composition of food products can be affected by processing steps as well as the bioavailability of bio-active compounds from the final products. It is therefore important to know what the effects of processing steps are on the level and activity of bioactive components such as antioxidants in foods. With this information more accurate figures can be given for epidemiological work and also product development can be directed to consumer foods with an optimal level of bio-active compounds. As examples processing effects on antioxidants in apple, tea and brassica vegetables are presented here. The possibilities to predict the antioxidant activity of a food product from its compositional data has been tested for apple juice and tea.

2 MEASUREMENT OF ANTIOXIDANT ACTIVITY

Lipid peroxidation was induced in male rat liver microsomes by ascorbic acid and Fe^{2+}. Peroxidation products were determined in the thiobarbituric acid assay. The inhibition of lipid peroxidation gives an indication of the antioxidant activity. For each component or food extract the inhibition of the oxidation has been determined for a range of concentrations. The activity is expressed as IC_{50}: the concentration of an individual compound or food extract at which the oxidation is inhibited for 50%. This value can be determined from the inhibition vs. concentration curves by a fitting procedure. The assay has been optimised and made suitable for a large number of samples by using microtiter plates and an ELISA reader[2].

3 EFFECTS OF PROCESSING

3.1 Apple juice production

Flavonoid content in apple juice is only 5-10% of the content of the apples used to produce the juice[3]. The conventional process of making juice from apples was investigated for the losses of flavonoids. It was found that the pressing process resulted in the largest losses due to binding of the flavonols to the pressing cake. Alternative processes have been shown to be possible in order to significantly improve the level of flavonoids in the final product. An example is the addition of alcohol (methanol, ethanol or isopropanol) to the pulp before pressing and thereby extracting the flavonols from the insoluble particles (Figure 1).

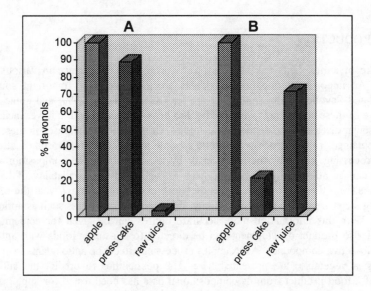

Figure 1 *Effect of processing of apple juice on the partitioning of total flavonols over the juice and the press cake (A: conventional pressing; B: alcoholic extraction before pressing)*

The antioxidant activity of a juice produced with an alcoholic extraction process was >10 times higher than a conventional produced juice. In order to develop an apple juice with a high antioxidant activity attention has to be given to sensory properties of the juice. Too high polyphenol content is known to be responsible for astringent taste, so there will be a maximum in the level of these compounds.

3.2 Tea infusion

Flavonoid content in tea varies widely among types and brands. This has a marked effect on the antioxidant activity of tea. Green tea flavonoids consist mainly of catechins and flavonol glycosides. During the manufacturing process of black tea a part of the catechins is converted to theaflavins and thearubigens by enzymatic oxidation. The infusion process of black tea was studied[4]. The effects of infusion time and temperature on the composition and antioxidant activity of the tea extract was determined (Figure 2). The effect of time on the antioxidant activity of the tea shows that during 'normal' extraction times of 3-5 minutes, the extract contains around 50-75% of the equilibrium activity. The infusion temperature does not have a large effect on the activity of the extract between 60 and 100 °C.

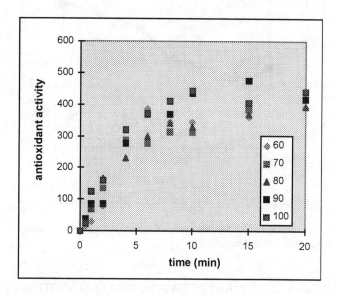

Figure 2 *Effect of infusion temperature and time on the antioxidant activity of the tea extract (antioxidant activity is expressed as the dilution factor of the tea to give a 50% inhibition in the peroxidation assay)*

The compositional data show an effect of temperature and time: at higher temperature more theaflavin is found in the extract and less catechins. At short extraction times the relative contribution of the flavonol glycosides is higher due to their faster extraction rates.

3.3 *Brassica* processing

Brassica vegetables are well known for their high levels of bio-active glucosinolates[5]. Glucosinolates and especially their breakdown products are potential anticarcinogenic compounds due to the induction of Phase II enzymes that are involved in detoxification processes. The level of glucosinolates can be strongly affected by processing.

Extracts of *brassica* vegetables also have an antioxidant activity. This activity is not due to glucosinolates but most likely to other compounds like flavonols, vitamins etc. Interesting results have been found for the effect of processing on antioxidant activity. After cutting white cabbage and exposing it to air an initial small decrease followed by a 50% increase in the antioxidant activity is observed (Figure 3). This observation has not yet been coupled to compositional changes in the cabbage.

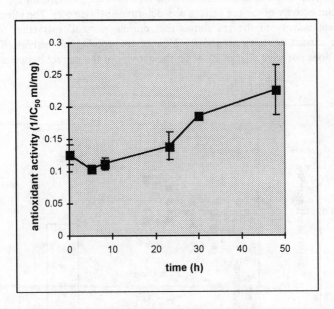

Figure 3 *Effect of cutting and exposing to air of white cabbage on the antioxidant activity of extracts of the cabbage.*

4 PREDICTING ANTIOXIDANT ACTIVITY FROM COMPOSITION

From the chemical analysis of food products and antioxidant activities of individual compounds it is theoretically possible to calculate the antioxidant activity of a product. This implies however that no synergistic/antagonistic or other matrix effects play a role and that all compounds with antioxidant activity are known and detectable[2,4]. For black tea and apple juice these calculations were performed using the antioxidant activity (expressed as $1/IC_{50}$) of the individual components or the complete mixture (equation 1).

$$\frac{\sum_{i=1}^{n} C_i}{IC_{50,mixture}} = \sum_{i=1}^{n} \frac{C_i}{IC_{50,i}} = \text{dilution factor of the product to obtain 50\% inhibition} \qquad (1)$$

$$C_i = \text{concentration of component i.}$$

In Figure 4 the contribution of known and detected antioxidants, as calculated based upon their concentration and specific activity with equation (1), is shown as a percentage of the measured total antioxidant activity.

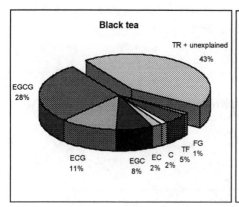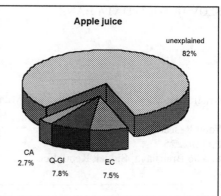

Figure 4 *Contribution of known antioxidants present in tea and apple juice (obtained by alcoholic extraction) to the total antioxidant activity of these drinks. (C: catechin, CA: chlorogenic acid, EC: epicatechin, ECG: epicatechingallate, EGC: epigallocatechin, EGCG: epigallocatechingallate, FG: flavonol glycosides, Q-Gl: quercetin glycosides, TF: theaflavins, TR: thearubigens).*

In black tea the majority of the measured activity can be explained by the contribution of known compounds. The measured antioxidants in apple juice only explain some 20% of the total activity.

5 CONCLUSIONS

It was shown that the antioxidant activity of food (extracts) can be assessed conveniently by a lipid peroxidation assay in microtiter plates. As a result of the processing of foods large effects on the antioxidant activity have been shown. Both a decrease as well as an increase in the antioxidant activity of the final product are possible. Attempts to relate the total antioxidant activity of food products to the compositional data show that sometimes matrix effects or other interactions are important. For some products (e.g. tea) a good match between the predicted total activity and measured total activity is found while in other products (apple juice) only 20% of the activity can be explained.

References

1. M.G.L. Hertog, P.C.H. Hollman, M.B. Katan and D. Kromhout, *Nutr. Cancer*, 1993, **20**, 21.
2. A.A. van der Sluis, M. Dekker and W.M.F. Jongen, 1998, *in preparation.*
3. A.A. van der Sluis, M. Dekker and W.M.F. Jongen, *Cancer Letters*, 1997, **114**, 107.
4. M. Dekker, A.A. van der Sluis, R. Verkerk, A.R. Linnemann and W.M.F. Jongen, 1998, *in preparation.*
5. R. Verkerk, M. Dekker and W.M.F. Jongen, 'Natural Toxicants in Food', Sheffield Academic Press, London, 1998, Chapter 3, p. 29.

SLOVAK FOOD DATA BANK

Kristína Holčíková, Eva Simonová and Milan Kováč

Food Research Institute
P.O.Box 25
820 06 Bratislava, Slovak Republic

1 INTRODUCTION

A knowledge of the chemical composition of consumed foods are an important tool required to promote an improvement in the health status of the population, to formulate health and dietary recommendations, and to meet consumers demands for information. Food composition information is currently available primarily in the form of printed tables which, while valuable, are limited in the amount and type of information that they can hold. Although food composition tables remain the major source of food composition data, they cannot accommodate the ever-increasing volume of data or meet the needs of users with specialised information requirements. Increasingly, these tables are intended to provide information for use by both the food industry and consumers and their associations. Given the current trends in domestic and international marketing, it is necessary to include a very large number of cooked dishes in these tables. This, in itself, justifies the use of computerised methods. Food composition databases are an important source of information in most countries. They are a modern computerised solution to the problem of access to these data. They not only improve the speed and accuracy of data use, but also extend the ways in which the data can be used.

The Slovak Food Data Bank (FDB) plays a very important role in giving information not only on the composition of foods, but also on other food characteristics. FDB is held at the Food Research Institute (FRI), Bratislava. Its computerised databases extend the ways in which the data can be utilised, including nutrition modelling among others. Slovak FDB can be compared with data banks of developed countries. According to suggestions of the FAO Food and Nutrition Division, FRI Bratislava will serve as the Subregional Data Centre for the network on food composition in Central and Eastern European Countries.

2 CHARACTERISATION OF THE SLOVAK FOOD DATA BANK (FDB)

Slovak Food Data Bank contains databases of different food characteristics and software for their management and application.

2.1 Databases of the Slovak Food Data Bank

- Physical properties of foods
- Chemical composition, nutritive and energy value of primary foods, food products and dishes
- Natural toxicants and antinutritive matters
- Food consumption in Slovakia and abroad
- Recommended daily allowances for different countries and categories of inhabitants
- Nutritive losses and gains
- Energy requirement of some working and sports activities

All databases are regularly supplemented and actualised

2.2 Software of the Slovak Food Data Bank

DMS – Data Management Software
ALIMENTA version 0.3
- Calculation of chemical composition and energy value of food products, meals and dishes considering mass changes and nutritive losses
- Simulation of calculated composition of foods according to principles of healthy nutrition
- Comparison of food composition with recommended daily allowances
- Evaluation of nutrient intake (percent of RDA)
- Calculation of actual price
- Monitoring of catering (individuals and groups) during the definite time
- Evaluation of diet composition
- Creating of dietary models according to selected criteria

2.3 Observed Parameters (Food Components)

Parameters (Food Components) stored in the database of Chemical composition, nutritive and energy value of primary foods, food products and dishes are:
dry matter, **proteins** and other nitrogenous matters (animal and plant proteins, gluten, amino acids, purine matters), total **lipids**, individual fatty acids, cholesterol, total **saccharides**, individual saccharides, starch, fibre complex, organic acids, **mineral matters,** total and individual elements, **vitamins**, **energy** value total and its structure.

3 APPLICATION OF THE SLOVAK FOOD DATA BANK (FDB)

Uses of food composition data have grown. In view of increased uses, the type of data has changed. Furthermore, there is considerable interest in the role of components other than traditional nutrients in the maintenance of health. Both health policy formulation and health research deal wih disaggregated consumption of specific food components by certain groups of the population, requiring updated composition data. In addition to changes in available data, foods have changed. The global distribution of foods, fresh as well as processed, simple and multi-component, stimulates the demand for new data. In addition, the abundance of new foods and the frequency of new entries into the world market require continuous effort to review, re-analyze, and update composition databases.

The health status of the Slovak population is unsatisfactory. Incorrect nutrition is considered to be one of the basic reasons. To achieve changes in diet habits towards the health nutrition and to increase the individual health responsibility of the whole population is the main task of the National Health Support Program in Slovakia. According to this, there is a need to increase the basic consumption of fruits and vegetables in Slovakia. Recent consumption of fruits is 68 kg per capita per year and vegetables 107 kg. According to recommendations it should increase to 75-80 kg for fruits and 120-125 kg for vegetables. Outputs from the Slovak FDB are very important for the education of the population. Special brochures (e.g. Cholesterol in foods) and the new edition of Food Composition Tables are printed regularly.

Review of applications of the Slovak FDB:

- Output data sets
- Balance calculation of composition of new food products and prepared meals
- Food labelling
- Basis for food tables
- Nutritional evaluation of food consumption
- Food quality evaluation
- Human nutrition management
- Modeling of diet
- Plans for prevention of mass illness
- Basis for food legislation
- Education sphere
- Users software for personal computers

4 INTERNATIONAL COOPERATION

Building up the food composition database is a time and money consuming process. Food composition data of comparable quality can be shared amongst countries, which in turn greatly assists cross-border trade and allows for more efficient use of resources. Therefore, **regional collaboration** is a possible way to reduce costs but still meet the needs for accurate food composition data.

Slovak FDB is a member of EUROFOODS (European network of Food Data System). FAO Food and Nutrition Division nominated the Food Research Institute, Bratislava, Slovakia as the Food Composition Database Centre for Eastern and Central European countries CEECFOODS. In this capacity, the Food Research Institute will pursue and promote sub/interregional food composition work including data interchange. Countries will share data on foods for inclusion in each database.

THE ECONOMY OF GOOD VEGETABLE PRODUCTION IN OPEN CULTIVATION

M. Stenberg

Pyhäjärvi Institute
Ruukinpuisto
27500 KAUTTUA
FINLAND

1 INTRODUCTION

The costs of vegetable production include material costs, labor costs and capital costs. The material costs amount to about 19 - 36 %, the labor costs to about 30 - 58 % and the capital costs to about 12 - 25 % of all costs in vegetable production in Finland. Today it is more and more important to reduce all these costs. This project has two main purposes: the first one is to find out the potential differences between three various farming methods, and the second one is to improve the economy of vegetable production on the basis of the results of the project. The farming methods studied are conventional farming, integrated production and organic farming. The vegetables under study are carrot, onion, gherkin, white cabbage, pea and beet. In this paper I will discuss carrot and onion production in detail. During this project we try to find out the costs and labor differences between the different farming methods and vegetables on the farm level. We started by asking voluntary farmers to take part in this project during the growing season 1997. The farmers recorded their labor hours and variable costs in a book made by researcher. At the beginning of 1998 the farmers returned the completed books to the researcher.

In figure 1 the columns illustrate the costs of carrot seeds, fertilizers and pesticides in Finnish marks. The two first columns have been drawn according to Lassheikki (1994).
The next three columns have been drawn according to the results of this project.The third column represents conventional farming, the fourth one integrated production and the last one organic farming. Figure 1 shows that nowadays farmers spend smaller amounts on seeds, fertilizers and pesticides than before.

In figure 2 the columns illustrate the costs of onion sets, fertilizers and pesticides in Finnish marks. The two first columns have been drawn according to Lassheikki (1994).
The next three columns have been drawn according to the results of this project. The third column represents conventional farming, the fourth one integrated production and the last one organic farming. Figure 2 shows that nowadays farmers spend smaller amounts on onion sets, fertilizers and pesticides than before.

Summary

In this project the reply percentage was 82. There were 40 voluntary farmers who recorded in a book their labor hours and variable costs for the production of the vegetables mentioned above. Some farmers gave information about several vegetables. In general, in Finland the largest cost items in vegetable production in open cultivation are seed and fertilizer costs, but this varies considerably according to the vegetables. Table 1 shows that in pea production the largest cost item was seeds, and in onion production the largest cost item was onion sets, which made up about half the costs. In gherkin production plastic and fleecy film costs were the largest cost items. The use of different materials (quantity and costs) varied considerably according to individual farmer. A quite small number of farmers took part for each vegetable, therefore we could not find any clear trend in the use of materials in conventional farming in comparison to the use of materials in integrated production.

Labor hours mainly consisted of harvesting and preparing vegetables for sale.

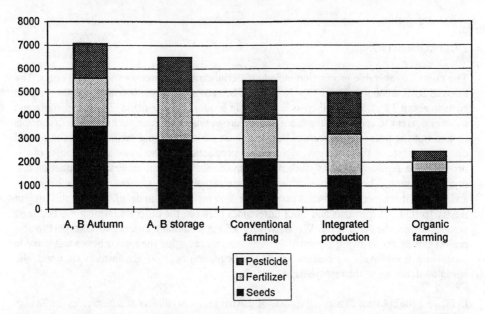

Figure 1 Seed, pesticide and fertilizer costs per hectare in carrot production in Finland. The first two columns have been drawn according to Lassheikki, the next three ones according to the results of this project.

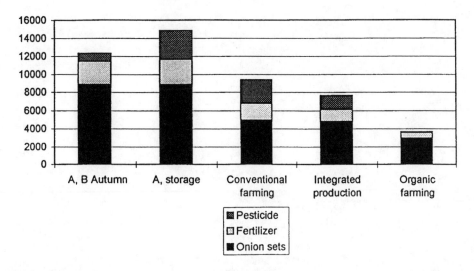

Figure 2 Onion sets, pesticide and fertilizer costs per hectare in onion production in Finland. The first two columns have been drawn according to Lassheikki, the next three ones according to the results of this project.

Table 1 The largest variable cost items (percentages) by vegetables and production methods. 1 = conventional production, integrated production = IP and 2 = organic farming.

	Pea*	Carrot			Onion			Beet		Gherkin		White cabbage	
	IP	1	IP	2	1	IP	2	1	IP	1	IP	1	2
Seeds	53	15	22	19				25	20	23	12	19	
Onion sets					30	47	49						
Plant										5	14		34
Fertilizer	20	12	27	4	11	13	10	14	19	20	20	22	
Pesticide	27	12	27	5	16	15	2	16	24	5	2	12	7
Plastic/fleecy film				2						33	22		6
Grading and packing		36		10	7	4	8	27	18			17	7
Transportation costs		22	16	16	19	8	15	15	18	10	10	26	16
Rents for machinery			1			4	16	3		3	2		10
Other		3	5	46	17	9	0		1	1	18	4	20
Total	100	100	100	100	100	100	100	100	100	100	100	100	100

References

1. Lassheikki, K. 1994. *Puutarhayritysten tuotantokustannusten seurantamallit*. Forssa. 138 p.

Session II Consumer Attitudes to Improving Crop and Food Quality

PUBLIC UNDERSTANDING OF FOOD SCIENCE AND CONSUMER ACCEPTANCE OF NOVEL PRODUCTS

L. J. Frewer, S. Miles, C. Howard and R. Shepherd

Department of Consumer Sciences
Institute of Food Research
Earley Gate
Reading, RG6 6BZ, UK

1 DEFINING PRODUCT QUALITY

It is important to understand the psychology that underpins consumer beliefs about food quality if improvements are to be made to the food chain. Consumer beliefs about quality attributes of different food products may be broadly classified as either "intrinsic" or "extrinsic". Sloof, Tijskens and Wilkinson [1] identify intrinsic properties as relating to factors such as flavour, quality, and appearance, whereas extrinsic properties may be identified as factors relating to brand name, price and packaging. It is proposed that this definition of "extrinsic" be extended to embrace other salient consumer perceptions, attitudes, and knowledge which determines the psychology of food choice.

Of particular relevance to food choice decision making are the perceptions of risk and benefit associated with consuming a particular product. Perceived risks may include identifying what people believe about dietary and nutritional issues (such as high levels of fat consumption, or patterns of dietary intake which exclude particular nutrients), or understanding perceptions that people have about the inherent dangers associated with the processes used in agriculture and food manufacturing.

The notion that uncertainty (in terms of both the probabilities associated with hazard occurrence and the consequences of hazard occurrence) has been important in understanding peoples responses to risk has been recognised in the literature for several decades. Both economic and cognitive psychology have utilised different types of psychological model in attempting to understand reactions to uncertainties.

Prospect theory [2] predicts that people tend to prefer decision making under conditions of certainty as opposed to uncertainty. People are not necessarily risk adverse, but are certainly loss adverse. They are only willing to take risks when they think it appropriate.

One of the findings that has consistently emerged from research on consumer preferences for the novel foods resulting from the application of emerging technologies is that much negativity is focused on this question of whether there is a benefit associated with applying a potentially risky technology. Differential accruement of the risks and benefits resulting from novel technologies is particularly important in determining consumer acceptance of resulting products. If there are consumer advantages, or benefits to the environment, associated with novel products, then some degree of risk-taking is tolerable. If the benefits accrue to industry, but the potential risks accrue to the consumer or the environment, application of innovative new technologies is associated with consumer negativity [3,4]. Negative consumer responses are also associated with any perceptions that animal welfare is compromised.

2 PUBLIC PERCEPTIONS OF GENETIC ENGINEERING : AN OVERVIEW

There has been a great deal of research into the deconstruction of public perceptions of risk associated with the new biotechnology, in particular genetic modification [5]. Some of this research has been at the level of individual countries (for example, see [6 7]), some has been at a cross-cultural level [8 9]. A more detailed overview of the European picture is provided elsewhere [10]. In summary, however, foods produced with novel processing technologies, whether these technologies are at the level of agricultural practice or manufacturing, are generally perceived to be relatively high in risk compared to the perceived benefits which are associated with them. Indeed, these relative perceptions are very similar to the way that people perceive the risks and benefits of nuclear technology [11]. Who is perceived to benefit from the application of the technology is also important. Perceptions that the equity of distribution of both risks and benefits has not accrued equally across all population groups, (for example, if transgenic development is perceived to be directed towards industrial profit, public acceptance is likely to be lower than if the consumer benefits [12]. Ethical issues cannot be dissociated from those of perceived risk and benefit [13 14], and it is essential that these concerns be addressed within the context of both regulation and release of GMOs, and strategic development of the technology. For UK populations, ethical concerns are greatest for applications involving transgenic animals, particularly if the application is for the purpose of food production [15]. Genetic modification of animals is more acceptable if it is seen to apply to a medical (as opposed to food-related) context [16].

Genetic modification cannot be regarded as a "unitary" technology in consumer acceptance terms, and consumer perceptions are likely to be focused by exposure to specific applications. Acceptance of different applications is increased if perceptions of benefit resulting from the application are perceived to offset perceptions of risk. The application of semi-structured interview techniques derived from already existing techniques such as laddering has shown that public concerns tend to be focused on higher order issues such as ethical problems in the case of genetic modification, and exhibit similarities with other technological hazards such as pesticides. Concerns about some other food-related hazards such as Salmonella tend to be much more related to concrete consequences of the hazard for human health [17].

3 PUBLIC UNDERSTANDING OF FOOD SCIENCE: IMPLICATIONS

The traditional view of the public as a body of "uninformed" individuals who are unable to make rational decisions about novel foods and the science of food production has long been superseded by awareness that the purpose of science communication is to create an informed public, who are able to make decisions about consuming novel products based on the best scientific information available. Increased knowledge may not increase acceptance. Rather views may further polarize with increased understanding [18]. If people already have crystallized views supporting or opposing a particular technology, they tend to select information from the environment which supports these views, and other information which opposes these views is cognitively "discarded". For undecided individuals, information including some form of reference to the scientific uncertainties inherent in the risk assessment process is likely to result in both increased acceptance of genetic modification, but also increased trust in the source providing the information [19].

4 THE MEDIA AND PUBLIC UNDERSTANDING OF FOOD SCIENCE

It is important to distinguish between risk presented within a chronic context, and risk as a "crisis". There has been much debate as to whether the media sets the agenda for public debate, or merely reflects the wider public discourse about risk and associated public concerns. Nor is it known whether media reporting of risk has a strong influence on public risk perceptions, or whether risk presented in a "crisis" context in the media has a differential impact from that presented as "chronic". Most members of the public obtain their risk information from the media. The extent and content of media risk reporting about the Chernobyl accident, and other associated potential hazards such as nuclear power in the UK, was analysed during the months of March, April and May of 1996. Most of the risk reporting during this period was about BSE, which appeared in the media in a "crisis" context in March of the same year. The results of the analysis indicate that BSE coverage amplified public risk perceptions, but that the effects were relatively short lived. Reporting about the 10th Anniversary of Chernobyl was reassuring in that the location of the accident was reported as being remote from the UK, and the responsibility of foreign risk regulators, and was not associated with national coverage of the risks of domestic nuclear power. The type of newspaper was important in determining coverage of different hazards. Local newspapers tended to be more concerned about domestic nuclear power, and report content reflects a local economic agenda as well as one concerned with risk-related issues [20].

5 CONCLUSIONS

It is clearly important that the public be provided with the best information available regarding the risks and benefits of scientific developments in food production. In addition, information about the science underpinning food production is also important, although improved public understanding will not automatically lead to improved acceptance. The social context of technologies must also be considered in the wider debate about the future strategic development of food science.

6 REFERENCES

1. M. Sloof, L. M. M. Tijskens, & E. C. Wilkinson, *Trends in Food Science and Technology*, 1996, **71**, 165-171.
2. D. Kahneman and A. Tversky, *Econometrica*, 1997, **47**, 263-291.
3. L. J. Frewer, C. Howard, D. Hedderley, and R. Shepherd, *Food Quality and Preference*, 1997, **8**, 271-280.
4. R. Deliza, A. Rosenthal, D. Hedderley, H. J. H. MacFie, and L. J. Frewer, in preparation.
5. B. Zechendorf, *Bio/Technology* **12**, September 1994, 870-875.
6. A. Hamstra, in *Biotechnology In Public. A Review Of Recent Research*, ed. J. Durant, Science Museum, London, 1992, 42-51.
7. W. J. M. Heijs, C. J. H. Midden, and R. A. J. Drabbe, Biotechnology: Attitudes and Influencing Factors, Eindhoven University of Technology, Eindhoven, 1993.
8. J. Durant, G. Gaskell and M. Bauer, Proceedings, 8th European Congress on Biotechnology, August 17th-21st Budapest, Hungary 1997. 9. A. Saba, A. Moles and L. J. Frewer, *Journal of Nutrition and Food Science*, 1998, **28**, 19-30.
10. L. J. Frewer and R. Shepherd (1998), in S. Roller (ed.) Genetic engineering for the Food industry: A strategy for food quality improvement, Blackie Academic, New York, 27-46.
11. L. J. Frewer, C. Howard, and R. Shepherd, *Journal of Risk Research*, in press.

12. C. T. Foreman, (1990) in NABC report 4, Animal Biotechnology: Opportunities and Challenges, ed. J. Fessenden Macdonald, National Agricultural Biotechnology Council, Ithaca, NY, 1990, 121-126.

13. D. Boulter,*Critical Reviews in Plant Sciences*, 1997, **16**, 231-251.

14. R. Straughan and M. Reiss, Improving Nature? The Science and Ethics of Genetic Engineering, Cambridge University Press, Cambridge, 1996.

15. L. J. Frewer and R. Shepherd, *Agriculture and Human Values*, **12**, 1995, 48-57.

16. L. J. Frewer, C. Howard and R. Shepherd, *Science, Technology and Human Values*, 1997, **22**, 98-124.

17. S. Miles and L. Frewer, in preparation.

18. G. Evans and J. Durant, *Public Understanding of Scienc*e, 1995, **4**, 57-74.

19. L.J. Frewer, C. Howard and R. Shepherd (In press), Agriculture and Human Values.

20. L. J. Frewer, E. Campion, D. Hedderley and S. Miles, Media reporting of risk in the UK. An analysis of UK media reporting surrounding the 10th Anniversary of the Chernobyl Accident, Report to the European Commission, 1997.

CONSUMER QUALITY PERCEPTION AND ATTITUDE TOWARDS QUALITY CONTROL OF VEGETABLES FROM FIELD TO TABLE

W. Verbeke, X. Gellynck, and J. Viaene

Department Agricultural Economics
University of Ghent
B-9000 Gent, Belgium

1. INTRODUCTION

Today's consumer attitude and behaviour are increasingly driven by quality, safety and health consciousness. From the producer viewpoint, quality control has evolved from an efficiency challenge to a tremendous opportunity by building competitive advantages through pursuing relationships based on an integrated chain approach with quality guarantees. To succeed in today's competitive agri-food marketplace, two options are available: organise production more efficiently and work more consumer oriented in order to meet consumer requirements[1]. During recent years, several concepts like Supply Chain Management[2,3,4], Efficient Consumer Response[5,6], Value-Added Partnerships[7], Total Quality Management[8,9], and Integrated Quality Management[10], have been introduced. All these concepts share the objectives of adding value to the entire chain, of realising competitive advantages and a better performance of the chain through increased responsiveness to consumer needs, wants and demands.

The research at hand addresses questions related to collecting valuable information at consumer level, since this is the prerequisite for the practical application of the aforementioned concepts by industries like e.g. the vegetable industry. This paper focuses on assessing both the quality perception of vegetables and the attitude of consumers to quality control throughout the vegetable production chain. The paper first addresses the research objectives and related methodology. The major part of the paper concerns the presentation and discussion of the results. Finally, in the conclusion section, key factors for a successful performance of the vegetable chain from field to table are identified.

2. RESEARCH OBJECTIVES AND METHODOLOGY

The research methodology is based on primary exploratory (qualitative) and conclusive (quantitative) consumer research. The qualitative research consisted of five focus group discussions with six to eight respondents. The objectives were to gain preliminary insights into consumer attitude, perception and behaviour towards vegetable consumption. Additionally, insights were gained in consumer information requirements (chain perception) concerning the vegetable chain and potential topics for communication. The qualitative research has been organised during March 1997.

Based on the information gathered from the qualitative research, hypotheses and key attention topics for quantitative conclusive research have been drawn. The quantitative primary data were gathered through a sample survey research. The research approach dealt with administering pre-tested formal questionnaires during personal in home interviews, led by trained field workers. The questionnaire comprised issues such as general consumer behaviour and attitude towards vegetable consumption and consumer's chain perception of and interest in different processes within the processed vegetable chain. The target population of the survey consisted of people living in Belgium, aged between 15 and 65 years, who are the main responsible person for purchasing vegetables within their household. The respondents were selected by means of non-probability quota sampling. Quota towards age and place of living have been established. The overall sample size is set at N=500 respondents, equally split up between Flanders (northern part of Belgium) and Wallonia (southern part). This sample size allows to meet the minimum sample size rules suggested[11]. The field work was realised during October and November 1997. After coding and editing the questionnaires, the collected data were analysed by means of the statistical package SPSS.

3. RESULTS

The qualitative research[12] revealed major consumer concerns related to the use of pesticides during the early stages in the production process and to vitamin and taste losses during the latter. Consumer interests in vegetable production and the processes from field to table included soil cultivation, seed choice, growth process, harvesting, processing and vegetable preparation at-home. The results of the quantitative research that are relevant to this paper, include vegetable quality perception and consumer attitude to quality control in the vegetable chain.

3.1 Consumer Perception of Vegetables

The perception of fresh, frozen, canned and glass vegetables is assessed by using a pick any scaling technique[13]. The respondents have been asked to choose for each type of vegetable the most relevant attribute out of a list of attributes, resulting from the qualitative research. The selection of an attribute by the respondents means that the attribute is highly associated with the product discussed. The attributes have been split up in to core product attributes, augmented product attributes, product benefits, situation factors and image components.

Fresh vegetables are most associated with the core product attributes healthy, tasty and natural. The perceived benefit of fresh vegetables consumption is that these provide people with necessary vitamins and minerals. Fresh vegetables are served in meals prepared for the whole family and most appreciated by the real or 'would-be-real' family mother and connoisseurs. No statistically significant differences are found in the perception of fresh vegetables between 'addicted' fresh vegetable consumers and people who frequently consume frozen, glass or canned vegetables.

Of all processed vegetables, frozen vegetables are best perceived in terms of vitamin and mineral content. This attribute is significantly more mentioned by consumers who frequently use frozen vegetable. These consumers moreover perceive frozen vegetables as easy and fast to prepare, providing variety and ideal for active and modern people. An association with the attribute 'industrial' also comes about.

Canned vegetables are perceived as cheap and industrial. Other associations deal with convenience, speed and variety. Canned vegetables have the image of being mainly the preferred vegetable of single people.

Glass vegetables have a less industrial and cheap, but more traditional image than canned vegetables. Apart from offering variety, important product benefits deal with a good presentation and with the fact that the quality of the product can be judged through seeing. A very similar response pattern for glass and canned is found for situation and image component.

The quantitative results about perception and image of vegetables fully confirm the findings of the qualitative research. Frozen vegetables are clearly the best and preferred alternative to fresh. This can be explained by the fact that consumers are nowadays more familiar with the freezing preservation technique, rather than with the sterilisation technique that was commonly applied in Belgian households some decades ago.

3.2 Consumer Attitude to Quality Control

During the qualitative research, vegetable consumers are asked to describe the ideal vegetable production process from the real beginning till the prepared vegetable as it is presented on a plate. The consumer concerns related to the six main steps (soil cultivation, seed choice, growth process, harvesting, processing and vegetable preparation at-home) in the production process from field to table are quantitatively assessed by using a pick any scale. For each step in the production process, relevant attributes are selected based on the qualitative research. These attributes are labelled as 'must be' and 'should not be', which respectively means that the ideal production process 'must' take care of specific practices and 'avoid' other ones . The attributes are presented to the respondents, who are asked to indicate the most relevant attribute for their imagined ideal vegetable production process.

All steps are almost attached equal importance by the consumers. Nevertheless, it is perceived that slightly more importance is attached to those processes the consumer is most familiar with through own vegetable production: tillage or soil cultivation, vegetable preparation at-home and harvesting.

According to the soil, consumers especially stress that the soil must be pure and carefully cultivated. Treatments with pesticides and to a lower extent with chemical fertilisers are not accepted for the ideal vegetable production process. Soil purity is significantly more mentioned by frequent glass and canned vegetable consumers.

The seed for the ideal vegetable production must be of premium quality and should not be treated or coated with pesticides. Remarkable is also that about a quarter of the respondents indicate that the seed should not be genetically engineered.

The growth process should be under permanent control. Again reserves against the use of pesticides are on top of the bill, together with the rejection of any kind of spraying on the vegetables.

The harvesting process must be done at the right moment. This attribute is further specified through mentioning that harvesting should especially not be done too late. Vegetables should further be harvested in a selective manner and care should be given to avoiding damaging the vegetables during the harvesting process.

During the processing of vegetables, special attention must be paid to careful washing the vegetables and to strictly limit the storage period between harvesting and processing. This concern refers to concerns about potential vitamin and taste losses due to long storage. The frequent users of canned and glass vegetables attach significantly more importance to 'not supplying vitamins' during the processing of the vegetables.

During the preparation, losses of taste, vitamins and minerals should be avoided. This topic is significantly more stressed by frequent users of frozen and fresh vegetables. The ideal vegetable must finally be appreciated by all family members as it is served.

The identified topics from a consumer viewpoint for the ideal vegetable chain reveal key factors for a successful improvement of the processes in the vegetable chain. This includes both key attention points for quality control throughout the chain and relevant topics for communication around chains and processes from field to table. It is obvious that some topics related to the production process of vegetables, such as manual labour or avoiding the use of machinery, hardly can be realised in practice. A great majority of the indicated issues for ideal vegetable production are however perfectly feasible in practice.

4. CONCLUSIONS

Recent changes in the working conditions of the agri-food business create opportunities for the development of competitive advantages through co-operation in chains[14]. These changes deal with new wants at consumer, retail and production level.

The qualitative and quantitative primary research described in this paper, indicate that the vegetable consumer is confronted with a dilemma. The consumer approaches vegetable consumption emotionally. Fresh vegetables are perceived as the best product in terms of health, quality, nutrition and naturalness. However, the preparation of fresh vegetables is increasingly considered as too time consuming, especially by fully employed people. Under these circumstances, the consumer looks for an alternative in the form of processed vegetables: frozen, canned or glass. To justify this choice to other people such as family members, the consumer looks for rational support. A great deal of this rational support is identified by the elements provided during the description of the chain perception in terms of the ideal vegetable production process. The search of consumers for rational support offers a tremendous opportunity for chains that manage first, to guarantee an integrated quality, as requested and defined by the consumer, and second, to work out the realised consumer driven chain improvements as an effective communication tool. It is up to the processing vegetable chain to translate these opportunities into adapted processes and institutional adjustments. The objective in doing so is to realise and implement the consumer's ideal vegetable, with clearly guarantees covering all the way throughout the chain, from field to the table.

References

1. K. Grunert, 'Research on agri-chain competence and consumer behaviour', Paper presented at EU-workshop on Agri-Chain Competence: Learning from other chains, 's-Hertogenbosch, the Netherlands, p. 23, 1996.
2. G. Evans, M. Naim and D. Towill, *Logistics Information Management*, 1993, **6**, 15.
3. C. Harland, *Int. Journal of Production Planning Control*, 1995, **6**, 209.

4. D. S-Bridge, *Farm Management*, 1996, **9**, 357.
5. A. Van der Laan, *Food Personality*, 1994, Sept., 8.
6. B. Wierenga, 'Competing for the future in the agricultural and food channel', In: B. Wierenga, A. van Tilburg, K. Grunert, J. Steenkamp and M. Wedel (eds.), 'Agricultural Marketing and Consumer Behavior in a Changing World', Kluwer Academic Publishers, Norwell, Chapter 2, p. 31, 1997.
7. M. Porter, 'The Competitive Advantage of Nations', MacMillan Press, London, p. 855, 1990.
8. J. Ross, 'Total Quality Management: Text, Cases and Readings', St-Lucie Press, Delray Beach, p. 325, 1993.
9. J. Cortada, 'TQM for Sales and Marketing Management', McGraw-Hill, New York, p. 232, 1993.
10. J. Viaene, X. Gellynck and W. Verbeke, 'Integrated Quality Management applied to the Processing Vegetables Industry', In: R. Shewfelt and B. Bruckner (eds.), 'Fruit and Vegetable Quality: An Integrated View', Technomic Publishing, Lancaster, forthcoming 1998.
11. S. Sudman, 'Applied Sampling', Academic Press, New York, p. 252, 1976.
12. J. Viaene and X. Gellynck, 'Consumer perception of integrated quality management for vegetables in Belgium', Paper presented at: International Conference of Fruit and Vegetable Quality, Potsdam, May 1997, p.11.
13. P. Van Kenhove, 'A comparison between the pick any method of scaling and the semantic differential', Working paper 95/10, University of Ghent, Department of Marketing, Gent, p. 14, 1995.
14. J. Viaene, W. Verbeke and X. Gellynck, 'Chain behaviour and chain reversal of the processed vegetables chain', In: J. Trienekens and P. Zuurbier (eds.), 'Proceedings of the Third International Conference on Chain Management in Agri-Business and the Food Industry', Wageningen Agricultural University Press, Wageningen, forthcoming 1998.

AN INTERACTIVE NEURAL NETWORK FOR ANALYSING THE FOOD CONSUMER BEHAVIOUR STABILITY

D. Thiel

E.N.I.T.I.A.A.
Department SMAD
Rue de la Géraudière, BP 82225
44322 Nantes Cedex 03 (France)

To improve the quality of food and the consumer behaviour knowledge, some research tackled an identification and evaluation of the perceived consumer attributes. For instance, for Claudian[1], the food behaviour factors which correspond to the attributes that we assign to a product perception, come from organic, psychological, and social origins. According to Bodenstedt[2], these factors take into account the food, the consumer, society's characteristics and the abiotic environment. They also depend on the subject's situation, on their context and finally, on the presentation of the product[3]. Moreover, every subject has its individual story which comes from learning which is conditioned by personal, family, social and cultural experiences[4].

In the objective to show the food attitude diversity, we will comment on two types of apple tasting, the first is stemming from global preference tests and the second, from sensorial evaluation results of a panel of judges.

1.THE BEHAVIOUR INSTABILITY AND THE EXISTENCE OF ATTRACTORS

Firstly, we based our investigation on global preference data from 139 consumers who have tasted four apple varieties. We discretized the values into three classes 0, 1 and 2 which respectively correspond to a low, medium or very good global preference. These data showed large food preference diversity by the number of possible « configurations » (for paper size limitations, we do not present the data here). Nevertheless, it emerges a certain number of identical preference classes that I called «collective preference attractors» (we draw here a parallel with the fixed points, limit cycles and chaotic attractor notions which can describe the non-linear system stability). These first observations bring us precisely to our research objective which consists of trying to explain the behaviour diversity and stability.

Secondly, sensorial data on eleven varieties of dessert apples were presented to a panel of 14 trained judges and 17 attributes from the flavour category were tasted in duplicate. It must been noted that we have not statistically operated these data. Our objective was only to show the existence of different types of attractors in each judge behaviour. For example, some collective attractors show almost unanimous judge behaviours faced with some particular attributes. This could be explained by their insensitivity to some products or by a collective memory of these particular attributes. We also observed that some judges had the same evaluations of the different product attributes, which show an other type of attractor in this human «measure instrument». Finally, other classes revealed «unstable» measures.

These two experiences empirically showed the complexity and the instability of the behaviours. In a second step, with a more theoretical point of view, we will present the usual models from cognitivist inspiration.

2. THE COGNITIVIST MODELS OF ATTITUDE FORMING

2.1. The compensatory models

One of the first compensatory models shows the consumer decision process as the choice among different proposals for that which gives the higher satisfaction (cf. Edwards[5,6]). The evaluation of the different options is calculated by : $U_c = \Sigma_i (PS_i . U_i)$, where PS_i corresponds to the subjective probability that for the chosen option of the consumer C, is linked the occurrence of the event i, and U_i the utility given by this event. According to Edwards and Rosenberg's[7] models, Fishbein's[8] approach also rests on an additive linear representation of the attitude concept - here it is a global evaluation which is formed by local evaluation aggregation. The equation of the attitude A_o is : $A_o = \Sigma_j (C_j . V_j)$ where C_j corresponds to the probability that option O has, or has not, the attribute j and V_j is the consumer weighing of this attribute. In Bass and Talarzyk's[9] model, the global evaluation of each option is given by a similar linear equation which compensates for the local evaluations : $A_o = \Sigma_k (P_k . A_{ko})$ where P_k corresponds to the appreciation criteria k weighting of the interviewed consumer; while A_o and A_{ko} give the global evaluation and the local evaluation of the option O over k, respectively.

Fishbein's model is certainly the most important model, in the literature of food choice, it is shown that this model and its variants (Fishbein and Ajzen[10]), are commonly used. For example, Shepherd[11,12,13] presents some applications of these models to food choice and validates them by different experimentations. He also proposes some extensions of the theory of reasoned action by considering the person's self-identity which may influence behaviour independently of his or her attitudes. Bagozzi[14], in his critical review of these different attitude process models, introduces other dimensions for attitudes such as : the necessity to introduce the process factors; the role of the goals; the introduction of non-reasoned determinants; the nature of the motivational component; and the halo effect. The last effect is particularly studied in psychosociology as a cognitive phenomenon. Wilkie[15] considers that the halo effect takes place when a consumer overvalues all the characteristics of a product which he likes (see an example of measurement of the halo effect in Stein and Nemeroff[16]). Beckwith and Lehmann[17] proposed on the individual level to distinguish the cognitive effect (perception → emotion) from the halo effect (emotion → perception). Their model shows the relationship between these two effects and quantifies the effects of the emotional on the perceptual, in parallel with Fishbein's extension of his first attitude process representation by integration of the social norms (influence of membership and reference groups on the individual perceptions), Beckwith and Lehmann's model gives for each product which was tested by the consumers, the following relation : $A_p = \Sigma_j \omega_j B_{pj} + \gamma A^*_p + u_0$, with : $B_{pj} = \beta_j A_p + \gamma_j B^*_{pj} + u_j$, for j = 1 to n (number of attributes), where A_p corresponds to the consumer attitude for the product p and A^*_p the average attitude of the consumers; B_{pj} indicates the perception of the product p by the attribute j and B^*_{pj} the average perception of the consumers; ω_j the weights of the attribute j and γ the weight of the average attitude, β_j the importance of the attitude for the product p in the perception B_{pj}, and γ_j the importance of the average perception B^*_{pj} in the perception B_{pj}. Finally, u_0 and u_j represent some random differences.

2.2. Critics of these models

On reviewing psycho-sociological and marketing literature about the theme of the representation modes of the attitude forming process (Bagozzi[14] ; Aurifeille[18]), an absence of dynamic aspects can be observed. For example, the identification of the internal composition law of the local evaluations to addition, is extremely open to criticism. Many experiences in psychology and psychosociology, show that the order of processing the attributes influences the final result. To illustrate these aims, the process of consumer attitude forming for a given product which is derived from Beckwith and Lehmann's model, was represented in our previous paper (Thiel and Robert[19]) which showed, according to the order in which the consumer considers the attributes of a given product, that his attitude progressively changes along the time, ending up in very different appreciations.

From this necessary dynamical vision of the attitude forming, we propose a connectionist approach based on Beckwith and Lehmann's model.

3. PROPOSITION OF A CONNECTIONIST MODEL OF ATTITUDE FORMING

To progress from cognitivism to connectionism, we change from an interpretation of the cognition facts and phenomena which initially concerns the *logic* (of the *discontinuum*), to an interpretation concerning the *dynamic* and the *topologic* (of the *continuum*). We propose to implement an automata or interactive neural network (cf. Mc Culloch and Pitts[20]) of Beckwith and Lehmann's model (see figure 1). The particulars of an automata network are that it may be defined, in a general way, as a (large) set of cells (finite automata), locally interconnected, which can evolve at discrete time steps through mutual interactions. Formally, an automata network can be described as a mapping F from S^n into itself, where S is a finite space (state space). The network is then made up of n interconnected cells. F defines the connection structure : cell i receive a connection from j if F_i depends on the jth variable (where F_i is the ith component of mapping F). A state of the network is a vector x in S^n.

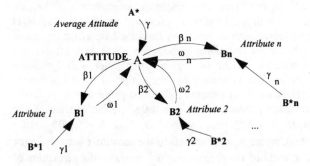

Figure 1 *The connectionist model*

In our application, Beckwith and Lehmann's model is transforming into a network where the automata correspond to n variables A, A*, and all B_j and $B*_j$, which will be discretized. The connections between the different automata which define the mapping F, have the following weights : ω_j, β_j and γ. The valuations of these arrows correspond for instance, to the connection weights of a scale which ranges through -3 to +3 (semantic differentiator going from "a little influence" to "very influent" : -3, -2, -1, +1, +2, +3). This network was constructed according to the causality relationships between $(B_j, A*) \rightarrow$ A and $(A, B*_j) \rightarrow B_j$. The figure 1 shows the principle of the structure of this automata (or neural) network.

The dynamics on the network are then defined through a rule (here given by Beckwith and Lehmann's equations) that transforms any vector x in S^n into a vector y in S^n. For example, the parallel iteration rule is defined as : $y = F(x)$ and can be interpreted as follows : at each time step, each automaton computes its next state by using its mapping F_i on the current state x. As the state space S is here finite, it is demonstrated that all trajectories will be periodic, limit cycles or fixed points. Here, we make a parallel with the previous attractors in consumer attitudes.

Finally, we will present the principle of our applied research program with the objective to continue to validate this approach.

4. CONCLUSION

Our previous experimentation results of consumer perceptions have already confirmed the validity of Beckwith and Lehmann's model and our theoretical hypothesis which consists of proving the sequentiality in the attitude forming process. The second step of this research consists in validating the connectionist modelling and in studying the attitude dynamics by showing different attractors according to the previous observations of apple consumption. Our difficulty in putting this research formalisation into practice will consist in rationally explaining which attributes are perceived important by the consumer for a given product and to quantify their respective weights. For the last ten years, a lot of work has focused on the study of food choice and on the factors which contribute to its determinism. It is also usually assumed that the number of beliefs or attributes consumers used when forming an evaluation or attitude is limited (see for instance Grünert[21]). The problem is then to be able to define and to measure these attributes and weights which are taken into consideration during the food choice.

In operational terms, the marketing and quality managers may gauge the different attributes weights by changing some elements of the mix or of the production and crop, with the objective of bringing the consumers nearer to the interesting attractors. In spite of the poorness of current rational knowledge of the human attitudes, it is nevertheless possible to demonstrate that this type of neuromimetic model is very robust.

References

1. J. Claudian and J. Trémolieres, *L'Univers de la Psychologie*, Ed. Lidis, Paris, 1978.
2. J-M. Bernard, P. Salles and C. Thouvenot, *Les journées de l'Anvie : pratiques et comportements alimentaires,* 1993, 11 juin, 5.

3. S. Issanchou and J. Hossenlopp, in *Plaisir et préférences alimentaires*, ed. CNERNA - CNRS, Polytechnica, Paris, 1992, 49.
4. P. Rozin, *Social learning-Psychological and Biological Perspectives*, eds. T.R. Zentall and B.G.J. Galef, Lawrence Erlbaum Associates, Hillsdale, New Jersey, 1988, 165.
5. W. Edwards, *Psychological Bulletin*, 1954, **51**, 380.
6. W. Edwards, *Conflicting objectives in decision*, eds. D.E Bell, R. L. Keeney & H. Raiffa, Chichester, Wiley, 1977.
7. M. J. Rosenberg, *Journal of Abnormal and Social Psychology*, 1956, **53**, November, 367.
8. M. Fishbein, *Human Relations*, 1963, **16**, 233.
9. F. M. Bass and W. Talarzyk, *Journal of Marketing Research*, 1972, **9**, 63.
10. M. Fishbein and I. Ajzen, *Belief, Attitude, Intention and Behavior : An Introduction to Theory and Research*, Addison-Wesley, Reading, Mass., 1975.
11. R. Shepherd, *Food Acceptability*, eds. Thomson D.M.H., Uni. of Reading, UK, Elsevier Applied Science, 1988, 253.
12. R. Shepherd and P. Sparks, *Measurement of Food Preferences*, eds. H.J.H. MacFie and D.M.H. Thomson Blackie Academic and Professional, 1994, 202.
13. R. Shepherd, *Measurements of consumer attitudes : EU Project AIR-CAT meeting*, As-Trykk, Norway, 1995, **1** (1), 11.
14. R.P. Bagozzi, *Recherche et Applications en Marketing*, 1989, IV, **2**/89, 61.
15. P.L. Wilkie, *Consumer Behavior*, New York, John Wiley & Sons, 1990, chap. 11.
16. R. I. Stein and C.J. Nemeroff, *Personality and Social Psychology Bulletin*, 1995, **21** (5), 480-490 (error report of the original article In *PSPB*, **22** (2), pp. 222).
17. N.E. Beckwith and O.R Lehmann, *Journal of Marketing Research*, 1975, **12**, 265.
18. J-M. Aurifeille, *Recherche et Applications en Marketing*, 1991, VI, **4**, 59.
19. D. Thiel and Ph. Robert-Demontrond, *Food Quality and Preference*, 1997, **8**, 5/6, 429.
20. W.S. Mc Culloch and W. Pitts, *Bulletin of mathematical biophysics*, 1943, **5**, 115.
21. K. Grünert, *Journal of Marketing*, 1996, **60**, 88.

Session III Sustainable Production

INTEGRATED AND ORGANIC PRODUCTION, ROUTES TO SUSTAINABLE QUALITY PRODUCTION

Peter Esbjerg

Department of Ecology and Molecular Biology
The Royal Veterinary and Agricultural University
Thorvaldsensvej 40
DK-1871 Frederiksberg C, Denmark

1. INTRODUCTION

While not many years ago quality of fresh market products was primarily a matter of appearance and freshness the picture is now becoming more complicated and sometimes even contradictory. Many comsumers are increasingly concerned about health aspects of the food products eg. whether fruit and vegetables are free of pesticide residues or at least below a certain minimum. On top of this environmental influence of the particuar production methodology is a sign of quality attracting increasing interest. The reason is, that more and more people realize that the primary production may pose a threat to natural life particularly at high production intensity based on intensive use of fertilizers and pesticides.

Though it is difficult to precisely define sustainability, it is relatively easy to separate certain production ways in terms of sustainability or the lack of it. If a production of fruit or vegetables implies twice as many chemical treatments as another production method and both lead to equivalent net income, then the second method is likely to be the more sustainable because of less ongoing pressure on non target organisms as well as reduced risk of polluting soil and water resources.

When mentioned as above sustainability refers to ecological sustainability which will be the sustainabilty focus of this presentation. Social and particularly economic sustainability may be locally relevant in developing countries. It is, however, becoming increasingly clear that economic sustainability of modern fruit and vegetable production will depend on ecological sustainabilty. This aspect is not only due to the enviromental interests but also to the fact that meeting the health aspects eg. connected to pesticides goes hand in hand with improving sustainability.

2. INTEGRATED AND ORGANIC PRODUCTION AS RELATED TO SUSTAINABILITY AND FRUIT AND VEGETABLE PRODUCTION

Long before the sustainability terminology had emerged organic farming had introduced a concept of sustainability by focussing on cycling of matter in accordance with the cycling in natural ecosystems. Later eg. the demonstated leaching of nitrogen and pesticides to groundwater as well as other side effects show that the prohibition of easily soluble nutrients and pesticides has in this respect brought organic production to the better side of sustainability.

Also Integrated Pest Management (IPM), which is the the first step towards Integrated Production / Farming (IP / IF), had brought in elements of sustainability before this terminology appeared. This is reflected by the definition of Integrated Control (the European version of IPM). Integrated control is a control of pests and diseases employing all methods consistent with economical, ecological and toxicological requirements, while giving priority to natural limiting factors and economic damage thresholds.[1] While organic production has from the beginning included both common field crops and high value crops as fruit and vegetables, IP has mostly been introduced within the high value crops. This has among others been supported by inputs from working groups (WG's) of IOBC/WPRS, The International Organization for Biological and Integrated Control of noxious insects and plants, Western Palaearctic Region Section. Furthermore this organization has produced general guidelines to IP (Commission IP-Guidelines, 1993) and more crop specific IP-guidelines eg. for apple production.[2,3]

2.1 From IPM to IP

Though environmental concern was part of the creation of the original IPM concept it has to be remembered that the other focal part was the need identified in USA for economical improvement of insect control in particular.[8] IPM was meant to become the manual for farmers, realizing that the contradiction was not as big as often believed.

In Europe early IPM interests focussed on fruit and vegetables because of the high treatment frequency and the importance of residues in products meant for fresh consumption. Carrots in Denmark may serve as an example of what is the result of a model IPM project[6] supplemented with results from several IOBC/WPRS WG's. For the project which started in 1978 carrot was selected as a model crop for the above mentioned reasons, the treatment frequency against the two major insect pests (larvae of carrot fly, Psila rosae, (CF) and cutworms, Agrotis segetum, (AS) being in the range of 6-14 per crop.

At present the following IPM components are recommended/demanded as part of IP:

4 year crop rotation (an old recommendation against CF)
Avoid carrots close to hedges and similar lee (an old recommendation, project)
Varieties with some resistance to CF (project and WG on insect resistant vegetables)
Late sowing to avoid host recognition by generation of CF (project)
Monitoring of CF with yellow sticky traps and use of "thumb rule control thresholds" (project and WG on Field Vegetables)
Monitoring of AS moths with pheromone traps and use of dynamic control thresholds (project, WG on pheromones, [5])
Control of AS larvae with synthetic pyrethroid, (virus) or by means of irrigation (project)
Chemical control of CF larvae (in some countries no recommended chemical is permitted)
Early harvesting and marketing of slightly infested areas before CF larvae start mining and starting of harvesting from edges (project)

Already at the end of the project the decrease in treatment frequency was in the range of 50-60%[5,7] and especially in the case of cutworm attack the level of injury after treatment was strongly reduced due to improved timing of treatments.

As indicated by the example of carrots it is generally accepted that IPM is related to crop level in space and not season in time. This may, however leave out the originally intended preventive effects of crop rotation (cf. carrot example) or of none crop elements to promote natural enemies eg. in fruit orchards. This problem was recognized early by founders of IP, among others Dr. Hans Steiner who also paid much attention to the need of preserving and promoting natural enemies in perennial non-crop rotation systems as eg. apple orchards and other fruit plantations. These ideas have been further developed and brought into a managerial frame by the IOBC/WPRS Commission on IP-Guidelines. From the section, Definitions and

Objectives, of the basic document[3] the following extracts of particular interest for the present area may be cited:

INTEGRATED PRODUCTION IS A FARMING SYSTEM WHICH:
INTEGRATES NATURAL RESOURCES AND REGULATION MECHANISMS INTO FARMING ACTIVITIES TO ACHIEVE MAXIMUM REPLACEMENT OF OFF-FARM INPUTS
These objectives address the basic intentions of a sustainable agriculture re
IP IS APPLIED HOLISTICALLY
IP is not a mere combination of IPM with additional elements such as fertilizers and agronomic measures to enhance their effectiveness. On the contrary it relies on ecosystem regulation
THE ENTIRE FARM IS THE UNIT OF IP IMPLEMENTATION
IP is a system approach focussing on the entire farm as the basic unit, IP practised on isolated individual areas of the farm is not compatible with the holistic approach
STABLE AGROECOSYSTEMS ARE TO BE MAINTAINED AS KEY COMPONENTS OF IP
NUTRIENT CYCLES ARE TO BE BALANCED AND LOSSES MINIMISED
IPM IS THE BASIS FOR DECISION MAKING IN CROP PROTECTION
Decisions about the necessity of control measures must rely on the most advanced tools such as prognostic methods and scientifically verified threshold aspects.
BIOLOGICAL DIVERSITY IS TO BE SUPPORTED
PRODUCT QUALITY MUST BE EVALUATED BY ECOLOGICAL   PARAMETERS OF THE PRODUCTION SYSTEMS AS WELL AS BY THE   USUAL EXTERNAL AND INTERNAL QUALITY PARAMETERS
Commodities produced under strict IP regulation do not only exhibit measurable external and intrinsic parameters but also meet the requirements of the ecological evaluation for the production processes. Hence a certification testifying to the achievements of the producers is a prerequisite for the IP-label

The last part cited concerning quality and labelling represents the most interesting change in the development from IPM to IP. With this it is clear that from the IPM concept being directed to farmers attitude to the use of control/management a leap has been made into a concept with a strong element of marketing hopefully appealing to consumers via special labelling.

2.2 Quality of products and of special labels

In case of labels as guarantee for quality of products the crucial question is how sharply defined the promise of quality is and to which degree it is controlled. In the case of fruit and vegetables the conventional products do not promise anything in particular. The products compete mainly on price and appearance and control may be according to EU and national standards on grades and acceptable level of residues both being checked by sampling. In contrast to this organically produced products are often sold in separate shops or under special labels. A convenient example may be the Danish Ø-label, which is only for registered organic producers being subjected to governmental inspection down to farm level. The main guarantee in the case of plant products is that no easily soluble fertilizers and/or pesticides have been used in the production. As this promise is clear and the control fairly easy (yes/no), the labelling signals a more healthy product and also a responsibility to the environment.
IP falls in between conventional and organic production. While the fertilizer inputs have

not been paid too much attention there are many examples in particular from fruit and vegetable research where, as in the case of the earlier mentioned carrot-example[6,7] the pesticide treatments have been reduced by 50% or more. However, it is questionable, at least to the critical consumer, whether this applies to practical farming. Neither is it clear to the consumer what the quality means. The label on the particular packet guarantees improved sustainability on average but in terms of personal health aspects he/she might have got a bad unit. To some degree this problem of IP relates to the few rigid demands (like 4 year crop rotation in annual crops and use of locally adapted cultivars in fruit production, cf. Strict Rule or Prohibition[3], and the many recommendations which are not easily controlled.
At the level of a particular vegetable crop it may also be a problem for the research side to deliver solid and precise enough guidance (e.g. as concerns control thresholds). For the growers it may be difficult to make the change to IP, among others because of too limited land resources. This problem is very relevant e.g. in the case of cabbage growing which is often highly specialized and intensive on very small pieces of land.

Looking at the above problem from the practical side shows that relatively few organizations and companies have approached IOBC/WPRS to obtain an approved label. However, many have utilized both the general IP-guidelines[3] and the crop specific guidelines (eg. for apples[3]) as background for developing their own labels. Such IP products are marketed in a very visible way in several European countries, while eg. in Denmark they are, despite govermental recognition and control, mostly placed among conventional products in contrast to Ø-labelled products being shelved separately and very visible in the majority of supermarkets.

2.3 Discussion

Following the above the key question is what the future of IP- and organically produced fruit and vegetables will be. In this sensitive area of high value products it appears that ORG products have most to offer. Though it has to be remembered how large a share of the production is still conventional and how much the average quality, particularly in terms of sustainability, may be improved if IP became the larger part. The practical relevance of this may also be seen in the light of many growers being capable of a gradual change to IP, while Organic production may at the beginning be a kind of production with no economical "safety net" because of yet lacking training and no control method to limit the sometimes severe infestations which may occur in the transitional phase. After some years of experience growers with the necessary capacity are likely to go further to Organic production which in terms of farm practise may be regarded the ultimate of IP but which in its dogmatic form offers possibility of special labelling and better prices accordingly. The whole change towards IP and further to Organic production may be promoted through research backing and development of sharper IP labelling standards with more consumer appeal. Most obviously research is especially needed within :
- Breeding for resistance towards diseases in fruit trees.
- Development of proper and operational control thresholds (based on economic damage thresholds / Economic Injury Levels).
- Development of more biological control methods.
- Establishment of a sharper distinction between "normal" pesticides and pesticides which may be permitted in IP.

As regards the development of better over-all standards of the best proposal for a central body appears to be EU, which will have the possibility of orienting its subsidizing towards environmental friendliness so the amount given per hectare to the grower will increase according to limitations in pesticide and fertilizer use (and maybe also with reduction of tillage and energy consumption in the long term). A rigid but consumer understandable procedure would be the use of pesticides only according to prescriptions by authorized but independant IP-advisors. With such limitations also fertilizer use is likely to drop to a lower level, particularly in vegetable production.

In conclusion the combination of IP and Organic production has a lot to offer as regards quality including sustainability. Organic products obviously have most to offer to the individual producer but IP has presumably most to offer in terms of over-all sustainability simply because of the much larger area which may be rapidly involved. Should this later on push further towards Organic production even more will be achieved in terms of quality.

References

1. L. Brader, *5th IOBC/WPRS Symp.*, 1975, 9-16.

2. J.V. Cross and E. Dickler, *IOBC/WPRS Bull.*, 1994, XVII,9.

3. A. El Titi, E.F. Boller and J.P. Gendrier, *IOBC/WPRS Bull.*,1993, XVI,1: 1-96.

4. P. Esbjerg, *Proc. 4. Danish Plant Prot. Conf.*,1987: 187-198.

5. P. Esbjerg, *Grøn Viden, Havebrug,* 1996, 96: 1-8.

6. P. Esbjerg, J. Jørgensen, J.K. Nielsen, H, Philipsen, O. Zethner and L. Øgaard, *Tidsskr. Planteavl,* 1983, 87: 303-355.

7. P. Esbjerg, H. Philipsen and A. Percy-Smith, *IOBC/WPRS Bull.*, 1988, XI,1: 14-25.

8. R.B. Norgaard, *Annual Rev. Entomology,* 1976, 21: 45-60.

LOW NITROGEN FERTILIZATION MAINTAINS THE INTERNAL QUALITY OF SWEDES

A-M Evers[1], E. Ketoja[2], M. Hägg[3], S. Plaami[3], U. Häkkinen[3] and R. Pessala[4]

[1]Department of Plant Production, P. O. Box 27, 00014 University of Helsinki, Finland
Agricultural Research Centre of Finland (MTT)
[2] Data and Information Services, 31600 Jokioinen, Finland
[3] Laboratory of Chemistry, 31600 Jokioinen, Finland
[4] Plant Production Research, Horticulture, 21500 Piikkiö, Finland

1 INTRODUCTION

The swede (*Brassica napus* L.) is one of the traditionally cultivated plants in Northern Europe. Environmental problems have forced agricultural and horticultural producers and researchers to critically evaluate production systems and their effects on crop quality and the environment. The Agricultural Research Centre of Finland in 1993 began a research programme to create instructions for new, modified production systems. This study is part of this research programme.

The purpose of this study was to determine whether an old variety, Simo, can be grown with low nitrogen rates and still maintain good internal quality, and simultaneously if nitrogen efficiency could be improved with sufficient irrigation without reducing crop quantity. Most previous internal quality studies with horticultural crops have focused on optimal and high nitrogen levels[1, 2, 3] whereas we focused on low nitrogen levels. The present recommendation for modern swede varieties (sandy clay, preplant clover) is 90 kg N/ha and we used 90, 60 and 30 kg N/ha.The quantity and internal quality aspects studied were growth (total yield and dry matter content), compounds closely related to carbohydrate metabolism (sugars, dietary fibre, vitamin C), nitrate, and 1994 vitamins (thiamine and riboflavine).

2 MATERIALS AND METHODS

The field experiment was conducted at the unit of Plant Production Research of MTT in Piikkiö, Southwest Finland (60°23'N, 22°33'E) in 1993 and 1994. In both years the trial was set up according to a standard split-plot design where the main-plot treatments, irrigation / no irrigation, were in a randomized complete block design with four blocks and the split-plot treatments, three nitrogen levels, were randomized within each main plot. The Simo swede was sown in a greenhouse at the begining of May and planted outside on June 8, 1993 and May 30-31, 1994.

The experiments were on sandy clay with good P, K, Ca and Mg contents, pH 7.3 and the preplant in 1993 red clover and 1994 pH 6.5 and preplant red clover - timothy grass. Before planting all the plots received placement fertilization with granular compound PK-fertilizer (2N, 7P, 17K, with trace elements) 400 kg/ha. To ensure the nitrogen levels ammonium nitrate (27.5N) as 30-60-90 kg N/ha was spread on the plots by hand and cultivated in the topsoil by harrow. All the plots were irrigated to 20 mm before planting to guarantee rooting and the start of growth. Irrigations for irrigated treatments were performed in 1993 in the second and third week of July at 30 mm each, thereafter no irrigation was done because of heavy rains at the end of July and August. The irrigations in 1994 were performed during the last two weeks of July and the second week of August at 24mm. The need for irrigation was evaluated with a soil moisture tensiometer (Model 2710, Soilmoisture Equipment Corporation, Santa Barbara, CA) and irrigation was started when soil water potential had decreased to -1 bar. The harvest and the samples for internal quality analyses were taken on the 22-24th August. Plant protection measures were kept at a minimum. The seeds were treated with biological Streptomyces fungicide. Weeds were sprayed before planting with trifluralin, the rest of the weeds were weeded out manually. The insects on the plot were sprayed with pyrethrum, mevinfos and malation, and mercaptodimetur against slugs was used once. Plant analysis are reported in (4).

2.1 Statistical methods

On each subplot several response variables were observed representing growth, substances closely related to carbohydrate metabolism, nitrate, and vitamins. Typically, the responses of the same quantity and quality aspects measured on the same plot were intercorrelated. Therefore, they were analysed by means of the multivariate analysis of variance (MANOVA) for the split-plot design, which takes the intercorrelations into account by examining the effects of the treatments on all relevant response variables simultaneously. If the multivariate tests were statistically significant the corresponding univariate analyses were examined in order to locate those response variables which contributed most to the results of the multivariate tests.

3 RESULTS

The study of the weather conditions showed that irrigation was relevant only in 1994 due to heavy rains at the end of July and in August 1993. Therefore, data for 1993 and 1994 were modelled separately. Checking of the models by diagnostic methods did not indicate any cross departures from the assumptions underlying the models.

3.1 Growth

The responses, yield and dry matter content, were used to represent the crop growth. In 1993 according to MANOVA swede growth was affected by nitrogen level ($F_{4,22}$=9.57, P < 0.001) and this effect was not dependent on irrigation ($F_{4,22}$=2.10, P=0.12 for irrigation x nitrogen interaction). Furthermore, the effect of irrigation per se was not statistically significant ($F_{2,2}$=0.73, P=0.58). Judging by the univariate analyses, both the swede yield and the dry matter content contributed to the significant nitrogen level effect. At the nitrogen levels 30, 60 and 90 kg/ha the mean yields were 408 (SEM 21.7), 499 (SEM 23.0) and 577 kg/100m^2 (SEM 21.7 kg/100m^2), respectively, and the mean dry matter contents were 11.3, 11.2 and 10.5% (SEM 0.18 %), respectively. When these mean profiles of yield and dry matter content over the levels of nitrogen were modelled by first and second-degree orthogonal polynomials for both of them the first-degree polynomial (linear trend) was statistically significant (P < 0.01).

In 1994 swede growth was affected by both factors, nitrogen ($F_{4,22}$=6.42, P < 0.005) and irrigation ($F_{2,2}$=77.52, P=0.01) but the multivariate test gave no evidence of interaction between them ($F_{4,22}$=1.11, P=0.38). On grounds of univariate analyses irrigation increased the total yield and decreased the dry matter content. The mean yield increased from 483 to 608 kg/100m^2 (SEM 30.5), i.e. 125 kg/100m^2 (95% CI: +10, +240) and the mean dry matter content decreased from 13 to 12% (SEM 0.18%, 95% CI for the difference +0.4, +1.6). Since the 95% confidence intervals did not include zero, they imply that the observed differences in yield and dry matter content were statistically significant at the 5% level. The mean total yields over the three levels of nitrogen were 496, 583 and 557 kg/100m^2 (SEM 30.3 kg/100m^2) and the corresponding mean dry matter contents were 13.3, 12.3 and 11.9 % (SEM 0.20%). For total yield the quadratic trend of nitrogen was statistically significant (P=0.05) and for dry matter content the decreasing linear trend (P < 0.001).

3.2 Substances closely related to carbohydrate metabolism

In 1993, according to MANOVA, both the irrigation ($F_{3,1}$=1107.67, P=0.02) and the nitrogen level ($F_{6,20}$=5.18, P < 0.005) effected substances closely related to carbohydrate metabolism (vitamin C, dietary fibre, total sugar), but the interaction between the factors was not statistically significant ($F_{6,20}$=1.51, P=0.22). On the basis of the univariate results the difference between the irrigated and unirrigated group seemed to originate mainly from the total sugar content, which increased from 5.1 to 5.4% (SEM 0.06%), i.e. by 0.3% (95% CI: +0.1, + 0.6) due to irrigation. With irrigation the means for glucose, fructose, and sucrose were 2.73%, 1.79%, and 0.87%, respectively, and without irrigation 2.56%, 1.62%, and 0.91% (SEM 0.03%, 0.03%, 0.04%), respectively. Consequently, the univariate results of the total sugar content reflect the increasing effect of irrigation on glucose and fructose, as the mean sucrose content was approximately the same in both irrigated and unirrigated groups. The increase of nitrogen level from 30 kg/ha to 90 kg/ha also increased the total sugar content in a linear fashion, but decreased dietary fibre and vitamin C. Of the different sugars glucose accounted for most of the significant nitrogen effect on the total sugar content.

In 1994, irrigation effected compounds closely related to carbohydrate metabolism through interaction with the nitrogen level ($F_{6,20}$=7.92, P < 0.001 in MANOVA). Inspection of the univariate results showed that this was mainly due to the total sugar content, whose mean profile over the nitrogen levels with irrigation lay lower than without irrigation and had a different curvature (P < 0.005 for equality of the quadratic trends), the phenomenon was the same in the mean profiles of all sugars (Figure 1a and 1b). When averaging the total sugar content over all of the nitrogen levels the difference between the means (4.5 v. 3.9%, SEM 0.10%) of the unirrigated and irrigated group was 0.6% (95% CI: +0.3, +1.0). In addition, the quadratic trends of the mean profiles of the dietary fibre content with and without irrigation tended to differ (P=0.03). The increase in nitrogen level decreased the mean dietary fibre both with and without irrigation, but with irrigation the biggest decrease occurred between 30 and 60 kg/ha (means 3.4, 3.1 and 3.0%, SEM 0.08%), whereas without irrigation it occurred between 60 and 90 kg/ha (means 3.3, 3.2 and 3.0%, SEM 0.08%). The mean difference over the nitrogen levels between the irrigated and unirrigated group was 0.02% (95% CI: -0.17, +0.21). The increase of nitrogen level caused a decreasing trend in vitamin C, which was quadratic in shape (P=0.04) and parallel in irrigated and unirrigated groups (P=0.19 for equality of the quadratic trends). The means of the vitamin C content over the nitrogen levels were 42.0, 38.1, 38.0% (SEM 0.78%) and

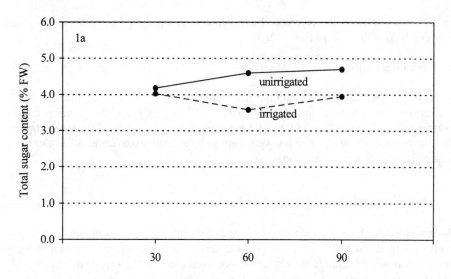

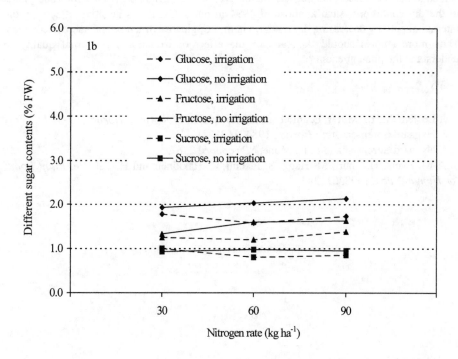

Figure 1. Irrigation-nitrogen interaction on swede total sugar (1a), and glucose, fructose and sucrose contents (1b) in 1994.

the difference between the unirrigated and irrigated group means (40.9 v. 37.9%, SEM 0.78%) was 3% (95% CI: +0.5, + 5.5).

3.3 Vitamins and nitrate

Irrigation and nitrogen rates did not have an effect on thiamine or riboflavine. All 24 riboflavine content measurements were 0.04 mg/100 g. Of the 24 thiamine content measurements 22 were either 0.06 or 0.07 mg/100g irrespective of the different treatments. The nitrate contents of the swedes were very low, in both years under 30 mg/kg fresh weight irrespective of the treatments.

4 DISCUSSION

The lowest nitrogen rates were found to give a slight advantage to swede internal quality, and the combination of 30 kg/ha nitrogen with irrigation gave a better yield than 90 kg/ha nitrogen without irrigation in the dry 1994 conditions (566 vs 485 kg/100 m², SEM 39.4). However, in the rainy 1993 conditions, the best yield level was achieved only by the highest nitrogen rate of 90 kg/ha. In both years the internal quality was slightly improved by the lowest nitrogen rates which increased swede dry matter, vitamin C, and dietary fibre contents, but decreased total sugar content. Even though irrigation is vital for good yields in the dry conditions, such as those of 1994 in our study, it has negative effects on the internal quality by decreasing dry matter, vitamin C, glucose, fructose and sucrose contents. Thus more studies should be made of the effect of irrigation on internal quality to understand the phenomenon.

References

1. J. Leclerc, M. L. Miller, E. Joliet and G. Rocquelin, *Biol. Agric. Hortic. ,*1991, **7**, 339.
2. J. Nygaard Sørensen, *Acta Hortic.*, 1984, **163**, 221
3. J. Nygaard Sørensen and L. H. Mune, *SP rapport*, **2**, 67.
4. A-M. Evers, H. Tuuri, M. Hägg, S. Plaami, U. Häkkinen and H. Talvitie, *Plant Foods for Human Nutr.*, 1998, **51**, 283

N UPTAKE BY CABBAGE, CARROT AND ONION

T. Salo
Crops and Soil
Plant Production Research, Agricultural Research Centre, FIN-31600 Jokioinen, Finland

1 INTRODUCTION

In sustainable production it is essential to adjust N fertilization according to N uptake by the crop[1]. In addition, the N available in crop residues must be considered as part of the N supply for the next crop[2]. The rate of N uptake of vegetables still warrants further study[3].

2 FIELD EXPERIMENTS

Field experiments on cabbage, carrot and onion to determine the amount and rate of N uptake at different yield levels were carried out during three years (Table 1). The plants were sampled four times during the growing season, and plant N concentration was determined by the macro-Kjeldahl method. Soil mineral N was also determined after harvest.

Table 1 *Experimental details.*

Experiment	N rate kg ha^{-1}	Plant density plants ha^{-1}	Planting date	Harvest
Cabbage 1993	0, 125,188, 250	67 000	25.05	07.09
Cabbage 1994	0, 80, 120, 160	44 000	01.06	07.09
Cabbage 1995	0, 160	50 000	16.06	03.10
Carrot 1993	0, 30, 70, 100	730 000	04.05	01.10
Carrot 1994	0, 30, 70, 100	785 000	06.05	30.09
Carrot 1995	0, 70	290 000	10.05	06.10
Onion 1993	0, 30, 70, 100	356 000	11.05	17.08
Onion 1994	0, 30, 70, 100	356 000	10.05	23.08
Onion 1995	0, 70	356 000	30.05	29.08

3 RESULTS

High yields, 80 t/ha for cabbage, 90 t/ha for carrot and 35-40 t/ha for onion, were obtained when the total crop N uptake was 300 kg/ha, 150 kg/ha and 120 kg/ha, respectively (Figure 1). Variation in yield and N uptake was highest with onion, whereas for carrot they were rather uniform each year. In cabbage almost 50% of total N was in crop residues, whereas in carrot and onion only about 30% of total N was in crop residues.

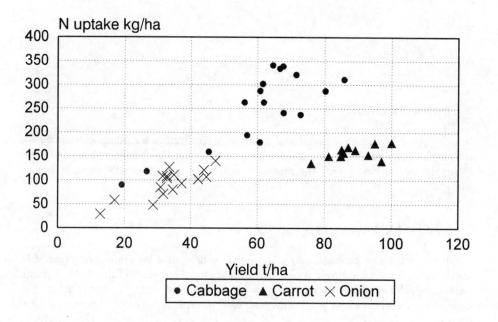

Figure 1 *Yield vs. total N at harvest in 1993-1995.*

N uptake from non-fertilized soil varied from 29 to 160 kg/ha, depending on the growing season and the crop. Cabbage and carrot efficiently utilised soil N,   usually taking up more than 100 kg/ha/year from non-fertilized soil. Onion, on the contrary, utilised relatively poorly soil N, usually less than 50 kg/ha/year from non-fertilized soil.

The rate of N uptake was low with all crops in early summer (Figure 2). After one month, N uptake increased in cabbage and onion. This uptake continued until harvest, i.e. mid-August for onion and early September for cabbage. N uptake by carrot started rapidly just two months after sowing, but continued until harvest at the end of September.

After harvest soil mineral N content was generally low, i.e. below 25 kg/ha at the depth of 0-60 cm. Onion was an exception with poor growth in 1994, when soil mineral N after the highest N rate was 68 kg/ha at the depth of 0-60 cm after harvest.

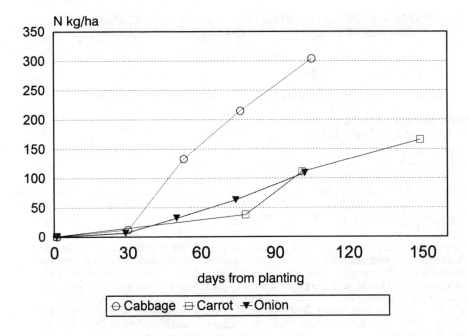

Figure 2 *Rate of N uptake with cabbage, carrot and onion averaged in 1993-1995.*

The vegetables differed widely in their requirement of N, and thus in their potential to cause losses of N. Especially the N requirement of cabbage was high and the crop residues contained large amounts of N. Thus cabbage requires careful management to keep N losses low. When results obtained in these experiments are compared with the N fertilizer recommendations applied in Finland[4], it can be concluded that the recommendations correspond to the actual N uptakes measured.

References

1. P.E.Bacon, 'Nitrogen Fertilization in the Environment', Marcel Dekker Inc., NewYork,1995, p. 295.

2. C.R.Rahn, L.V.Vaidyanathan and C.D.Paterson, *Aspects Appl. Biol.* 1992, 30, 263.

3. C.Gysi, *Acta Hort.* 1996, 428, 253.

4. Soil Testing Laboratory of Finland, 'Soil analysis and application in horticulture' (in Finnish), Mikkeli, 1997, p. 9.

BRASSICA VEGETABLES: RELATIONSHIP BETWEEN CHEMICAL COMPOSITION AND *IN VITRO* CALCIUM AVAILABILITY.

M. Lucarini, R. Canali, M. Cappelloni, G. Di Lullo and G.Lombardi-Boccia

Istituto Nazionale della Nutrizione
Via Ardeatina 546, Roma - Italy

1 INTRODUCTION

Calcium absorption from vegetables is generally considered low because they contain substances (phytate, oxalate, dietary fibre components) which bind calcium in unabsorbable compounds. Studies have clearly shown that foods with high concentrations of oxalic acid and phytic acid strongly reduce calcium availability (1,2,3). Dietary fibre is another food constituent interacting with minerals with consequence on their absorption. Among dietary fibre components, uronic acids have been demonstrated to bind calcium (4). Most literature data on calcium availability derives from *in vivo* studies, little work has been done to evaluate *in vitro* the availability of calcium especially from vegetable foods (3,5).

In this study measures of dialysable (D), soluble (S), ionic dialysable (ID) and ionic soluble (IS) calcium were used as indicators of the potential calcium availability. Because the availability of minerals from single food or when included in a meal can vary greatly, we also studied the effect of including *brassica* vegetables in composite dishes prepared following traditional italian recipes (macaroni and broccoli, macaroni and cauliflower).

2 MATERIALS AND METHODS

Brassica vegetables (broccoli, cauliflower, cabbage, kale) and macaroni were purchased locally. Vegetables were pressure cooked for 8min in deionized water (150ml). Composite dishes were prepared by boiling 80g of macaroni in 1lt of deionized water with 2g NaCl, macaroni were then strained and mixed with cooked broccoli or cauliflower (110g, raw weight). Samples were freeze-dried before subsequent analysis.

Calcium analysis was performed by Atomic Absorption Spectrometry on a Varian SpectrAA 400. Dietary fibre and uronic acids were determined by the method of Englyst & Cummings (6). Organic acids were extracted in mild acidic conditions and analysed by HPLC. Dialysable calcium (D) was assessed by using the *in vitro* method of Miller et al. (7). Soluble calcium (S) was determined in the retentates after *in vitro* digestion. Ionic-dialysable calcium (ID) and ionic-soluble calcium (IS) were determined by a calcium selective electrode (Model 93-20, Orion, Boston, MA). Calcium content of dialysates and retentates was expressed as percentage of the total calcium.

3 RESULTS AND DISCUSSION

All the selected vegetables are naturally rich in calcium and organic acids (Tab.1). The amounts of organic acids of composite dishes were strictly dependent on the vegetable ingredients. *Brassica* vegetables were similar in dietary fibre content and composition. All of them contained high amounts of uronic acids in the soluble fibre fraction, representing about 50% of this fraction. Total dietary fibre content of composite dishes was slightly higher than that of *brassica* vegetables, reflecting a higher content of the soluble fibre fraction (data not shown). Nevertheless the uronic acids content of the soluble fraction was approximately half the amount determined in *brassica* vegetables, representing only 24.4% of this fraction. This was dependent on the contribution of macaroni to the dietary fibre composition of the composite dishes.

Table 1 *Calcium, organic acids, dietary fibre and uronic acids content in cooked brassica vegetables and composite dishes (fresh weight)*

Food source	Total Ca mg/100g	Citric ac. mg/100g	Malic ac. mg/100g	D.fibre g/100g	Uronic ac* g/100g
Vegetables					
Broccoli	35.3±2.1	15.0±2.8	65.5±10.1	2.3±0.4	0.58±0.2
Cauliflower	20.6±1.8	64.0±4.2	74.0±9.8	2.4±0.1	0.60±0.1
Cabbage, green	22.8±1.9	49.0±8.0	68.0±10.8	2.5±0.8	0.59±0.1
Kale	34.2±2.2	21.1±1.9	41.0±2.7	2.3±0.1	0.45±0.2
Composite dishes					
Macaroni and broccoli	37.5±2.1	7.8±3.2	34.2±7.4	2.6±0.1	0.33±0.04
Macaroni and caulifl.	32.5±1.8	41.9±6.4	38.2±9.0	2.6±0.1	0.33±0.01

Each value represents the Mean±SD of triplicates.

*Soluble fibre fraction

In *brassica* vegetables approximately 25% of the total calcium was dialysable (D) and only about 7% of it was in ionic form (ID) (Tab. 2). This finding indicated that most of the calcium released during the *in vitro* digestion was bound to low molecular weight compounds. The similar concentration found for both bound (D and S) and ionic (ID and IS) calcium in both side of dialysis membrane indicated that during the *in vitro* digestion the dialysis equilibrium was reached.

The addition of cereal products to vegetables (composite dishes) lowered significantly ($p<0.001$) only calcium dialysability (D) compared to broccoli and cauliflower (Tab. 2). Unlike vegetables, in composite dishes soluble calcium (S) was significantly higher ($p<0.001$) than dialysable calcium (D). This finding indicated that not all soluble complexes containing calcium were dialysable, and that calcium was likely bound to soluble complexes of molecular weight higher than the cut-off of the dialysis membrane. Same finding was reported in a previous *in vitro* study (3) dealing with calcium dialysability from beans. This was probably due to differences in the chemical forms of calcium between seeds (legume, cereal) and green vegetables.

Brassica vegetables are essentially phytate- and oxalate-free vegetables, therefore dietary fibre components and organic acids are the constituents that could mainly influence calcium availability. James et al. (4) showed that the binding by the non-cellulosic fraction of dietary fibre will reduce the absorption of calcium from small-intestine. In *in vitro* experiments, in which pectin degradation does not occur, we expected a negative effect on calcium availability. This did not occur. In *brassica* vegetables therefore the uronic acids content, though high, seems not to impair calcium availability. Probably a high percentage of uronic acids of *brassica* vegetables are methyl-esterified; therefore the level of calcium dialysability found in this study might be the consequence that in these vegetables little calcium was bound to uronic acids. The presence in dialysates of higher amounts of bound calcium compared to free ionic calcium (Tab.2), lead to the idea that most of the calcium was bound to low molecular

Table 2 *In vitroDialysable (D), Ionic-Dialysable (ID), Soluble (S) and Ionic-Soluble (IS) calcium in cooked brassica vegetables and composite dishes (fresh basis).*

Food source	D Ca	ID Ca	S Ca	IS Ca
	%	%	%	%
Vegetables				
Broccoli	22.9±1.3	8.1±0.9	27.5±4.4	6.9±1.3
Cauliflower	23.4±1.6	8.0±2.2	27.1±6.1	8.8±2.5
Cabbage, green	24.8±0.8	5.7±1.2	28.5±1.3	4.7±0.5
Kale	28.9±1.4	6.9±1.1	39.7±7.3	10.9±3.2
Composite dishes				
Macaroni and broccoli	18.7±0.3	10.1±1.1	32.0±8.4	12.5±5.7
Macaroni and cauliflower	19.0±0.7	12.9±1.5	29.2±6.0	15.0±1.9

Each value represents the Mean±SD of triplicates.

weight complexes. Organic acids in *brassica* vegetables (Tab. 1) could likely bind calcium and, consequently, might be responsible for the high calcium dialysability showed by these vegetables. Composite dishes differed from vegetables essentially in the percentage of dialysable calcium (D). This difference in calcium dialysability was likely dependent on the presence in composite dishes of wheat constituents affecting the bioavailability of calcium. Among these phytate might be the most involved factor. Macaroni generally contains about 2.6 mg/g of phytate (8), a factor for which the negative effect on calcium bioavailability is well known (2,3). The levels of the potential calcium availability from *brassica* vegetables found in this study are lower than those reported for kale and broccoli *in vivo* studies (9-10). However the *in vitro* results clearly cannot account for the calcium released by the microbial fermentation; this absorbable calcium might be responsible for the observed discrepancies with *in vivo* results.

In conclusion our findings suggest that *brassica* vegetables can be regarded as a good source of available calcium. As well, the consumption of these vegetables together with foods of low calcium availability (eg.macaroni) contributes to maintaining the potential calcium availability at quite a good level.

References

1. C.M. Weaver, B.R. Martin, J.S. Ebner, C.A. Krueger *J. Nutr.* ,1987, **117**, 1903-6.
2. R.P. Heaney, C. Weaver & M.L. Fitsimmons *Am. J. Clin. Nutr.*, 1991, **53**, 745-7.
3. G. Lombardi-Boccia, M. Lucarini, G. Di Lullo, A. Ferrari, E. Del Puppo, E. Carnovale. *Food Chem,.* 1998, **61**:167-71
4. W.P.T. James, W.J. Branch & D.A.T. Southgate *The Lancet* , 1978, 638-9.
5. O. Rejkdal & K. Lee *J. Food Sci.* 1991, **56** (3): 864-6.
6. H.N. Englyst & J.H. Cummings *J. Assoc. Off. Analyt. Chem.*, 1988, **71**, 808-14.
7. D.D. Miller, B. Schricker & A. Cederblad *Am. J. Clin. Nutr.* ,1981, **34**, 2248-56
8. B.F. Harland & D. Oberleas *World Rev. Nutr. Diet.,* 1987, **52**: 235-59.
9. R.P. Heaney & C. Weaver *Am. J. Clin. Nutr.*, 1990, **51**, 656-7.
10. R.P. Heaney *J.Food Sci.* 1993, **59**(6), 1378-80.

IPM IS AN ESSENTIAL PART OF SUSTAINABLE PRODUCTION AND QUALITY MANAGEMENT OF FIELD VEGETABLES

K. Tiilikkala

Agricultural Research Centre of Finland (MTT)
FIN-31600 Jokioinen, Finland.

1 INTRODUCTION

Factors affecting the production of high-quality vegetables were studied in a research programme (VIVI) carried out by the Agricultural Research Centre of Finland (MTT) in 1993-1997. Integrated pest management (IPM) formed an essential part of the programme, as farmers and the vegetable processing industry regard IPM as an important factor in quality management. Another goal was to speed up the development of integrated farming practices (IP) and ecological farming systems in vegetable production. The examples presented in this paper illustrate the diversity of factors affecting food quality and the quality management of field vegetable production.

1.1 What is IPM?

IPM is an interdisciplinary approach to reducing crop losses and quality risks through the use of optimum combinations of pest control techniques. IPM encompasses the goals of agricultural productivity, environmental sustainability and cost effectiveness. IPM has arisen from a need to avoid the problems of pest resistance build-up (leading to pest resurgence), secondary pest outbreaks, adverse effect on human health, high cost of pesticide control and environmental degradation caused by excessive and inappropriate use of chemical pesticides. The approach aims at enabling farmers to make decisions on crop protection on the basis of valid information concerning their agro-ecosystems.

With its emphasis on making the best use of local and human resources, IPM encourages, wherever appropriate, the use of natural control mechanisms (for instance pest predators) and "traditional" pest management techniques. However, the adoption of practical alternatives to chemical methods of control may be constrained by lack of technical solutions, the lack of resources, or socio-economic and other factors. Where such constraints are severe, an optimal IPM control package may include selective chemical treatments used in combination with alternative nonchemical control techniques.

1.2 Why IPM ?

There is a great global need for greater food security, and effective pest control is required to enable sufficient food supply in many areas of the world. In LDC's IPM is not only a part of the quality management of food production but also a prerequisite for an acceptable quality of human life. The agribusiness is well aware that food quality can be impaired by pests: insects, plant pathogens, nematodes and many other organisms using food crops as their hosts and plant tissue as a source of energy and nutrients. A pest attack or infection can change the shape, colour and structure of vegetables, fruits and berries in a way that renders the final product unsaleable to consumers. Taste and food quality, determined by a vast number of natural chemicals or biochemicals can also be altered by pests and diseases. Pesticides play a special role in consumers' estimations of food quality. The most important is the fear of pesticide residues. Their second concern is the quality of the production system and the potential side effects of pesticides on nontarget organisms and the environment. Both of these quality risks can be minimised by means of IPM.

Farmers and the food industry face economic risks in terms of losses in yield and quality which must be solved if the economy of the production chains is to be sustained. Proper food quality through economically and ecologically sound farming hardly can be achieved without well-implemented IPM.

2 COMPONENTS OF IPM

On the farm level, the composition of IPM depends on the needs of quality management and the set of organisms relevant to the particular agroecosysstem. The key pests should be identified and the factors affecting their capacity to cause yield losses and quality risks should be predicted at the production planning phase. Within MTT's VIVI Programme many kinds of methods were developed and tested to help farmers improve their production systems and quality management through sophisticated IPM decision making.

2.1 Forecasting and warnings

Insect development and activity often depends on temperature, and thus records of accumulated heat units can be used to predict pest activity and migration to the fields. Modern information technology allows efficient collection and analysis of weather data as well as a rapid information exchange between research teams and end users of the information.

In the VIVI Programme activities of the carrot fly, *Psila rosae*, and the cabbage root fly, *Delia brassicae*, were forecast on the basis of meteorological data provided by the Finnish Meteorological Institute. Air temperatures at 35 automatic weather stations were analysed using GIS (geographical information system) tools, and the predicted activities of the pests were displayed as thematic maps on the AGRONET (*http://www.mtt.fi/ksl/ajankohtaista*). The forecasts were validated by means of pest monitoring data stored in the MTT database.

2.1.1 Carrot fly forecast. In the summer of 1997, the flight of overwintered adults of the bivoltine type of *P. rosae* (southern Finland) started during the 23rd week of the calendar year and peaked during calendar week 24. Preliminary validation of the forecasts suggested that the threshold temperature sum (effective temperature sum, ETS) should be 255 DD5 for the start of the flight and 355 DD5 for the peak. The forecast concerning the second flight was in line with the monitoring data. The flight started when 800 DD5 had accumulated and peaked at 860 DD5.

Forecasts of the activity of the univoltine type (northern Finland) of the carrot fly were precise enough to allow farmers to be informed about the need to initiate monitoring and control at the level of the individual field. In practice, the same ETS values can be used for predicting the activities of both types of the carrot fly.

2.1.2 Cabbage fly forecast. The forecasts concerning the first flight of the cabbage root fly, *D. radicum*, coincided relatively well with the first egg-laying period. In the forecast, 80 DD5 was used as the threshold ETS value for the start of the flight and 150 DD5 for the peak. Regarding the second flight, the forecast was about two weeks early compared with the monitoring data. The preliminary threshold ETS values (600 DD5 for the start and 750 DD5 for the peak) should be adjusted after final analysis of the monitoring data. The main problem in forecasting cabbage fly activity in Finland is the occurrence of *Delia floralis*. The flight activity of *D. floralis* normally peaks between the two peaks of *D. radicum*. In many regions in Finland, both species may occur in the same fields.

GIS was found to be a powerful tool for making and presenting forecasts and for analysing monitoring data. AGRONET/Internet services provided a rapid and facile means of delivering the information to farmers and extension services. GIS also has the potential to produce more sophisticated simulation models for forecasting pest activity, provided that relevant weather data are available at a reasonable price.

2.1.3 Use of insect radar. Insect pests can sometimes migrate thousands of kilometres in one day, and farmers should also be aware of such risks. In the VIVI Programme use of insect radar was tested in 1995 and 1997 when the diamondback moth (DBM), *Plutella xylostella,* migrated to Finland. During the last week of May in 1995, fields in Finland were filled by actively moving moths. Catches in yellow sticky traps showed that the moths seen flying everywhere were DBM adults. It was subsequently found that a suction trap catch from the same week in Helsinki included DBMs. An analysis of entomological radar information revealed that an enormous swarm of insects had been driven by southern wind from Estonia to Finland on May 26. It being early in the year, this migration was very exceptional. Yield losses caused by the DBM were severe in Finland as well as in the northern parts of Sweden. The first generation of DBM developed on cruciferous weeds and the second generation on cultivated crucifers. A total of four generations was possible during the growing season, which is very rare in Finland. In 1997 DBM migrated to Finland on June 9. This time the radar information from the University of Helsinki was disseminated through MTT's plant protection service to farmers so rapidly that yield losses and quality problems could be avoided with biological control methods.

2.2 Monitoring and threshold values

Field-specific monitoring of pests is one of the most important aspects of IPM because population densities can vary widely among different fields. Risks to yields or quality also depend on the variety grown, and thus any control method should be based on field-specific monitoring of pests. In practice, it may not be easy to spot small insects, and it may be even more difficult to count the number of the key species or their natural enemies.

2.2.1 Use of traps. One goal of the VIVI Programme was to develop monitoring methods and practices for vegetable farmers with a view to the use insecticides without unexpected yield losses or quality problems. One of the test organisms was the carrot psyllid, *Trioza apicalis,* which is the most dangerous insect pest of carrots in Finland. Population densities of the psyllid vary substantially annually and among different cropping areas. Development of a monitoring system has been essential because of the variable need for control. Catch-it^R yellow sticky traps were used for trapping the psyllid. Threshold values were assessed in a field experiment established in 1994 and 1995. In the experiment, carrot plots were covered with an insect net for 13 different time periods. Calculation of threshold values was based on counting weekly trapped psyllids and correlating these with carrot yields during the netting experiments. Results from both years showed distinct yield losses during any experimental week when the average value of 1 psyllid/trap/week was exceeded. In both years, the population density of the psyllid remained above the threshold for about 3-4 weeks. Two or three applications of pyrethroids controlled the insect well. Control with the net was excellent when the timing of the netting was based on data provided by the monitoring system.

It was also found that yellow sticky traps can be used for monitoring the carrot fly, *Psila rosae,* the diamondback moth and cabbage flies (*Delia radicum* and *D. floralis*). The presence of many beneficial species can also be monitored with yellow sticky traps. Blue traps are used in monitoring the activity of *Lygus* bugs. Cabbage fly poses the most important quality risk in cabbage production; its control strategy should be based on counting eggs from soil samples.

2.2.2 Data management. In many cases it is impossible to give an exact threshold value applicable in any location for deciding upon control. This means that decision-making on the farm level should be based on effective use of monitoring data gathered over several years. In the VIVI Programme, data collection and data management were improved by GIS. Field-specific data collected from different production chains by regional extension services and private companies have been stored in an MTT database and will be used for further development of threshold values and IPM strategies.

2.3 Pest management

IPM should become an integral element of the whole farming system and not confined to a series of treatments during the growing season. The most important elements of IPM are crop rotation, use of resistant varieties and use of healthy plant material. Crop rotation must be planned to cover many years and concern the whole farm or even larger areas. Results of the VIVI Programme showed that the most important insect pest of cabbages, *Delia radicum*, was able to develop and propagate on oil-seed rape at the beginning of the growing season, increasing the risk of fly attack on cabbage fields during the second flight in August. In such situations, IPM should be planned in collaboration among many farmers and as part of quality management of production chains.

Consumer expectations of minimised use of chemical insecticides was found to be realistic in many production systems. Use of insect nets for control of the carrot psyllid gave excellent results without any chemical control. Use of insecticides in swede production could be avoided by delayed sowing in areas where only one species of *Delia* flies occurred. The most positive results by biological control methods were achieved in 1997 when DBM migrated to Finland. Proper use *Bacillus thuringiensis* prevented yield losses and quality faults very effectively. In a cool climate and in agricultural areas with high biodiversity of the surrounding landscape, biological control seemed to be a useful option.

3 IPM AS PART OF QUALITY MANAGEMENT

The VIVI Programme found that many kinds of information are needed for quality management of field vegetables. It was also found that collaboration among different research teams is very important. "Food science" should not be reserved for teams focusing on food processing , storage or other factors that determine food quality at the end of the chain. Quality management should start at the breeding process and extend to consumer feedback on the entire production chain. IPM can play a special role in chain-specific quality management in a wide range of production systems.

TOWARDS SUSTAINABLE PRODUCTION OF PROTEIN-RICH FOODS

Anita Linnemann[1], Dolf Swaving Dijkstra[1], Francesca O'Kane[1], Maarten Koornneef[2], Tiny van Boekel[1] and Wim Jongen[1]

Food Science Group[1] and Laboratory of Genetics[2]
Wageningen Agricultural University
P.O. Box 8129, 6700 EV Wageningen, The Netherlands

1 INTRODUCTION

Development of sustainable food production requires a reduction in the strain that present production systems pose on the environment. Intensive animal husbandry for meat production draws particularly heavily on the environment by the generation of unintentional emissions and surplus manure, and the excessive use of energy, space and raw materials. Novel Protein Foods (NPFs) from plants are a promising alternative that can reduce the disturbing impact of food production on the environment.

This project is based on the outcome of an interdepartmental research programme called Sustainable Technological Development (STD). STD explores the opportunities to ensure that within the next 40 years the Dutch society becomes 20 times more environment-friendly in the areas of food, transport, housing, water use and chemistry. One of the areas studied as part of the food topic, was the future development of protein products which can take over the dietary and cultural role currently played by meat[1].

2 OBJECTIVE

This project aims at the facilitation of the development of NPFs by an integral chain approach. This approach will be used to link technological know-how to options for improvement of primary production systems, and to identify opportunities for further reductions of the strain imposed on the environment by the production of protein-rich foods.

3 RESEARCH APPROACH

Analysis of potentially suitable sources of plant proteins for the Dutch situation. Sources of plant proteins are screened on production levels, possibilities to recover the proteins and the functional properties of these proteins. The crops under investigation are:
- the legumes pea (*Pisum sativum*) and lupin (*Lupinus* spp.) with estimated protein yields of about 1,250 and 2,000 kg/ha, respectively.
- the leafy crops grass (*Lolium* spp.) and alfalfa (*Medicago sativa*), both with an estimated protein yield of 2,500 kg/ha (with about 25% rubisco).

- the (pseudo)cereals triticale and quinoa (*Chenopodium quinoa*) with protein yields of approximately 900 and 650 kg/ha, respectively.
- two industrial crops that yield protein as a waste product after processing, namely the starchcrop potato (*Solanum tuberosum*) and the oilcrop rapeseed (*Brassica napus*) with protein yields of about 900 and 850 kg/ha, respectively.

Listing of technological requirements for proteins used in food production. Information is collected on the following topics:
- functional properties (e.g. solubility, gelation, foaming capacity, emulsifying properties)
- nutritional value (protein content, anti-nutritional components, amino acid composition, health-protecting components like secondary plant metabolites)
- technological possibilities (recovery of the protein, behaviour during storage, possible applications, opportunities for modification by e.g. traditional breeding methods, genetic modification and process technology)
- cultivation (yield per ha, stability of yields in relation to environmental factors, demand for inputs like pesticides and plant nutrients, place in crop rotation, production of waste streams)

Selection of the most promising protein sources. A limited number of protein sources will be selected for further research.

Chemical and physical characterisation of selected protein sources. Information on chemical and physical characteristics will be collected from literature, and, if necessary, completed with own experiments.

Construction of production chains. For each of the selected protein sources a process scheme will be drawn up for the production from raw material to a food ingredient.

Life cycle analysis (LCA) of production chains. LCAs will be performed to determine the impact of the proposed production chains on the environment.

Formulation of options for the production of protein-rich foods. The results of the LCAs will be used to assess further possibilities for the reduction of environmental impact.

Ranking of protein sources and products. The selected protein sources will be ranked on the basis of their productivity and the quality of the obtained proteins.

4 PRELIMINARY RESULTS

Protein isolates from Lupinus angustifolius. Research has been initiated to optimise procedures for the preparation of protein isolates from *Lupinus angustifolius*. This should eventually result in protein isolates which can compete with soya protein isolates, for inclusion in a variety of human food products. The optimisation of the isolation procedure initially focuses on the overall protein yields, the composition and the functional properties of the protein isolates. Soya is used as a reference during this research to compare this lupin species with the already frequently used soya protein.

Table 1 *Total Protein Yields and Protein Contents of Isolates derived from toasted and Non-toasted Flours of Lupinus Angustifolius and from Toasted-Defatted Soya Flour* (The flours were obtained from the company L.I. Franck in Twello [NL.])

Protein Isolate	Protein Yield/ % of total protein	Protein Content/ %* dry weight
Soya (toasted/defatted)	24	91
Lupin (toasted)	29	84
Lupin (non-toasted)	65	81

*Protein content was calculated using conversion factors of 5.52 and 5.40 for soya and lupin, respectively[2].

The isolates were prepared via alkaline extraction of the protein in Tris-HCl buffer (100 mM, pH 8.0, 1 h, at room temperature) at a ratio of 1:10 (flour:buffer, w/v), followed by acid precipitation (pH 4.8, 1 h, 4 °C). The precipitated protein was collected by centrifugation at 8000 g for 30 min at 10 °C followed by washing with acetate buffer (5 mM, pH 4.8). Thereafter the protein contents (Table 1) and the protein solubility profiles were determined (Figure 1).

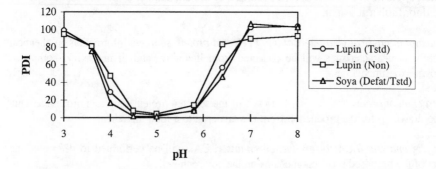

Figure 1 *Protein solubility profiles of protein isolates derived from toasted and non-toasted flours of Lupinus angustifolius and from toasted-defatted soya flour*

The results show that this extraction procedure is suitable for the preparation of protein isolates from non-toasted flour of *Lupinus angustifolius*. A protein isolate with a high solubility above pH 6.4 was obtained with a yield of 65%. Further research is on the way to characterise the extracted proteins in more detail.

References

1. Quist, O. de Kuijer, A. de Haan, H. Linsen, H. Hermans and I. Larsen, Restructuring meat consumption: Novel Protein Foods in 2035. Paper for the Greening of Industry, Heidelberg, November 24-27, 1996.
2. J. Mossé, *J. Agric. Food Chem.*, 1990, **38**, 18.

MANAGEMENT OF BRASSICA CROP RESIDUES TO DECREASE N LEACHING, WEEDS AND CABBAGE FLIES

T. Salo[1], K. Tiilikkala[2] and M. Aaltonen[3]
[1] Crops and Soil   [2] Plant Protection
Plant Production Research, Agricultural Research Centre, FIN-31600 Jokioinen, Finland

[3] Häme Research Station, Agricultural Research Centre, FIN-36600 Pälkäne, Finland

1 INTRODUCTION

The amount of N in Brassica crop residues often exceeds 100 kg/ha[1,2]. Mineralisation and leaching of this N can be controlled by proper timing of residue management and use of catch crops. Apart from N leaching, timing of residue management also affects the occurrence of pests and weeds.

2 FIELD EXPERIMENTS

Management of Brassica crop residues and use of catch crops were studied in 1993-1997 at Häme Research Station. Inorganic soil N was measured to estimate N leaching. N uptake by catch crops was determined to estimate the effectiveness of catch crops in northern conditions. The number of pupae of cabbage flies was recorded in three springs during the experiment.

3 RESULTS

Cauliflower was harvested in August, and incorporation of crop residues in August resulted in fast mineralisation of N. The catch crops used, i.e. rye and Italian ryegrass, were able to take up 15-30 kg/ha N until November. If rainfall was high during autumn, a considerable amount of N was leached below the depth of 60 cm (Figure 1). Cabbage was harvested in September, and incorporation in September reduced the mineralisation of N (Figure 2). Management practices did not affect the number of pupae of cabbage flies or weeds. The number of pupae was quite low and the number of weeds was dependent on the seed bank formed in the soil before the experiment.

In Finnish conditions the management practice to reduce N leaching most effectively seems to be postponing the incorporation until the end of September when soil temperature has decreased. In the case of a large number of weeds requiring incorporation in August, catch crops can reduce the leaching of N.

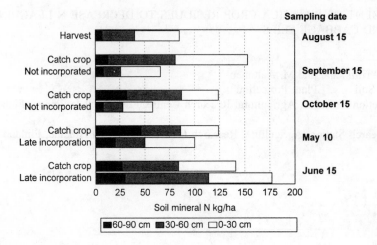

Figure 1 *Effect of catch crop and timing of incorporation on soil mineral N at the depth of 0-90 cm after cauliflower. Amount of soil mineral N averaged over 1993-1996.*

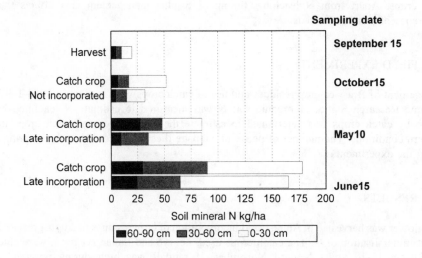

Figure 2 *Effect of catch crop and timing of incorporation on soil mineral N at the depth of 0-90 cm after white cabbage. Amount of soil mineral N averaged over 1993-1996.*

References

1. A.P. Everaarts, *Acta Hort.* 1993, 339, 149.
2. C.R.Rahn, L.V.Vaidyanathan and C.D.Paterson, *Aspects Appl. Biol.* 1992, 30, 263.

ASSESSMENT OF THE QUALITY OF POTATOES FROM ORGANIC FARMING

V. Schulzová, J. Hajšlová, J. Guziur and J. Velíšek

Department of Food Chemistry and Analysis
Institute of Chemical Technology
Prague
Czech Republic

1 INTRODUCTION

The increase in the demand for products of organic farming is distinct not only in Czech Republic (CZ) but also in foreign trades and generally it can be considered as a trend in agriculture all over the world.

The aim of the presented work is to evaluate the influence of different (conventional vs. organic) ways of cultivation on the quality of potatoes. Potatoes (*Solanum tuberosum*) which represent an important component of many diets are known as a good source of carbohydrate, protein, minerals (namely phosphorus, potassium and calcium), and moreover, they provide a large portion of daily intake of vitamin C[1]. On the other hand they also contain natural toxins - glycoalkaloids ("natural pesticides") that may play an important role in protective mechanisms of plants against a variety of diseases[2,3].

1.1 Description of Samples

Eight common varieties of potatoes (Rosara, Koruna, Krystala, Christa, Krasa, Monalisa, Karin, Rosella) have been cultivated by organic as well as by conventional modes of farming at two studied localities in CZ (1 - Jindrichuv Hradec, 2 - Vodnany) in 1996 and 1997.

Several parameters related to the nutritional and sensory quality of potatoes were determined to recognise the potential relationship between the value of parameters examined and the way of cultivation. For result determination the following analytical methods were used: HPLC/DAD (ascorbic acid, chlorogenic acid, glycoalkaloids), ICP/MS (metals), isotachophoretic (nitrates), polarimetric (starch), etc. For result assessment statistical methods were employed.

2 RESULTS AND DISCUSSION

2.1 Agricultural Parameters of Potatoes

2.1.1 Yields per Hectare. Potato yields from conventional farming (avg. 52.9 t/ha) were about twice as high as those from organic farming (avg. 26.5 t/ha), differences were higher in 1996.

2.1.2 Distribution Size of Potato Tubers. The average size of potatoes from organic farming was lower than that from conventional farming.

2.1.3 Dry Weight. Dry weight of organically grown potatoes was slightly higher in most varieties (on average by 9%).

2.1.4 Content of Starch. Content of starch in fresh weight was approx. higher by 15% in organically grown potatoes.

2.2 Content of Metals and Nitrates

2.2.1 Content of Metals Mn, Fe, Co, Ni, Cu, Hg, Zn, As, Se, Pb, Cd. In most varieties of organically grown potato higher content of Mn, Fe, Co, Cu, Zn, Se and Ni and in most of conventionally grown potato varieties higher content of Cd were found. In both localities these differences were statistically significant. The content of As, Co, Ni and Fe was considerably higher in potatoes at the locality 2, obviously due to their higher content in soil. There were no differences in metal contents between the studied years. In no case the content of metals was well below the hygienic limit and the levels of Hg and Pb were below the detection limit.

2.2.2 Determination of Nitrates. The average content of nitrates was lower in organically grown potatoes (111 mg/kg - 1996, 96 mg/kg - 1997, range 19 - 209 mg/kg) compared with conventionally grown potatoes (148 mg/kg - 1996, 219 mg/kg - 1997, range 28 - 398 mg/kg). Some of samples thus exceeded the hygienic limit (200 mg/kg) in CZ. Higher contents of nitrates were found in conventionally grown potatoes as well as in potatoes from locality 2 and in 1997 corresponded to 4 times higher content in soil - 307 mg/kg. The differences of nitrate content in organically and conventionally grown potatoes were statistically significant.

2.3 Toxic Glycoalkaloids (α-Solanin and α-Chaconin)

The average content of glycoalkaloids (GA) α-solanin and α-chaconin in fresh potatoes was in 1996 in organically grown potatoes 66 mg/kg and in conventionally grown potatoes 55 mg/kg. The total GA content ranged from 26 to 245 mg/kg, the highest content of GA (245 mg/kg) was found in the variety Karin (exceeded the hygienic limit in CZ - 200 mg/kg). No significant differences of GA levels were recognised between the tubers of the same variety. The content of GA is mainly dependent on the variety of potatoes and storage conditions during post harvest period.

2.4 Ascorbic Acid

The average content of the ascorbic acid (AA) in fresh potatoes was in organically grown potatoes 38 mg/kg (1996) and 103 mg/kg (1997) and in conventionally grown potatoes 32 mg/kg (1996) and 97 mg/kg (1997). The total AA content ranged in 1996 from 9 to 115 mg/kg fresh potatoes (3 months after harvest) and in 1997 from 16 to 138 mg/kg fresh potatoes (1 month after harvest). There were found no statistical differences between organically and conventionally cultivated tubers.

2.5 Factors Related to Enzymatic Browning

2.5.1 Chlorogenic Acid. This compound is a major representative of phenolics, which are precursors of brown pigments originating in the process of enzymatic and non enzymatic browning[4,5,6]. The average content of chlorogenic acid (CA) in organically grown potatoes was 127 mg/kg (26-166) and in conventionally grown potatoes 86 mg/kg (19-135) in 1996. Higher amounts of CA acid were found in organically cultivated tubers.

2.5.2 Activity of Enzymes and Rate of Fresh Tubers Browning. The intensity of enzymatic browning in fresh tubers corresponds to the activity of polyphenoloxidases, which oxidise o-diphenols to o-chinons[6]. The average activity of enzymes and the rate of fresh tubers browning were higher in organically grown potatoes in both localities. Statistically significant increase was marked in varieties Monalisa and Rosara.

2.6 Sensory Analysis (Profile of Cooked Tubers)

An important aspect of the comparison of examined potato groups is the organoleptic quality. Differences in sensory profiles of cooked tubers originating from organic farming and those from conventional farming were identified.

2.7 Statistical Evaluation of Acquired Data

Acquired data were evaluated by the multivariate statistical methods - linear (stepwise) discriminate analysis (LDA).

Samples were classified into exclusive groups according to the way of cultivation. By the LDA method 100% of the total sample set could be correctly stratified in both years. The content of metals especially Cu, As and Fe was the most important variables for the discrimination into two groups. Another very good discriminating variable was the distribution size of potato tubers (lower then 4 cm and higher then 6 cm). According to the sensory parameters of cooked potatoes 78 % (1996) and 68% (1997) of the total sample set were grouped correctly. As regards the parameters characterising the nutritional quality of tubers only nitrates and chlorogenic acid were able to be used for a good predication of the way of cultivation.

According to the locality as well as according to the crop year and to the variety 100% of the set of raw data can be correctly classified.

3 CONCLUSION

No general conclusions on a better quality of potatoes from organic farming, sometimes declared, can be drawn from the data set available hitherto. To eliminate the influence of climatic conditions in particular years on the variability in tuber composition, a long-term experiment has to be carried out.

References

1. J. Dlouhý, Doctor's Thesis, Inst. för Växtodling, Uppsala, 1981.
2. M. Friedman, J. R. Raynburn and J. A. Bantle, *J. Food Chem. Toxicology*, 1992, **40**, 1617.
3. P. Slanina, *Var Foda*, 1990, **43**, 1.
4. M. Friedman, *J. Agric. Food Chem.*, 1997, **45**, 1523.
5. L. Dao and M. Friedman, *J. Agric. Food Chem.*, 1994, **42**, 635.
6. M. S. Ramamurthy, B. Maiti, P. Thomas and P. M. Nair, *J. Agric. Food Chem.*, 1992, **40**, 569.

EFFECT OF CROP ROTATION ON STORAGE DISEASES OF CARROT

T. Suojala and R. Tahvonen

Agricultural Research Centre of Finland (MTT), Plant Production Research, Horticulture
Toivonlinnantie 518, FIN-21500 Piikkiö, Finland

1 INTRODUCTION

Fungal diseases can spoil a large proportion of carrots during storage. Two of the major storage diseases, licorice rot (*Mycocentrospora acerina*) and Sclerotinia rot (*Sclerotinia sclerotiorum*), have a wide spectrum of host plants and they are able to survive in soil for years.[1] Good crop rotation is therefore essential in minimizing the storage losses due to diseases. The objective of this study was to determine the effect of crop rotation on the occurrence of storage diseases on vegetable farms. Emphasis was laid on the frequency of the carrot cultivation during previous 4-5 years.

2 MATERIALS AND METHODS

The effects of precrops on the storability of carrot were studied in farm experiments. Nine farms in 1995 (cv. Fontana F_1, Bejo Zaden, The Netherlands) and 15 farms (cvs. Fontana and Panther F_1, Sluis and Groot, The Netherlands) in 1996 were included in the study. Carrots were harvested on three dates in 1995 and on four dates in 1996. Storage losses (weight loss and proportion of carrots infected by storage diseases) were analysed three times during the storage. The proportion of all diseased carrots and carrots infected by licorice rot (means of all harvest dates and storage times) were analysed by the analysis of variance. Fields with or without a history of carrot cultivation during the previous 4-5 years were compared.

3 RESULTS

In 1995, the six fields that had no carrot in crop rotation produced carrots with a significantly better storability than fields with earlier carrot production (Table 1). The storage disease rates were 49 % and 24 % in fields with and without a history of carrot cultivation, respectively. The incidences of licorice rot were 26 % and 6 %, respectively.

The trend was the same in 1996 (Table 2): in cultivar Fontana (6 farms), the rate of infection was 43 % for carrots grown on fields with earlier carrot production and 17 % for those grown on fields with no history of carrot cultivation. The rates of licorice rot were 32 % and 17 %, respectively. In cultivar Panther (9 farms), the overall rates of diseases were

17 % and 13 % and those of licorice rot 6 % and 1 %, respectively. Increasing carrot production from one year to two years during the previous four years increased the occurrence of all diseases and especially licorice rot.

4 DISCUSSION

Repeated carrot cultivation on the same field increases the storage losses of carrot. Experimental fields had produced carrots in 0-3 of the previous 4-5 years. The results indicate that even one year of carrot production in the near history may cause a clear increase in storage losses. Cultivation history of the field should be known over a longer period in order to fully understand the occurrence of diseases. Furthermore, the presence of weeds that maintain the pathogens should be registered.

References

1. A.L. Snowdon, 'Color atlas of post-harvest diseases & disorders of fruits and vegetables', CRC Press, Aylesbury, 1992, Vol. 2. 416 p.

Table 1 *Precrops and average percentages of carrots (cv. Fontana) infected by all storage diseases or licorice rot in 1995.*

Year					% of carrots infected by	
1990	1991	1992	1993	1994	all diseases	licorice rot
grain/grass	grain/grass	grain/grass	grass	cabbage	21	5
sugarbeet	sugarbeet	cucumber	fallow/potato	barley	20	6
oats/red beet	barley	turnip rape	oats	barley	27	1
fallow	fallow	fallow	fallow	oats	26	14
oats/red beet	potato	fallow	**carrot**	**carrot**	30	12
carrot	red beet	barley	barley/beet	barley	41	20
potato	oats	wheat	**carrot**	wheat	52	27
carrot	**carrot**	red beet	fallow/potato	**carrot**	59	34
oats/red beet	barley	onion	onion	**carrot**	63	38
Contrast:					p	p
carrot vs. no carrot					0.0001	0.0001

Table 2 *Precrops and average percentages of carrots infected by all storage diseases or licorice rot in 1996.*

Year					% of carrots infected by	
1991	1992	1993	1994	1995	all diseases	licorice rot
cv. Panther						
	barley	barley	turnip rape	barley	4	0
	grass	grass	oats	11	1	
barley	barley	potato/barley	barley	potato/barley	13	1
	barley	**carrot**	barley	barley	13	2
	oats	oats	**carrot**	**carrot**	18	12
	oats	oats	oats	**carrot**	18	1
	carrot	**carrot**	barley	red beet	19	9
		barley	barley	turnip rape	23	1
		oats	**carrot**	**carrot**	29	24
Contrast:					p	p
carrot vs. no carrot					0.0002	0.0001
1 year carrot vs. 2 years carrot					0.0701	0.0001
cv. Fontana						
	barley	barley	turnip rape	oats	14	3
barley	barley	oats	barley	fallow	15	4
oats	oats	oats	barley	oats	19	9
potato/beet	potato/beet	potato/beet	potato/beet	potato	20	11
carrot	red beet	red beet	red beet	barley	40	33
swede	**carrot**	**carrot**	(not known)	**carrot**	46	31
Contrast:					p	p
carrot vs. no carrot					0.0001	0.0001

Session IV Effects of Post-harvest Practice on Quality

GESSI: A GENERIC ENZYME SYSTEM ON STIMULATION AND INACTIVATION DURING STORAGE AND PROCESSING

L.M.M. Tijskens, M.L.A.T.M. Hertog and C. Van Dijk

ATO-DLO, P.O.Box 17,
6700 AA, Wageningen, the Netherlands

1 INTRODUCTION

The phenomena, observed in living plant parts as response to their environment, either natural or artificial, are so numerous, scientists have already been studying the behaviour of food as long as science exists. Nature has almost unlimited resources of chemical and biochemical compounds, and can play with different levels of those compounds to achieve its goals: prolonging the life span and increasing the distance a plant can proliferate. The number of processes that nature uses to reach this overwhelming variability in observed phenomena, however, is in a generic sense rather limited: with for example only the thirty-odd chromosomes present in humans, a uniqueness among all men, living and deceased, is obtained. The same generic approach can be used to describe the phenomena observed with the quality of our food.

2 ENZYMES IN LIVING PLANTS

During the life span of agricultural foods, all kinds of enzymes are active. These enzymes serve the plant to stay alive by fulfilling the natural physiological purpose that plant part has for the plant as a whole. Even during processing operations, like blanching and sterilisation, these enzymes may play a major part in the observed changes in what human like to call foods quality.

During research, conducted in the framework of a EU-AIR program on textural behaviour during processing, a model was developed that describes the changing level of pectin methyl esterase (PE), polygalacturonase (PG) and peroxidase (POD) in several products (peaches, potatoes, carrots, green beans) and the effect of blanching temperatures on the activity (heat denaturation) and exerted action of those enzymes. The observed behaviour and exerted action were very different for these enzymes and for the different products. This overall mechanism, however, has been applied not only to the aforementioned enzymes and products, but also to completely different enzyme systems like lipase, lipoxygenase and aroma forming enzymes in a number of products like rapeseed, green beans and bell peppers.

3 ENZYME ACTIVITY

When assaying enzyme activities, usually the amount of substrate converted into some product is measured at some standard temperature and expressed as activity. What really is measured is the rate of conversion at that temperature. If we reflect on the Michaelis Menten equation (eq. 1), we see that the activity of an enzyme, expressed in such a way, is a combination of the specific activity at that temperature and the (unknown and un-measurable) concentration of the enzyme. The specific activity will depend on temperature, presumably according to Arrhenius' law. What this combination will do at higher temperatures, where a first order exponential inactivation or denaturation of the enzyme occurs is shown in **Figure 1**.

So, in the remainder of this paper, we have to remember that all activities are expressed as

$$\frac{\partial S}{\partial t} = - \frac{k_s \, Enz \, S}{K_m + S} \qquad \text{Eq. 1}$$

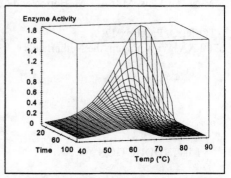

***Figure 1** Denaturation of an arbitrary enzyme*

the activity, at a standard constant temperature, of the enzyme remaining after a heat treatment at different temperatures and times.

4 SPECIFIC MODEL FORMULATIONS

For POD in peaches, carrots and potatoes a bound and a soluble enzyme was found, both of which can be active[7]. The mechanism proposed is shown in eq. 2. Also for PE in peaches, carrots and potatoes a bound and a soluble enzyme was found, again both active [4,6,8]. The same mechanism as proposed for POD is applicable (eq. 2).

The same mechanism was found to be applicable to PE in several cultivars of green beans (ATO, unpublished).

The most prominent aspect of the mechanism for PG in peaches[5] was the conversion from an inactive precursor into an active configuration (eq. 3). This could be regarded as a turnover normally present in enzyme systems of living plants.

$$POD_{bnd} \xrightarrow{k_c} POD_{sol}$$

$$POD_{bnd} \xrightarrow{k_{d,bnd}} POD_{na} \qquad \text{eq. 2}$$

$$POD_{sol} \xrightarrow{k_{d,sol}} POD_{na}$$

In each of these mechanisms denaturation (index d) occurs in an inactive or at least less active configuration, either by heat treatment or by senescence. The denaturation of Lipases was studied in rape seed oil as a function of heat treatment by steam or microwave [2]. Again, two iso-enzymes were found to exist, this time

$$PG_{pre} \xrightarrow{k_f} PG$$

$$PG \xrightarrow{k_d} PG_{na} \qquad \text{eq. 3}$$

without a conversion from the one species into the other. A third iso-enzyme resisted denaturation completely, thereby allowing for a residual activity after heat treatment. The enzyme system related to aroma development in bell peppers after heat also showed susceptibility to heat denaturation[1].

5 GENERIC MODEL FORMULATION

All these models, built for each combination of enzyme type and product, seem to be a special case of a generic underlying overall mechanism. This mechanism includes:
- generation of active species from a precursor
- conversion from one active configuration to another one (iso-enzymes)
- senescence induced loss of activity (natural turnover)
- heat induced denaturation of both active species

The generic mechanism is schematically presented in **Figure 2**. This mechanism can describe an almost infinite number of combinations of enzyme systems, living plant parts, and applied temperature scenarios. Although the apparent enzyme behaviour, measured or observed, is different for different combinations of products and enzymes and for different batches of products, the mechanism remains unchanged. Furthermore, the kinetic

$$Es_{pre} \xrightarrow{k_f} Es_1 \xrightarrow{k_{d1}} Es_{1, na}$$

$$k_{cf} \downarrow \uparrow k_{cb}$$

$$Es_2 \xrightarrow{k_{d2}} Es_{2, na}$$

Figure 2 Generic Model Formulation

parameters (the reaction rate constants at reference temperature and the energies of activation) seem to remain unchanged for different batches of products. They can (probably) be regarded as specific for a species or cultivar. This opens a wide alley to application of this model and the connected kinetic parameters for situations occurring in practice (Tijskens et al.[3]).

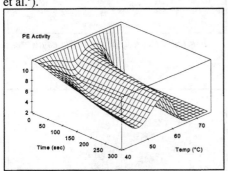

Figure 3 PE activity in peaches during blanching (season 1994)

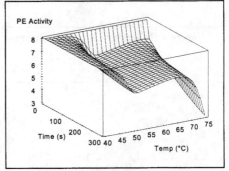

Figure 4 PE activity in peaches during blanching (season 1995)

In **Figure 3** to **Figure 10** some three dimensional examples are given to elucidate how differently the observation on enzyme activity can be in different circumstances of product, enzyme type, seasonal variation, all based on the same generic system and the species specific parameters. The parameters were obtained by analyses of appropriate sets of data, and based on the specific model formulations, derived for the combination of product and enzyme type. In Table 1 the value of the parameters, used in the graphs and the demonstration, are shown without further detail of analysis.

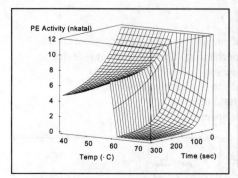

Figure 5 *PE activity in peaches during blanching (1994, combined analysis)*

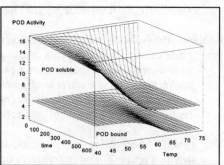

Figure 6 *POD activity, bound and soluble, in peaches during blanching*

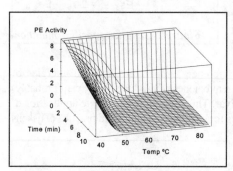

Figure 7 *PE activity in carrots during blanching*

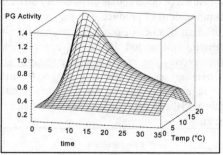

Figure 8 *PG activity in peaches during storage*

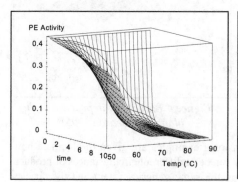

Figure 9 *PE activity in green beans (cv. Masai) during blanching*

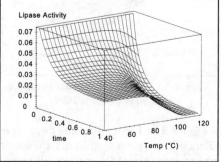

Figure 10 *Lipase activity in rapeseed oil during heating in steam and microwave*

6 ENZYME CLASSIFICATION

When iso-enzymes can be converted into one another the molecular mass may change (cleavage of the protein carrier chain, unbinding to matrix structure), the stereo configuration may change (internal cross links, hydrogen bonds or internal chelating), a change in shielding of the active site may occur (sometimes more, sometimes less), but the general structure of the active site remains roughly the same.

To stress the similarities in active sites of convertible iso-enzymes, the generally accepted classification of iso-enzymes, mainly based on molecular mass (kDaltons) and iso-electric focussing point, all of which are aimed at the properties of the protein carrier, should be extended to include a specification of the active site itself, irrespective of the protein chain they are tied to, and the magnitude and configuration of the protein chain itself. A classification of the obtained model parameters, especially the kinetic parameters connected to activity, conversion and denaturation, could serve as a start and a guideline for this distinction in iso-enzyme type.

An example of this approach is the lipoxygenase activity in green beans. The activity of the extracted enzyme was roughly 20 times lower as in the intact beans, either by change of stereo-configuration or by lesser accessability. The denaturation properties (rate constant of denaturation and energy of activation) were, however, in both cases exactly the same.

7 ACKNOWLEDGEMENT

This study was conducted within the framework of the EU AIR project AIR1-CT92-0278 on *The biochemistry and archestructure of fruit and vegetable tissue as quality predictors for optimising storage and processing regimes: Basic research leading to applicable models and rules*, partly financed by the European Union.

8 REFERENCES

1 Luning P.A., Tijskens L.M.M., Sakin S. and Roozen J.P., submitted *J. Sci. Food Agric.* Chapter 8 in: Luning P.A., 1995, PhD. Thesis Landbouwuniversiteit Wageningen, September, Wageningen, the Netherlands.

2 Ponne C.T., Möller A.C., Tijskens L.M.M., Bartels P.V. and Meijer M.M.T., *J. Agric. Food Chem.* 1996, **44**, 2818-2824

3 Tijskens L.M.M., Hertog, M.L.A.T.M. and van Dijk C., COST 915 & COPERNICUS/CIPA workshop *Food Quality Modelling*, (in press) June 1996, Leuven, B.

4 Tijskens L.M.M. and Rodis P.S., *Food Technol. & Biotechnol.* 1997, **35**, 45-50.

5 Tijskens L.M.M., Rodis P.S., Hertog, M.L.A.T.M., Kalantzi U. and Van Dijk C., *J. Food Eng.* 1998, **35**, 111-126.

6 Tijskens L.M.M., Rodis P.S., Hertog M.L.A.T.M., Proxenia N. and van Dijk C., submitted *J. Fd. Eng.*

7 Tijskens L.M.M., Rodis P.S., Hertog, M.L.A.T.M., Waldron K.W., Ingham L., Proxenia N. and Van Dijk C., *J. Food Eng.* 1997, **34**, 355-370

8 Tijskens L.M.M., Waldron K.W., Ng A., Ingham L. and Van Dijk C., *J. Food Eng.* 1997, **34**, 371-385

Gessi model
Parameter estimates

Product	Cultivar	Enz	Year	Tref	Ep0	Es10	Es20	Esfix	kfref	Ef/R	kcfref	Ecf/R	kcbref	Ecb/R	kd1ref	Ed1/R	kd2ref	Ed2/R
Carrot	Bak	PE	1994	60	0	1.473	12.5	0	0	0	1.64	14535	0	0	0.1352	0	0.04407	19318
Peach	Andross	PE	1994	60	0	8.5	1.4	2.07	0	0	1.2	57000	0	0	3.46E-01	20962	1.954E-03	41833
Peach	Andross	PE	1995	60	0	8	0	0	0	0	1.2	57000	0	0	0.008	52995	4.02E-06	50490
Peach	Andross	PE	1994	60	0	9.2	2.24	0	0	0	0.01027	8031	0	0	0.008	500	0.000391	68143
Peach	Andross	PE	1995	60	0	4	4.88	0	0	0	0.01027	8031	0	0		500	3.33E-07	68143
Potato	Bintje	PE	1994	60	0	1.823	5.5	0	0	0	0.255	10960	0	0	0.1357	0	0.06506	17809
Gr. Beans	Flotille	PE	1995	60	0.884	0.4315	0	0	0.00215	28518	0	0	0	0	0.00827	18941	0	0
Gr. Beans	Masai	PE	1995	60	0.003988	0.14892	0	0	0.001623	32938	0	0	0	0	0.179	25932	0	0
Peach	R Haven	PG	1994	10	0.3	8.37	0	0	0.0173	13646	0	0	0	0	0.269	5671	0	0
Peach	R Haven	PG	1995	10	12.52	122	0	0	0.0664	6266	0	0	0	0	0.1023	0	0	0
Peach	R Haven	PG	1994	10	2.276	0.235	0	0	0.0638	11711	0	0	0	0	0.1023	0	0	0
Peach	R Haven	PG	1995	10	40.91	18.17	0	0	0.0638	11711	0	0	0	0		0	0	0
Peach	Andross	POD	1994	60	0	3.5	10	0	0	0	0.0013	28541	0	0		14249	0.00844	18026
Carrot	Bak	POD	1994	60	0	40	0	0	0	0	0	0	0	0	9.6E-04	51034	0	0
Potato	Bintje	POD	1994	60	0	15	0	0	0	0	0	0	0	0	8E-07	26026	0	0
Gr. Beans	Masai	POD	1996	60	0	21.816	0	0	0	0	0	0	0	0	0.01059	5900	0	0
Rape Seed		Lip	1993	90	0	0.039	0.025	0.016	0	0	0	0	0	0	0.096	0	0.124	18041
Rape Seed Oil		Lip	1993	90	0	0.048	0.025	0.007	0	0	0	0	0	0	5.27	6140	0.46	0
Bell pepper	Hexenol	unk	1993	80	0	1	0	0	0	0	0	0	0	0	0.1957	0	0	0
Cucumber	El Trans	unk	1998	20	0	1	0	0	0	0	11.5	18500	1	0	0	0	0	0

A NOVEL APPROACH TO CONTROL CELL SEPARATION IN RELATION TO TEXTURE OF FRUITS AND VEGETABLES

K. W. Waldron, A. C. Smith, A. J. Parr, A. Ng and M. L. Parker

Institute of Food Research
Norwich Research Park
Colney Lane
Norwich NR4 7UA
UK

1 INTRODUCTION

During the past few decades, there has been considerable interest in controlling the textural quality of plant-based foods. As a result, there has been much research into the chemistry and biochemistry of plant cell walls, and on how processing in conjunction with biotechnology might be used to manipulate quality parameters. It is clear that a multidisciplinary approach is required, making use of expertise not only of biochemists, chemists and molecular biologists, but also of material scientists and investigators of sensory perception.

The way a plant tissue deforms during chewing depends on the forces of oral mastication[1], and the structural characteristics of the food. The latter will depend on contributions from the different levels of structure, and how they interact with one another. In Fig. 1, the cell-wall polymers comprise the lowest level of structure to be considered. Their arrangement and interactions within the cell wall, together with wall thickness and intrinsic structures (e.g. plasmodesmata) will largely determine wall mechanical properties[1]. Cells and their shape and size, will constitute the next level of structure; this is followed by their arrangement into tissues. At this level, intercellular spaces and turgor will also make contributions. Finally, the arrangement of tissues will affect the mechanical properties of the plant organ. Hence, modification of one or more of these interacting structural entities due to agronomic, physiological (e.g. maturation and ripening) or processing (post-harvest storage, mechanical manipulation, thermal treatments) events will influence the final "textural" character of the food.

2 CELL WALLS AND TEXTURE

When a fruit or vegetable tissue is eaten, its texture is strongly influenced by the way in which the cell walls break[2]. The firm crunchy texture of uncooked vegetables or unripe fruits is associated with tissue fracture involving rupture across the cell walls. Turgor and water mobility will have an important role here. In contrast, the cells of many over-cooked vegetables and over-ripe, mealy fruits separate easily during eating, giving the tissues a very soft texture due to loss of the tissue structure. Such separation results from the solubilisation of the wall polymers involved in cell adhesion. This is generally thought to be due to dissolution of the (pectic) polysaccharides of the middle lamella[2] which are perceived to be involved in cell adhesion. Heating-induced solubilisation of these polymers

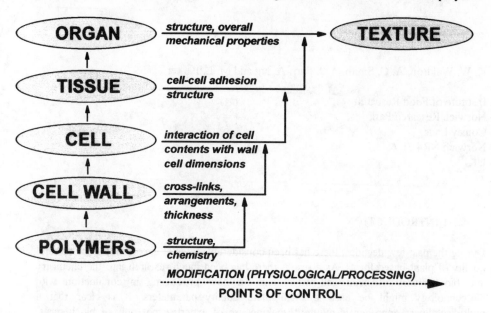

Figure 1 *Schematic representation of the levels of structure underlying the mechanical properties of plant tissues.*

probably results from a combination of depolymerisation by β-eliminative cleavage and chelation of divalent cations by endogenous organic acids[2]. Ripening-related solubilisation is biochemically mediated[3].

3 THERMAL STABILITY OF TEXTURE IN CHINESE WATER CHESTNUT

As part of the research programme on texture at the Institute of Food Research, we have investigated the basis of crispness in raw and cooked Chinese water chestnuts (CWC). The edible portion of the corm comprises thin-walled cells, full of starch, very similar to those in potato. However, unlike potato, CWC retains its firm and crunchy texture even after extensive cooking or even canning. This is due, largely, to the maintenance of cell adhesion so that fracture involves cell wall rupture. Our studies have indicated that the thermal stability of cell adhesion may be due to the presence of cell-wall ferulic acid (FA) dimers which cross-link the wall polymers to which they are esterified[1,4,5]. Other work on sugarbeet and beetroot have provided further evidence for the involvement of these phenolics in determining texture[6].

More recent investigations[7] have shown that there is a heterogeneity in the extractibility of the dimers, and fluorescence microscopy has demonstrated that the more alkali-stable phenolic components form an interesting pattern on the cell surface[4] (Fig. 2). These phenolic components are located predominantly at the edges of the cell faces. Not only does this image provide a direct visualisation of the number of neighbours that the cell

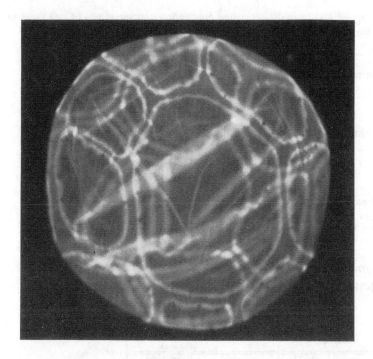

Figure 2 *Micrograph of the autofluorescence of a single cell of Chinese water chestnut after separation in hot, dilute alkali. The cell is approximately 70μm in diameter*

has, but it also gives some insight into how wall components are arranged to enhance cell adhesion. Because turgid cells will try to become spherical, cell separation will start at the edges of the cell faces - exactly where these more alkali-stable moieties are concentrated. The phenolic components in cell walls that exhibit these fluorescent zones have been extracted and analysed, and the two 8-8'- dimers (8,8'-diferulic acid and 8,8'-diferulic acid aryltetralyn form) are particularly prominent[7]. We have suggested that they may be particularly important in cell adhesion.

4 CONTROLLING TEXTURE WITH CELL-WALL PHENOLICS

How might these findings be used to enhance the textural quality of fruit and vegetable tissues? In most fruits and vegetables studied[8], phenolics are present only in very small amounts. However, the fact that they are present at all indicates that the metabolic pathways required are in place, and these may lend themselves to regulation. The main routes by which FA and related cross-links reach the cell wall are shown in Fig. 3[1]. An increase in diferulic acid cross-linking will require an increase in the FA (or other appropriate phenolic) components on cell-wall polysaccharides at the required locations in the wall (e.g. edges of cell faces). In addition, enzymes that are able to catalyse oxidative cross-linking (e.g. peroxidases), and those involved in the formation of H_2O_2, will also have to be present in these locations. It is likely that peroxidase enzymes are already present in most walls: indeed we have demonstrated that ionically-bound POD is located in the middle lamella in apple[9] and bean tissues, often concentrated at the edges of cell faces and in the plasmodesmata.

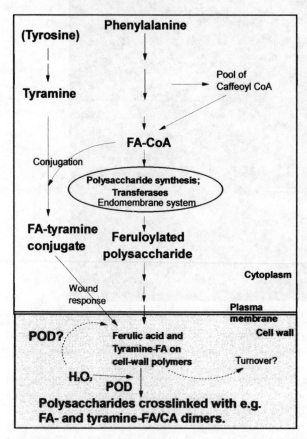

Figure 3 *Routes by which ferulic acid may cross-link cell-wall polymers. Abbreviations: FA - ferulic acid; POD - peroxidase; CoA - co-enzyme A.*

There appears to be little information concerning the location of peroxide-generating enzymes, although it would be logical to assume that these would also be present in locations where peroxidases are located. Therefore, an important regulatory step in determining the degree of diferulate cross-linking will be the levels of feruloylated polysaccharides in the cell wall. The factors which control the incorporation of phenolics into walls remain poorly understood. At present, it is difficult to identify the relative importance of the steps involved in feruloylation of cell-wall matrix polysaccharides (which probably occurs in the Golgi apparatus), and their insertion into the cell wall. However, it is likely that phenolic cross-linking will be governed, in part, by the availability of the phenolic moieties themselves. Perhaps the intracellular availability of FA may have a direct effect on the levels of phenolics in the cell wall[1].

5 GENETIC ENGINEERING

Genetic engineering may be used to complement biochemical studies in order to elucidate the factors which control phenolic incorporation into the cell wall. It will also provide a mechanism to improve product quality. Most of the genetic engineering on phenolic metabolism carried out so far has concentrated on the lignin fraction. However, in the light of these and other biochemical studies, a few possible control points in the production of feruloylated cell-wall polysaccharide may be proposed.

The enzyme phenylalanine ammonia lyase (PAL) will have considerable influence on wall phenolics by increasing flux of phenylpropanoid metabolism (Fig 3). However, a finer degree of control might be exerted by controlling the availability of FA-CoA, which is considered to be the most likely candidate to act as a substrate in the feruloylation of cell-wall polysaccharides. Manipulation of enzymes which control the direction of phenylpropanoid flux (e.g. into other alternative routes, such as that involving Caffeoyl-CoA) may provide an effective means. In addition, those routes which have been recently highlighted in the response of plants to wounding and infection, may also be exploited to enhance phenolic cross-linking in the cell wall. Of particular interest is the formation of feruloylated biogenic amines such as feruloyltyramine. This, and other moieties, may be produced during a wound response, inserted and cross-linked into the cell wall, thereby modulating cell-wall mechanical properties. Whilst little is known about the details of these mechanisms, exploiting the plant's response to wounding or infection may also provide a route to switching on the production of H_2O_2 necessary for the oxidative cross-linking of cell-wall phenolics.

6 ACKNOWLEDGMENTS

The above work was funded by the UK Biotechnology and Biological Sciences Research Council.

7 REFERENCES

1 K. W. Waldron, A. C. Smith, A. J. Parr, A. Ng and M. L. Parker. *Trends. Food Sci. Technol.*, 1997, **8**, 213.

2. J. P. Van-Buren. *J. Texture Stud.*, 1979, **10**, 1.

3. C. T. Brett and K. W. Waldron, 'Physiology and Biochemistry of Plant Cell Walls' Chapman and Hall, London, 1996.

4. M. L. Parker and K. W. Waldron. *J. Sci. Food Agric.*, 1995, **68**, 337.

5. A. J. Parr, K. W. Waldron, A. Ng and M. L. Parker. *J. Sci. Food Agric.*, 1996, **71**, 501.

6. K. W. Waldron, A. Ng, M. L. Parker and A. J. Parr, *J. Sci. Food Agric.* 1997, **74**, 221.

7. K. W. Waldron, A. C. Smith, A. Ng, M. L. Parker and A. J. Parr. Proceedings "Plant Biomechanics Vol. 1", eds. Jeronimidis, G. and Vincent, J.F.V., Univ. of Reading, 1997, 137.

8. A. J. Parr, A. Ng and K. W. Waldron. *J. Agric. Food Chem.*, 1997, **45**, 2468.

9. L. M. Ingham, M. L. Parker and K. W. Waldron. *Physiol. Plant.* 1998, 102, 93.

Optimal Quality Control in Potato Storage

G.J.C. Verdijck

Department of Process Control
Agrotechnological Research Institute (ATO-DLO)
P.O. Box 17, NL-6700 AA Wageningen, The Netherlands

1 INTRODUCTION

Improving the processing of agricultural products involves pushing the operating domain further to the limits. Interest in the performance of processing agricultural products is increased due to pressure exerted directly by consumers, indirectly by the economy and more directly by environmental driven considerations. Processing agricultural products differs from other processes with respect to the unidirectional dynamics and the limited ability of manipulating the processes. Other differences are the natural spreads of size, properties, quality and dynamic behaviour of the individual species.

Currently, control of post-harvest processes does not account for the particularities involved in processing agricultural products. Most industrial applications use PID, on/off and man-operated controllers. The complexity that can be captured by these structures is limited and not easily related to process models. Furthermore, the aim of these controllers is to keep process variables as air temperature and humidity as close as possible to some prespecified values. This means that product quality is a priority translated into specifications for the process variables. This approach results in a non-optimal processing state. Advanced (model based) process control is necessary to improve the processing of agricultural products. The use of Model Predictive Control (MPC) is thus an interesting option.

In this research the storage process of potatoes is studied. In the Netherlands storage is necessary because the food industry demands year-round supply of potatoes. Storage efficiency depends on costs of storage and product quality. Product quality is influenced by storage conditions while these conditions affect mechanisms as cold-induced sweetening, senescent sweetening, respiration, evaporation and sprouting. The objective in this research is to improve process control by incorporating product quality in the control problem. The research is part of the P-Watch development program of Tolsma Techniek b.v. in Emmeloord, The Netherlands.

In Section 2 the modelling of product and process is described. In Section 3 the controller methods and structure are analysed in more detail. The results using the proposed controller in a simulation study are discussed in Section 4. This paper ends with conclusions and aspects for further research.

2 MODELLING

The modelling consists of the parts product and process modelling. Product modelling describes product quality aspects on individual product level and process modelling describes the product environment and storage costs.

2.1 Product modelling

Cold induced and senescent sweetening cause sugar accumulation and these mechanisms are modelled using two enzyme concentrations.[1] The differential equations are

$$\dot{E}n_{cold} = -k_{dena} \quad En_{cold}$$
$$\dot{E}n_{sene} = k_{form} \quad En_{sene} \tag{1}$$
$$\dot{S} = k_{cold} \quad En_{cold} \quad Starch + k_{sene} \quad En_{sene} \quad Starch - k_{resp} \quad S \quad O_2$$

where En_i represents enzyme-concentration i, k_i is the reaction rate i that depends on temperature and S is the sugar content.

2.2 Process modelling

The storage facility consists of the potato stock, the air above the stock, the airshaft and the air channel. The potato stock is separated in an air phase and a product phase. Spatial effects are modelled by dividing the stock in vertical direction in layers and in horizontal direction in columns. In fact this reduces the partial differential equations (pde's) that result from mass and energy balances to ordinary differential equations (ode's).

Essential in process modelling are the assumptions that were made in the modelling process. They determine the validity and usability of the model. The most important assumptions made in this research are constant potato volume, uniform distribution of the potatoes in space, uniform air stream, constant specific heats (temperature independent), moisture and temperature profiles inside the potatoes can be neglected and heat conduction in air phase can be neglected.

The differential equations describing moisture content are deduced from mass balances for air and product phases. Temperature equations are deduced from energy balances.

2.2.1 Air phase. Air humidity in the product stock is described by

$$\varepsilon \rho_a A \frac{dD}{dt} = -\alpha GA\rho_a \frac{\partial D}{\partial x} + A_{sp} Ak_v \Delta P \qquad [kgm^{-1}s^{-1}]. \tag{2}$$

The first term of the right hand side (r.h.s.) in equation (2) represents the air stream. The second term represents the interaction between product and air phase. This is the amount of evaporation. Air temperature is described by

$$\varepsilon \rho_a A(Cp_a + D(x)Cp_{wd}) \frac{dT}{dt} = -\alpha GA\rho_a (Cp_a + D_{in}(x)Cp_{wd}) \frac{\partial T}{\partial x} + UA_{sp} A(\Theta - T)$$
$$\frac{\partial M_{1sp}}{\partial t} \frac{1}{\delta x} Cp_{wd}(x)(\Theta - T) \qquad [Js^{-1}]. \tag{3}$$

The first term of the r.h.s. represents the energy flow that is associated with the air stream. The second term is the heat exchange between product and air phases. The energy involved in the mass stream from product to air phase is represented by the third term.

The differential equations for the other parts in the storage facility resemble equations (2) and (3). A difference is the absence of interaction terms between product and air phase. Further, external ventilation introduces outside air temperature and humidity in modelling the airshaft.

2.2.2 *Product phase.* In the differential equation for product moisture content the evaporation is represented by the first term of the r.h.s. and the respiration by the second term.

$$(1-\varepsilon)A\,\delta x\rho_p\,\frac{dM_{sp}}{dt}=-A_{sp}A\,\delta xk_v\Delta P+c_1\rho_pA\,\delta x(1-\varepsilon)\qquad[kg\,s^{-1}].\qquad(4)$$

Product temperature is described by

$$(1-\varepsilon)A\,\delta x\rho_pCp_p(x)\frac{d\Theta}{dt}=H_{sr}A\,\delta x-UA_{sp}A\,\delta x(\Theta-T)-\frac{\partial M_{1sp}}{\partial t}\Delta H_v+$$
$$\lambda A\,\delta x\frac{\partial^2\Theta}{\partial x^2}+\lambda_hA\,\delta y\frac{\partial^2\Theta}{\partial y^2}\qquad[J\,s^{-1}].\qquad(5)$$

In (5) the aspects of respiration energy, heat exchange between air and product phase, evaporation energy and the second order effect of heat conduction in the product phase in horizontal and vertical directions are present.

The energy consumption is determined by the ventilation periods. The amount of ventilation depends on the heat exchange between the storage facility and the environment, on the energy production by respiration and on the desired storage conditions.

3 CONTROL ARCHITECTURE

A simple optimisation procedure determines the setpoints for product quality. This procedure is performed beforehand. Monitoring product quality during storage enables process control to follow the setpoints by acting upon changes to assure an optimal product quality when the potatoes are delivered.

Advanced (model-based) process control techniques as Model Predictive Control (MPC) have been applied with great success in process industry.[2,3] The basic component of MPC is a dynamic process model that predicts the process output for a time horizon for future adjustments of the input. An optimisation process incorporating constraints results in the optimal processing strategy. The obvious condition is the availability of suitable models as the performance of the MPC-controlled plant is a function of the model quality.

In this project a controller is developed that offers the possibility to incorporate quality aspects in the control problem and to optimise the process. This controller determines optimal storage conditions based on product quality and process efficiency. The most important manipulated value is the storage temperature, because of its direct influence on the mechanisms involved in sugar accumulation and on the evaporation leading to weight-loss. The controller predicts product quality and costs for a certain time-period using a simplified model of the product quality and of the storage facility. This simplified model is based on the simulation model derived in Section 2. The profits and costs related with sugar content, weight-loss and usage of energy are determined and used in the optimisation. Restrictions in conditions that can be realised in storing potatoes are due to weather conditions and aspects such as CO_2 concentration and freezing damage.

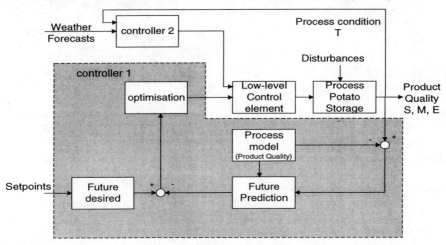

Figure 1 *Control structure for potato storage*

Constraints can be incorporated in the optimisation procedure. Controller 1 calculates the temperature setpoint for the storage process (Figure 1). A lower level controller has to realise this setpoint. This lower level controller uses several parameters. Using weather forecasts Controller 2 calculates the value for the parameter that represents the temperature difference between inside and outside temperature that is demanded for external ventilation. In other words, Controller 1 determines the need for cooling and Controller 2 the possibilities to do this.

In this section two controllers are introduced. Controller 1 in Figure 2 uses linear MPC techniques. This asks for linear models of the product quality aspects sugar content and weight-loss. Further a linear model is needed for the cost that can be effected. These models follow from linearisation of the equations (1) and (4) around a trajectory. If conditions change this linearisation must be performed again to assure a good performance of the controller.

4 SIMULATION RESULTS

The controller structure, which is discussed in Section 3, is tested in a simulation study using the non-linear model as the real process. The time period in the simulation was 34 weeks. A gradual temperature strategy is followed, because it increases storage profits (Figure 2a). Natural heat production is used to reach the desired temperature for delivery. The sugar content, which is a main product quality aspect, is plotted against its desired trajectory (Figure 2b). The desired trajectories are calculated with the non-linear models. The quality aspects follow their setpoints reasonably well. The next step in this research project is the implementation of the controller in an application for an existing storage facility.

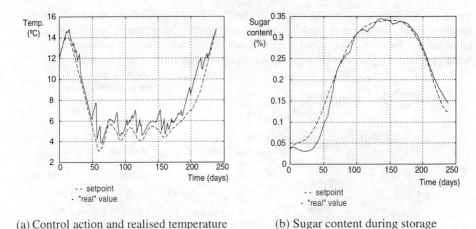

(a) Control action and realised temperature   (b) Sugar content during storage

Figure 2 *Simulation results: Sugar content and Temperature*

5 CONCLUSIONS AND FURTHER RESEARCH

Direct product quality control improves the processing of agricultural products by monitoring the quality aspects and by calculating the necessary control actions. As illustrated by the simulation results product quality stays close to its setpoints and the controller acts upon disturbances. Improving product quality leads to better products and less variety in quality. However, it is necessary to have accurate models available that describe the quality aspects. The use of non-linear MPC would probably further improve the control performance, because model accuracy is increased.

One of the main challenges in using MPC or similar control procedures in post-harvest and food processes lies in the nature of the processes. Modification of advanced control techniques is necessary to meet the particular characteristics and requirements of post-harvest processes. Important is the extension of techniques and ideas developed in this project to improve the processing of agricultural products. In further research a control methodology will be developed that optimises processing efficiency and product quality. For this development a generic model is essential. Different processes and products must be described with the same model structure to allow the use of the control methodology that is specifically developed for the improvement of processing agricultural products.

References

1. M.L.A.T.M. Hertog, L.M.M. Tijskens, P.S. Hak, *Postharvest Biology and Technology*, 1997, **10**, 67.
2. E.F. Camacho and C. Bordons, 'Model predictive control in the process industry', Springer, Berlin, 1995.
3. P.L. Lee, 'Nonlinear process control: applications of generic model control', Springer-Verlag, Berlin, 1995.

INTEGRATED TREATMENTS FOR APPLE SUPERFICIAL SCALD PREVENTION

M. Simčič and J. Hribar

University of Ljubljana, Biotechnical Faculty
Department of Food Science and Technology
Jamnikarjeva 101, Ljubljana, Slovenia

1 INTRODUCTION

Superficial scald is a postharvest physiological disorder of apples that has been recognised as a serious problem for as long as apples have been stored and marketed commercially. It is characterized by the browning of apple skin during storage. The biochemical basis of superficial scald was first suggested in the early seventies[1]. Alpha-farnesene, a terpene produced in hypodermic cells of apples appeared to migrate to the cuticle, undergoing slow oxidation to the conjugated triene hydroperoxides[2] (CTH). Harmful CTH and unstable free radicals accumulate in the apple skin and cause superficial scald. This disorder can be effectively controlled by the antioxidants diphenylamine (DPA) and ethoxyquine. Because of doubts about safety of these chemicals as food additives they are no longer acceptable as post harvest treatments.

For these reasons there is an increased interest in alternative treatments that can be applied for the superficial scald prevention. New methods to prevent scald are treatments with ethyl alcohol (EtOH) vapours[3,4] and prestorage heating [5,6].

In our research the DPA treatment was compared with ethanol, methanol and "apple aroma" vapour treatment on apples cv. 'Granny Smith'. The aim of this work is to find possible synergetic effects caused by these treatments on the scald control and integrate these results with different storage temperatures and harvest dates.

2 MATERIALS AND METHODS

Apples cultivar "Granny Smith" were picked in the Krško region at optimal harvest date 10.10. 1997 (I) and 10 days before optimal harvest date 30.9.1997 (II).
They were randomised in samples and exposed to several treatments:
TEMP - 30 apples of both picking dates were stored at 1°C, 4°C and 10 °C.

VAPOUR TREATMENTS: In all vapour treatments 20 apples from the first picking date (I) were treated in glass exicator (10 L).
All vapour treatments were performed in duplicate in the period between 27. 10. - 6. 11. 1997.

ETHANOL - 100 ml of 98% ethanol at 10°C and 20°C for 2 and 10 days
METHANOL - 100 ml of pure methanol at 10°C and 20°C for 2 and 10 days
AROMA - 150 ml of natural apple aroma (ETOL- Celje) at 10°C and 20°C for 2 and
        10 days
CONTROL - 100 ml of distilled water 10°C and 20°C for 2 and 10 days
ETHANOL (sil.) - apples were treated with 100 ml of ethanol mixed with silicagel at
                20°C for 2 days
DPA - apples were treated with 2000 ppm of Diphenylamine (DPA)

After the treatments apples were stored in normal atmosphere conditions at 1°C and 90% RH. The levels of hydroperoxides (CTH) and L* a* b* chromametric values of apple skin were measured during the storage.
Chromometric values were measured by MINOLTA chromometer CR-200b.
All the measurements were performed on each apple on the same signed point on the equatorial part of the fruit.

2.1 Measurements of hydroperoxides

Levels of hydroperoxides (CTH) were measured by the following method.
Six apples from each sample were peeled in the equatorial part. The cortex part of the peel was carefully removed by a sharp knife. On the opposite sides of the equatorial peel strip two circular segment (2,5 cm^2)of the cuticle were cut with a cork borer (diameter 1,8 cm).12 circular segments were put in 15 ml of hexane (Merck-for spectrophotometry) for 2 minutes.
Spectrophotometric measurements of clear hexane extract were performed between 200 and 300 nm. The highest peak at 268 nm was used for the calculation of the CTH value in A . 10^4 /cm^2.

After five and six months of storage fruit samples were transferred to 20 °C. One week later the incidence of superficial scald was recorded as a weighted Lurie index (LI) of scald intensity using the formula :

L.I. = [(% light scald * 1) + (% medium scald * 2) + (% severe scald * 4)] / 4

light scald ....under 10% of affected skin surface of fruit
medium scald................11% - 40% of affected skin surface of fruit
severe scald...................more than 40% of affected skin surface of fruit

The analysis was performed separately by 4 panellists.

3 RESULTS AND DISCUSSION

After 6 months of storage at different temperatures apples of two different harvest date has similar scald intensity. The apples stored at 4°C had low scald intensity but at the end of the storage the colour of 'Granny Smith' skin was yellow.

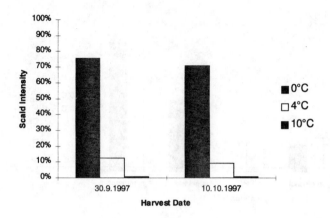

Figure 1 *Scald intensity for apples "Granny Smith" stored at different temperatures*

The CTH value of vapour treated apples were lower than the control. The most effective treatments for the reduction of CTH were ethanol and methanol at 20 °C.

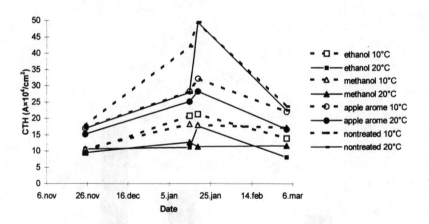

Figure 2 *Changes in hydroperoxide values in apples 'Granny Smith' after different vapour treatments*

Ethanol treatment for 10 days at 10°C was the most effective for the prevention of the superficial scald. The apples treated at 20°C had ethanol injury on the skin. Methanol at 10°C was less effective than ethanol. The aroma treatment reduced the scald intensity from 90% to 60%. The nontreated apples had very severe scald at both temperatures.

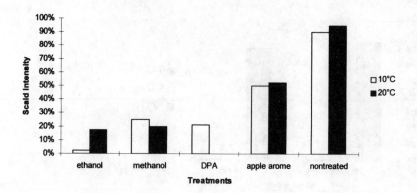

Figure 3 *Scald intensity in vapour treated apples 'Granny Smith' after 5 months of storagre compared with DPA treatment*

The two days vapour treatments were less effective than 10 days treatments. The combination of ethanol and silicagel had the largest effect on scald prevention .

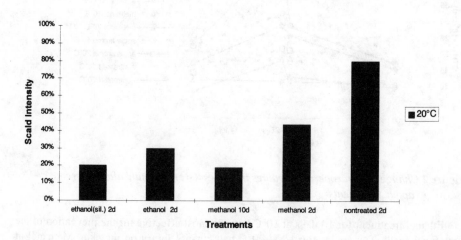

Figure 4 *Scald intensity in two days (2d) vapour treated 'Granny Smith' apples after 6 months of storage*

4 CONCLUSIONS

The ethanol treatments were effective in superficial scald prevention but they caused a high incidence of internal browning after two months of storage. The 10 day treatments at 20°C developed very pronounced internal browning after storage. In the case of methanol treatments severe internal breakdown were detected. The aroma treatment was the least effective in apple scald prevention but no internal disorders appeared after storage.

Ghahramani and Scott[7,8] demonstrated that exogenous ethanol vapour gave effective control of superficial scald. They achieve similar results with endogenous production of ethanol caused by low O_2 stress. The prevention of superficial scald by low O_2 may be partially affected due to ethanol production by these treatments. In our research there is evidence that methanol and aroma compounds have similar effects on superficial scald prevention. Further research is required to determine the mode of action of volatile compounds and to minimise the internal disorders caused by vapour treatments.

References

1. E. F. L. J. Anet, *J. of The Sci. of Food and Agricult.*, 1974, **25**, 299.
2. Z. Du and W.J. Bramlage, *J. Amer. Hort. Sci.*, 1993, **118**, 807.
3. R. Lo Scalzo and A. Testoni, In: Proceedings vol.2, International Control Atmosphere Reasearch Conference,Davis 1997, 5.
4. K.J. Scott, C.M.C. Yuen, F. Ghahram, *Postharvest Biol. Technol.*,1995, **6**, 201.
5. J.D. Klein and S. Lurie, *J. Amer. Hort. Sci.*, 1993, **115**, 265.
6. S. Lurie, D.J. Klein and R. Ben Aire, *J. Horticul. Sci.*,1990, **65**, 503.
7. F. Ghahramani and K.J. Scott, *Aust. J. Agric. Res.*, 1998, **49**, 199.
8. F. Ghahramani and K.J. Scott, *Aust. J. Agric. Res.*, 1998, **49**, 207.

THE EFFECT OF MATURITY AT HARVEST ON POTATO QUALITY DURING STORAGE

M.L.A.T.M. Hertog, L.M.M. Tijskens and P.S. Hak

ATO-DLO
P.O. Box 17
6700 AA Wageningen
The Netherlands

1 INTRODUCTION

The quality of potatoes for the frying industry is largely determined by the reducing sugar content. The more sugars accumulate during storage, the darker the frying colour. The moment of harvest strongly determines the extent of sweetening during storage. Besides the moment of harvest, there are all kinds of pre-harvest factors determining the maturity at harvest.

Hertog et al. [1] developed a mathematical model that describes sugar accumulation during storage as a function of time and temperature. The batch differences in cold susceptibility could largely be accounted for by a different state of maturity at harvest. When the degree of cold susceptibility can be determined at the start of storage seasons, the storage behaviour of specific batches can be predicted. Based on this prediction a batch devoted storage management can be realised taking into account the demands of the specific batch [4].

This article deals with recent data from storage experiments during the season 1995-1996 using Bintje and Agria potatoes from five subsequent harvests stored at 6 different constant temperatures. The experimental data are analysed with the available model as a function of time, temperature and moment of harvest. A method is described to estimate the degree of cold susceptibility at the start of storage.

2 SUGAR ACCUMULATION MODEL

Hertog et al. [1] developed a dynamic mathematical model, based on underlying physiological processes to describe, analyse and predict the storage behaviour of potato (*Solanum tuberosum* L.) tubers in terms of accumulation of reducing sugars. The model can be represented in a simplified way by Fig. 1. Two major processes are incorporated, cold induced sweetening and senescent sweetening. In both cases sugars are released from a large pool of starch.

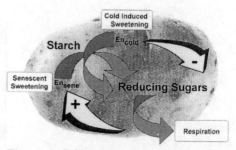

Figure 1 *Schematic representation of the sugar accumulation model for potatoes*

The sugars can be removed again by respiration. It is assumed that two different enzyme systems are involved for the two different processes. The enzyme system responsible for senescent sweetening (En_{sene}) is accumulating in time while the enzyme system responsible for cold induced sweetening (En_{cold}) is susceptible to denaturation. All process rates are assumed to depend on temperature.

The data, necessary for calibration and validation of the model, were gathered during long-term storage experiments over a wide range of storage temperatures for several seasons and cultivars. Although the model is based on a considerable simplification of the occurring physiological processes, it accounts for about 95% of the observed storage behaviour.

3 STATE OF MATURITY

Hertog et al. [1] applied the model to storage data coming from different seasons. The differences in storage behaviour between different seasons could be explained by the differences in length of the growth season. This obviously resulted in batches with different states of maturity.

During the development of tubers attached to the mother plant, the activity of several enzymes related to sugar metabolism change. Merlo et al. [3] reported for developing tubers a maximum curve for most enzyme activities. This observation of Merlo et al. implies that the initial value of En_{cold} ($En_{cold,0}$) at harvest will depend on the moment of harvest. Hertog et al. [1] therefore postulated the concept that the state of maturity at time of harvest determines storage behaviour through the initial amount of enzyme (or enzyme system) responsible for cold-induced sweetening, $En_{cold,0}$. The other model parameters are supposed to be cultivar specific and common for all different batches of the same cultivar.

This approach of forcing all maturity effects to $En_{cold,0}$ was successfully applied to data on Bintje and Saturna potatoes from different seasons [1] and to data on different cultivars and different harvests from one and the same season [2]. In this latter case $En_{cold,0}$ decreased almost linearly as a function of harvest time and thus of the related state of maturity.

4 EXTENDED DATA

The data, Hertog et al. analysed [2] on different cultivars and different harvests from one and the same season, were gathered at one temperature only (4°C). During the storage season 1995-1996 an extended set of data was collected to check if the approach also holds for an extended set of data in function of time, temperature and moment of harvest. We studied the storage behaviour of 5 batches for both Bintje and Agria potatoes coming from 5 subsequent harvests and stored them at 6 different constant temperatures. This data is now analysed in full detail.

4.1 Storage behaviour

Potatoes from Bintje and Agria were harvested 5 times at 2 week intervals starting at the 15th of August for Bintje and starting two weeks later for Agria. Agria was sampled two weeks later because it is known as a 'late' cultivar. Samples were stored at constant temperatures (4, 5, 6, 8, 11, 14 °C) and the accumulating reducing sugars (glucose and

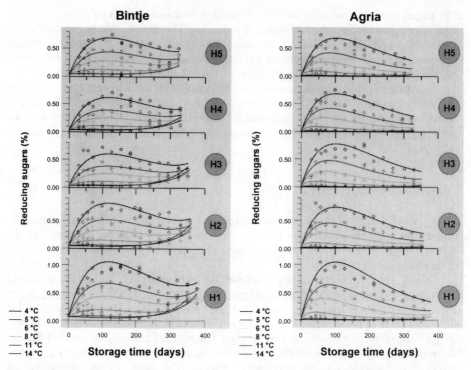

Figure 2 *Storage behaviour of Bintje potatoes as a function of time, temperature (4-14°C) and moment of harvest (H1-H5)*

Figure 3 *Storage behaviour of Agria potatoes as a function of time, temperature (4-14°C) and moment of harvest (H1-H5)*

fructose) were determined at 4 week intervals. The measured data are presented in Fig. 2 and 3 by symbols.

The first harvest of Bintje (H1, Fig. 2) shows the most pronounced sweetening. The maximum levels of reducing sugars reached around day 100, decrease for the subsequent harvests up to the third harvest (H3). No further reduction of the sugar accumulation is observed for the last two harvests (H4, H5). As the batch from the first harvest was stored for the longest time, senescent sweetening most clearly develops at the end of the storage season, especially at high temperatures (8-14°C). The batches from the other harvests of Bintje also develop senescent sweetening, but to a less extent. The pattern of reducing sugar accumulation as a function of time and

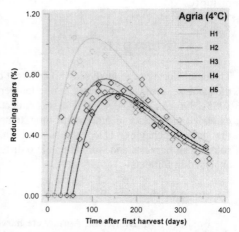

Figure 4 *The accumulation of reducing sugars of Agria potatoes at 4°C for the subsequent harvests (H1-H5)*

temperature is comparable for the five harvests.

Agria (Fig. 3) does not reveal any senescent sweetening. The effect of the moment of harvest on sugar accumulation is mainly limited to the first two harvests (H1, H2). No clear further decrease is observed with a further delay of the moment of harvest. Apparently, Agria was not that 'late' as expected. The effect of the moment of harvest on sugar accumulation of Agria is represented in more detail for 4°C in Fig. 4.

4.2 Initial levels of En_{cold}

The statistical analyses of the experimental data using the developed model resulted in an explained part of 92% for Bintje and 90% for Agria. The model results are presented in Fig. 2-4 by the solid lines. The model is indeed capable of describing the storage behaviour as a function of time, temperature and harvest (Fig. 2, 3). The observed differences between subsequent harvests could completely be described by the model parameter $En_{cold,0}$ which is a function of the moment of harvest (Fig. 5). The value of $En_{cold,0}$ for the subsequent harvests is expressed relative to the value of the first harvest. The general storage behaviour as described in section 4.1 is

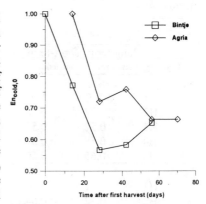

Figure 5 *The model parameter $En_{cold,0}$ as a function of harvest (expressed as time after first harvest) for Bintje and Agria*

confirmed by the values found for $En_{cold,0}$. For Bintje, $En_{cold,0}$ declines with the first three harvests while in case of Agria, $En_{cold,0}$ mainly declines between the first two harvests.

$En_{cold,0}$ may thus be considered as a general maturity dependent parameter determining the storage potential of a given cultivar. It is not very likely that En_{cold} can be identified as a specific enzyme. Probably En_{cold} must be seen as the reflection of a more complex metabolic mechanism. For practical purposes however, it is not necessary to identify En_{cold} as a single enzyme. It will be enough to correlate $En_{cold,0}$ with a physiological property of maturing potato tubers at the start of the storage season.

5 DETERMINING COLD SUSCEPTIBILITY

$En_{cold,0}$ can be estimated easily by storing a sample for 4 weeks at 2°C. Sugar accumulation will be induced and the cold susceptibility of the specific batch can be determined. Providing the cultivar specific model parameters are known, the parameter $En_{cold,0}$ can be estimated on the data from such short-term cold storage experiment.

This approach has been applied with the currently studied batches of Bintje and Agria potatoes from the 5 moments of harvest. At each harvest date a sample was stored at 2 °C (Fig. 6). Reducing sugars were sampled at 3-4 days intervals. This detailed sampling reveals an accumulation pattern slightly different from what was expected based on the rough model developed for the whole storage season. The data (Fig. 6) show a sigmoïdal type, while the model exhibits an exponential type of sugar accumulation. So, the duration of the cold storage experiment should be long enough to overcome this discrepancy. Table 1 contains the estimated values for $En_{cold,0}$ based on either the long-term storage data (Fig.

2, 3) or on the short-term storage data (Fig. 6). For Agria there is a slight difference between the two approaches. For Bintje, the estimates are comparable.

Table 1 *Estimates of $En_{cold,0}$ based on either long-term or short-term storage data*

	$En_{cold,0}$	
Harvest	long-term storage	short-term storage
Bintje		
H1	1	1.02
H2	0.77	0.78
H3	0.57	0.52
H4	0.58	0.57
H5	0.65	0.66
Agria		
H1	1	1.20
H2	0.72	0.64
H3	0.76	0.80
H4	0.66	0.53
H5	0.66	0.55

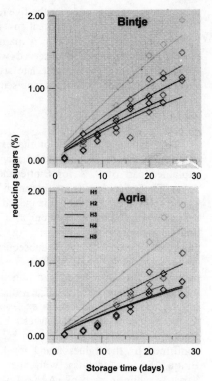

Figure 6 *Accumulation of reducing sugars during a short-term cold storage experiment (2 °C)*

6 PRACTICAL APPLICATION

When storage data on the accumulation of reducing sugars is available for a given potato cultivar the cultivar specific parameters of the model can be determined. The batch specific parameter $En_{cold,0}$ can be estimated at the start of the storage season, using a short term cold storage experiment. Subsequently, the storage behaviour of the specific batch can be predicted using the available model [1]. Based on this quality change model the warehouse management can be optimised taking into account the demands of the specific batch under storage [4]. In this way the quality of raw material for the frying industry can be guaranteed.

References

1. M.L.A.T.M. Hertog, L.M.M. Tijskens and P.S. Hak, *Postharvest Biology and Technology*, 1997, **10**, 67.
2. M.L.A.T.M. Hertog, B. Putz and L.M.M. Tijskens, *Potato Research*, 1997, **40**, 69.
3. L, Merlo, P. Geigenberger, M. Hajirezaei and M. Stitt, *Journal of Plant Physiology*, 1993, **142**, 392.
4. G.J.C. Verdijck, Lecture held at Agri-Food Quality II, Turku, Finland, 1988.

HEAT-INDUCED CHANGES IN CELL WALL POLYSACCHARIDES OF VEGETABLES IN RELATION TO TEXTURE

A. Ng and K.W. Waldron

Institute of Food Research,
Norwich Research Park,
Colney,
Norwich NR4 7UA,
United Kingdom

1 INTRODUCTION

The structure and textural properties of fruit and vegetable tissues are dependent, largely, on the cell wall[1]. Heat-induced softening of vegetables is due, mainly, to an increase in the ease of cell separation[2,3]. This is accompanied by an increase in the solubility of pectic polysaccharides, probably as a result of β-eliminative degradation[4,5]. To reduce an undesirable degree of softness resulted from processing, vegetables are often precooked at moderate temperature for a period of time, followed by cooking[6,7]. Investigations into pre-treatment of carrots or potatoes have shown that pre-treatments at 50 °C result in a reduction in the solubilisation of pectic polysaccharides[8,9]. The aim of this work was to compare the effect of texture-modifying pre-treatments on cell-wall chemistry of carrots (*Daucus carota* cv Amstrong) and potatoes (*Solanum tuberosum* cv. Bintje).

2 HEAT-INDUCED CHANGES IN CARROTS AND POTATOES

2.1 Thermal softening

Heating reduced the firmness of carrots and potatoes to approximately 20% of their original firmness[8,9]. Heat-induced tissue softening also involves cell separation. Preheating prior to heating of carrot resulted in a significantly higher firmness than heating alone. In contrast to carrots, preheated potatoes did not exhibit enhanced firmness after subsequent heating.

2.2 Carbohydrate Composition and the Degree of Methylesterification (DM).

2.2.1 *Fresh carrot and potato tissues.* The carbohydrate compositions of CWM from carrots were rich in pectic polysaccharides, as indicated by the levels of rhamnose, arabinose, galactose and uronic acid; they also contained relatively small amounts of xylose and mannose (Table 1). Compared to carrots, the carbohydrate composition of CWM of potatoes contained galactose as the major neutral sugar. The DM of the UA of carrots and potatoes were 66 and 53%, respectively.

2.2.2 *Sequential Extraction of CWM of fresh carrot and potato tissues.* The polysaccharides released by the sequential extraction[8,9] were predominantly pectic in nature

Table 1   *Carbohydrate composition of CWM of fresh carrots and potatoes*
   *(*results obtained from Ng and Waldron*[8,9])*

	Rha	Fuc	Carbohydrate (mol%)						Total	DM
			Ara	Xyl	Man	Gal	Glc	UA	µg mg[-1]	%
Carrots*	2	trace	7	2	3	9	42	34	601	66
Potatoes*	1	trace	6	3	1	35	28	26	870	53

as reported in Ng and Waldron[8,9]. Most of the extractable pectic polymers were solubilised by the water (WSP), CDTA-1 (CSP-1) and Na_2CO_3-1 (NSP-1) extraction, whilst relatively little was released by NaCl (SSP), CDTA-2 (CSP-2), Na_2CO_3-2 (NSP-2). The DM of the UA from water- and CDTA-1-soluble polysaccharides, and CDTA-insoluble residues (CIR) were similar. Similar results were obtained with potato.

2.3 Effect of Heating on Wall Chemistry

Heating of carrots and potatoes had a very great effect on the extractability of the cell wall polysaccharides. It resulted in an increase in total uronic acid-rich pectic polymers of WSP and SSP (approximately 2-fold). This was accompanied by a decrease in total uronic acid rich pectic polymers of CSP-1 and NSP-1 and insoluble residues uronide (approximately 15-20% to their control value) and, in particular, a large decrease in the levels of total pectic polymers solubilized by Na_2CO_3 (approximately 50-60% to their control value). This indicates that heating-induced changes, which are probably due to β-elimination[10,11], occur in most, if not all, pectic components throughout the cell wall. The DM of the CWM uronic acid component, extracts and RES were reduced by heating (approximately 15-20% of their control value). This decrease probably involved chemical de-esterification of methyl ester groups[11]. However, some de-esterification may have been caused by enhanced pectin methylesterase activity during the initial stages of heating of the tissues[12].

2.4 Effect of Pre-heating Following by Heating on Cell Chemistry

The effect of preheating prior to heating of carrots and potatoes was to reduce the effects of heating. In particular, pre-heating reduced the heating-related increase in WSP and SSP extracted uronide, and the decrease in CSP-, NSP- and residual uronide[8,9].

2.5 Vortex-induced Cell Separation (VICS) of tissues

Soaking of pre-heated and heated carrot tissues in CDTA overnight resulted in total VICS. Fresh carrot tissues did not undergo VICS after similar treatment. This indicates that the preheating effect was manifest through enhanced cell-cell adhesion[9], due to the increase in the thermal stability of calcium-cross-linked pectic polysaccharides. Our results are consistent with the hypothesis that pre-cooking de-esterifies cell-wall pectic eliminative degradation, and increases their potential for calcium-cross-linking[2]. Interestingly, preheated potato tissues did not exhibit enhanced firmness after subsequent heating even though, as in carrots, a preheating-induced increase in thermal stability was manifest in nearly all of the pectic polymer fractions of the cell wall, including the

CDTA-soluble components which might be expected to bind and cross-link with calcium (see section 2.3). However, soaking potato tissue with 3 mM of calcium prior to preheating followed by heating resulted in a 30% increase in firmness compared with the heated tissues. Hence, our results indicate that in Bintje potatoes, pre-heating can enhance the thermal stability of cell wall polymers, but a firming effect fails to occur due to a lack of calcium.

3 CONCLUSIONS

Heat-induced softening of carrot and potato tissues results in the solubilization of pectic polysaccharides; this is accompanied by the loss of uronic acid from all the wall fractions studied and is consistent with the general depolymerization of pectic polymers through β-elimination. Heating also involves cell separation[8], this is consistent with a heat-related weakening of cell-cell adhesion[13]. Pre-heating prior to heating results in a reduction in the DM of the pectic moieties and reduces heating-induced modifications to all pectic fractions.

Unlike carrots, Bintje potatoes do not exhibit a measurable firming effect if they are subjected to pre-heating followed by heating. This is in spite of the presence of CDTA-soluble polymers, the degradation of which was reduced by the pre-heating. However, a firming effect in Bintje potatoes could be induced if the tissues were soaked in dilute $CaCl_2$ prior to the preheating.

This is consistent with the hypothesis that the preheating enhancement of firmness in heated tissues is due to an increase in the thermal stability of calcium-cross-linked pectic polysaccharides, which are involved in cell-cell adhesion.

References

1. D. M. Klockeman, R. Pressey and J.J. Jen, *J. Food Biochem.*, 1991, **15**, 317.
2. J. P. S. Van-Buren, *J. Text. Stud.*, 1979, **10**, 1.
3. M. C. Jarvis and H. J. Duncan, *Potato Res.*, 1992, **35**, 83.
4. M. J. H. Keijbets, PhD Thesis, Agricultural University, 1974.
5. T. Sajjaanantakul, J. P. Van-Buren and D. L. Downing, *Carbohydr. Polym.*, 1993, **20**, 207.
6. C. Y. Chang, Tsai, Y. R. and Chang, W.H., *Food Chem.*, 1993, **48**, 145.
7. J. Schoch and J. L., U.S. Pat. 3 669 686, 1972.
8. A. Ng and K. W. Waldron, *J. Sci. Food. Agric.*, 1997, **73**, 502.
9. A. Ng and K. W. Waldron, *J. Agric. Food Chem.*, 1997, **45**, 3411.
10. T. Sajjaanantakul, J. P. Van-Buren and D. L. Downing, *J. Food Sci.*, 1989, **54** 1272.
11. L. C. Greve, R. N. McArdle, J. R. Gohlke and J. M. Labavitch, *J. Agric. Food Chem.*, 1994, **42**, 2900.
12. L. M. M. Tijskens, K.W. Waldron, A. Ng, L. Ingham and Van Dijk, *J. Food Eng.*, 1998, in press.
13. C.T. Brett and K.W. Waldron, The Physiology and Biochemistry of Plant Cell Walls (2nd edn). Chapman and Hall, London, UK, 1996.

THE QUALITY AND SHELF LIFE OF SOME NEW FINNISH STRAWBERRY VARIETIES

M. Mokkila[1], U. Häkkinen[2], M. Hägg[2], K. Randell[1] and E. Laurila[1]

[1] Technical Research Centre of Finland (VTT), Biotechnology and Food Research, P.O. Box 1500, FIN-02044 VTT, [2] Agricultural Research Centre of Finland (MTT), Laboratory of Food Chemistry, FIN-31600 Jokioinen

1 INTRODUCTION

The main strawberry varieties cultivated in Finland are 'Jonsok', 'Senga Sengana' and 'Zephyr'. The areas in which 'Senga Sengana' and 'Zephyr' are cultivated are decreasing and it looks as if new varieties are coming. Although the flavour of Finnish varieties is excellent the texture of the strawberries is soft they normally have very short shelf lives, of only one to three days. There is a need for new varieties possessing good flavour and long shelf lives but neither the quality nor shelf life of new varieties cultivated in Finland have been systematically studied.

Methods and possibilities for improving the post-harvest quality of strawberries were studied in a joint research project between VTT, MTT, domestic producers and Finnish packaging and refrigeration companies in the period 1995-1997. One of the tasks undertaken was the characterisation of the differences in quality and shelf life between the main current cultivars 'Jonsok' and 'Senga Sengana' and eight new varieties 'Bounty', 'Dania', 'Elsanta', 'Honeoye', 'Korona', 'Lambada', 'Nora' and 'Polka'.

2 MATERIALS AND METHODS

Experiments began in 1995 with only four varieties ('Bounty', 'Dania', 'Elsanta' 'Senga Sengana') and the number of varieties increased to nine in 1996 and to eight in 1997 (Table 1). It was not possible to source all the varieties being studied from a single location and strawberries were therefore obtained from five different cultivation sites in south and middle Finland. 'Elsanta' and 'Dania' were no longer studied in the 1997 season. 'Dania' was almost totally contaminated by mould and 'Elsanta' had been so seriously damaged by frost during the 1996-1997 winter that no fruit was available.

For all the experiments, strawberries were placed in cold storage as soon as possible after harvesting. In the 1997 season, pre-cooling was delayed for two to three hours and this had the effect of decreasing the shelf life of the strawberries. Fruit was delivered by refrigerated transport to VTT (located in Espoo) for sensory evaluation and to MTT (located in Jokioinen) for nutritional analysis. Strawberries were stored either at 2°C (1997) or 5°C (1995, 1996). The experiments were repeated one to three times during each season.

Differences between varieties were studied by determining the sensory and nutritional quality of strawberries (Table 2) after storage. In all seasons, quality was followed as a function of storage time for four to eight days from the harvesting time when the fully-ripe strawberries were picked. In addition, in the 1997 season the effect of the degree of ripeness on shelf life was studied. In the 1997 season the sensory quality of strawberries was also evaluated in two test series in which domestic trade and exportation of strawberries were simulated. In the domestic trade test strawberries were picked when

Table 1 *Strawberry varieties and harvesting places in experiments*

Variety	1995	1996	1997
Bounty	MTT's research station	MTT's research station	Farm A
Dania	MTT's research station	MTT's research station	
Elsanta	MTT's research station	MTT's research station	
Honeoye		Farm B	Farm C
Jonsok		Farm B	Farm C
Korona		Farm B	Farm C
Lambada			Farm C
Nora		Farm B	Farm C
Polka		Farm B	Farm D
Senga Sengana	MTT's research station	MTT's research station	Farm D

MTT's research station is located in Mikkeli, Farm A in Suonenjoki, Farm B in Hämeenkoski, Farm C in Leppävirta and Farm D in Suonenjoki.

Table 2 *Quality parameters used to measure the quality and shelf life of strawberries*

Quality property analysed	Determined as	Method used
Sensory quality	Appearance	Sensory evaluation by 2 trained judges according to the Karlsruhe quality scale from 1 to 9
	Texture and flavour	Sensory evaluation by 8-10 trained judges according to the Karlsruhe quality scale from 1 to 9 or by a round table discussion recorded by 3-5 assessors
	Mouldy berries	Counting
Nutritional quality	Vitamin C	HPLC (Speek et al[1], Hägg et al[2])
	Sugars (fructose, glucose, sucrose)	GLC (Li and Schuhmann[3], Haila et al[4])
	Organic acids (malic acid, citric acid)	GLC (Li and Schuhmann[3], Haila et al[4])

fully-ripe or 4/5 ripe and stored at 2°C over night after being picked. On the next day the storage temperature was raised to 20°C and the fall in quality during the day was monitored. In the exportation test strawberries were picked when 4/5 ripe and initially stored at 2°C for five days. After this they were stored at 20°C for two days and at 2°C during the night between these two days. The fall in quality was monitored.

3 RESULTS AND DISCUSSION

Significant differences were found in the shelf life and in the sensory and nutritional quality of the varieties at different harvesting weeks and seasons. The sensory quality of strawberries decreased from the beginning to the end of the 1995 and 1997 seasons, and was at its best in the middle of the 1996 season. The best levels of quality (Figure 1) and the longest shelf lives were achieved during 1996. The quantity of mouldy berries was lowest in the middle of each season. In the 1995 season, vitamin C (Figure 2), sugar and acid contents were higher than in the other seasons. Shelf life of strawberries could be improved by picking the fruit when it was 4/5 ripe, but the initial sensory quality was then less good than berries which were picked when fully-ripe.

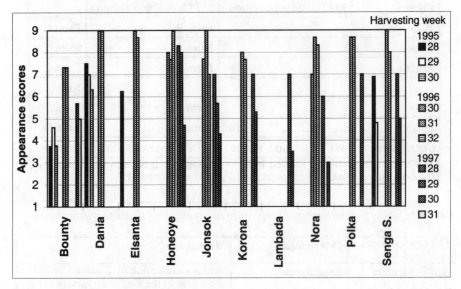

Figure 1 *Appearance of strawberry varieties. Scores: 1-3: unacceptable for sale, 4-6: acceptable, 7-9: excellent*

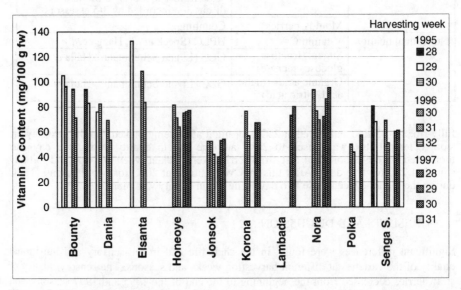

Figure 2 *Vitamin C content of strawberry varieties*

'Bounty' exhibited extremely good flavour and good texture but its shelf life proved to be only one to three days when picked as fully-ripe and only slightly better or no better than the shelf life of 'Senga Sengana'. In the domestic trade test the quality of fully-ripe 'Bounty' deteriorated when the fruit was stored at 20°C. Compared to many other varieties, the flavour of 'Bounty' was good when picked as partly-ripe and it would be reasonable to pick 'Bounty' as partly-ripe late in the season. 'Bounty' is suitable for exportation only very early in the season. 'Bounty' proved to be quite sensitive to moulding. The vitamin C content of 'Bounty' was one of the highest found, around 90 mg/100g fw.

The quality and shelf life of 'Dania' and 'Elsanta', picked as fully-ripe, was investigated in the 1995 and 1996 seasons. The texture of both varieties was evaluated as being too hard and their flavour was considered to be poor. Although the sugar content of 'Elsanta' was high and the flavour was evaluated as sweet in 1995, the intensity of the strawberry flavour was poor. In 1996 both varieties were evaluated as acidic and they had no sweetness. 'Elsanta' had a long shelf life and the fruit maintained an excellent appearance over four days. 'Dania' had nearly as long a shelf life as 'Elsanta'. 'Elsanta' did not prove to be sensitive to moulding but in 'Dania' samples many mouldy berries were observed. The vitamin C content of 'Elsanta' was the highest measured, over 100 mg/100g fw, and at its best in 1995 it was over 130 mg/100 g fw.

'Honeoye' proved to be the most promising new variety. Fruit had a very long shelf life of four to six days when picked as fully-ripe and was not sensitive to moulding. In the domestic trade test the quality of fully-ripe 'Honeoye' was well maintained during storage at 20°C. Flavour was also good especially in the early season but decreased in late season. In the exportation test 'Honeoye' maintained its quality best of all varieties assessed.

The flavour of 'Jonsok' was evaluated as being quite acidic and it very easily acquired off-flavours during storage. The quality of 'Jonsok' varied from season to season. The shelf life of 'Jonsok' also shortened significantly in late season and fruit quality was very poor if the strawberries were not efficiently pre-cooled. When 'Jonsok' was picked as partly-ripe its shelf life was extended but flavour deteriorated. 'Jonsok' proved to be quite sensitive to moulding. The vitamin C content of Jonsok was the lowest found, only some 50 mg/100 g fw.

'Korona' was evaluated as sweet and intensive in strawberry flavour. The outer layer of the berries was soft but the heart proved to be hard. The shelf life of 'Korona' varied between the 1996 and 1997 seasons. In 1996 the shelf life was short compared to all the other varieties and a substantial number of mouldy berries were found during storage in late season. In 1997 the shelf life was longer, two to five days, when fruit was picked as fully-ripe. It is also possible to pick 'Korona' also as partly-ripe because it was found that the flavour of the fruit develops and improves during storage.

The quality and shelf life of 'Lambada', picked as fully-ripe, was only investigated in the 1997 season. The flavour of 'Lambada' was excellent. It was evaluated as being very sweet, intensive in strawberry flavour and not acidic. The texture of the berries was very soft. Early in the season the shelf life was long, five days, but deteriorated dramatically in late season when it became unacceptable for sale on the day after picking.

The flavour of 'Nora' was sweet and 'like blueberries'. In 1996 the shelf life of 'Nora' was a little better than that of 'Jonsok's' but in 1997 the shelf life of 'Nora' was poor. On the other hand, 'Nora' did not appear to be sensitive to moulding. It would be reasonable to pick 'Nora' when almost-ripe, not fully-ripe, to maintain better quality. The vitamin C content of 'Nora' was quite high, approximately 80 mg/100g fw.

'Polka' had very good flavour when picked as fully-ripe. It was sweet, intensive in strawberry flavour and not too acidic. Early in the season the shelf life was quite good, around four days, but the quality of the fruit was not maintained in the domestic trade test. In the late season the shelf life was shorter. When picked partly-ripe, 'Polka' was evaluated as being poor in both flavour and texture. The berries were judged unripe, flavourless and hard. 'Polka' did not prove to be suitable for exportation. The vitamin C content of 'Polka' was as low as that of 'Jonsok', approximately 50 mg/100 g fw.

Finns like the flavour of 'Senga Sengana'. It is intensive in strawberry flavour, the flavour is appropriately sweet and acidic. The shelf life of 'Senga Sengana' was found to be between only one and two days when picked when fully-ripe. In the domestic trade test the quality of fully-ripe 'Senga Sengana' deteriorated when it was stored at 20°C. It would be reasonable to pick 'Senga Sengana', as with 'Nora', when nearly-ripe, not fully-ripe, in order to better maintain the quality of the fruit.

4 CONCLUSIONS

It was not possible to list the 10 strawberry varieties handled in this project into an order of superiority. Many varieties that had a long shelf life had quite poor flavour, while varieties that possessed good flavour had quite short shelf lives and exhibited an increasing number of mouldy berries during storage. If a long shelf life is required, for example in exportation, the work described here showed 'Honeoye' to be the most promising variety. 'Honeoye' have a very long shelf life, their flavour was quite good and their vitamin C content was also quite high.

The work done in this project emphasises the importance of efficient cooling as soon as possible after picking, particularly in the case of varieties which have short shelf lives.

References

1. A. Speek, J. Schrijver and W. Chreurs, J. Agr. Food Chem., 1984, **32**, 352-355.
2. M. Hägg, S. Ylikoski and J. Kumpulainen, J. Food Comp. Anal., 1995, **8**, 12-20.
3. B. Li and P. Schuhmann, J. Food Sci., 1980, **45**, 138-141.
4. K. Haila, J. Kumpulainen, U. Häkkinen and R. Tahvonen, Food Comp. Anal., 1992, **5**, 100-107.

QUALITY CHANGES IN MINIMALLY PROCESSED 'ROMAINE' LETTUCE AS AFFECTED BY SEVERAL TREATMENTS

F. Artés, J.A. Martínez and J.G. Marín

Postharvest and Refrigeration Laboratory. Food Science and Technology Department. CEBAS-CSIC. PO Box 4195. Murcia. E-30080. Spain.

1 INTRODUCTION

Production of minimally processed fruit and vegetables has dramatically increased in the recent years. Some volumes of packaged 'Iceberg' or 'Romaine' lettuce to be consumed as salad or together with other selected vegetables, such as carrot or cabbage, are very usual in all markets.

Minimally processed lettuce consists of a difficult product due to brown discoloration of the cut surfaces when it is stored for several days[1]. Another very usual difficulty to extend the shelf life of shredded lettuce for more than one week consists of the microbial flora evolution of lettuce pieces during storage. The presence or absence of both human pathogens and plant tissue spoilage organisms is important[2].

Various treatments to prevent browning and spoilage extending the shelf life have been proposed from the handling up to the storage conditions: modified atmospheres and/or low storage temperatures, the kind of packaging and the use of some chemical solutions in the washing water[3].

The aim of the present work was to study the influence of two temperatures, two type of films, the system of the atmosphere modification (passive or active) and the use of chlorine and citric acid on the quality of minimally processed 'Romaine' lettuce.

2 MATERIALS AND METHODS

Autumn heads of mature 'Romaine' lettuce grown under Mediterranean climate, were obtained from an open orchard located in Torre-Pacheco, Murcia (Spain). All heads were immediately transported 50 km by ventilated car to the Laboratory, and precooled by forced air to reach 10 °C in about 10 hours. The heads were kept at 10 °C and 90% RH until the next morning, when they were handled at this temperature, as in commercial practice, by hand-eliminating external leaves and stem up to about 50% of fresh weight.

Sound young leaves were cut at 10°C by using commercial blades to obtain a size ranging between 2-3 cm. wide and 10-12 cm^2. Shredded lettuces were washed at 10°C either with tap water or with water containing 50 ppm chlorine and 200 ppm citric acid (pH = 6.8). Subsequently the pieces were centrifuged. After centrifugation lettuce pieces were packed in passive or active (4% O_2 and 96% N_2) modified atmosphere packaging of selected 35 μm. thickness, microperforated oriented polypropylene (OPP) or standard OPP films antifog treated. The bags were stored for 8 days either at 1 or 5°C by simulating real conditions during transport or retail sale. Five packages containing 250 g lettuce pieces each were made for each treatment and temperature. The gas composition was measured throughout the storage period by using a Lippke CheckMaster 2+1 O_2 and CO_2 analyser.

The quality attributes and disorders were evaluated by a group of trained judges. These attributes were classified according to López-Gálvez et al.[1].

3 RESULTS AND DISCUSSION

3.1 Package Atmosphere

At the steady state a slight modification of the atmosphere within microperforated film bags was obtained (average 17% O_2 and 1% CO_2), while in nonperforated OPP the gas modification was higher (average 6% CO_2 and 6% O_2). The pieces of lettuce stored under higher levels of CO_2 showed lower overall visual quality and higher leaf edge and surface browning (Table 1). These results are in contrast with those reported by Hanza et al.[4] who confirmed that good modified atmospheres for shredded lettuce are established up to 15% of CO_2 (optimum around 10%), whereas O_2 should be more than 1%. This atmosphere reduced enzymatic browning.

According to the present results, probably the effects of the chemical treatment could be more important than gas composition in our experiment. Brecht[3] stated that chlorine treatment gave good results in reducing browning of shredded lettuce. Results of the present study confirm that the chemical treatment applied had been efficient in reducing leaf edge browning. On the other hand, active modification of the atmosphere did not seem to produce any improvement when compared to passive modification. These results agree with those previously reported by Aharoni *et al.*[5] who showed that gas flushing bags of 'Romaine' lettuce gave no improvements in lettuce quality over non-flushed bags when they were stored at 1°C for 9 to 17 days.

3.2 Quality Attributes

No differences were observed in texture and aroma among different treatments. Off-odour was only detected after opening the bags, and its disappeared quickly. These off-odours was mainly detected in standard OPP at 5°C.

3.3 Disorders and Microbial Development

Leaf edge browning was the most important disorder developed in almost all the bags. Packaging in microperforated OPP reduced this disorder but it was not enough to eliminate it. Leaf edge browning was responsible for the general loss of visual quality occurred at the end of the storage period. The highest value obtained was 6.1 relatively close to the limit of saleability (5), as shown in Table 1.

Russet spotting and brown stain were the other two main disorders detected. They are known as leaf surface browning, because they only appeared on the midribs of the leaf. Sometimes they became visible on green tissues in severe cases. Russet spotting was associated with ethylene accumulation (> 0.1 ppm) while brown stain with carbon dioxide accumulation (> 2%)[6]. In unusual cases, browned mechanical damages in the midribs can be also included in this disorder.

Table 1 *Gas composition within the packages, quality attributes and disorders of shredded 'Romaine' lettuce stored for eight days under different conditions.*

Parameter	1°C								5°C							
	standard OPP				microperforated OPP				standard OPP				microperforated OPP			
	passive		active[z]		passive		active		passive		active		passive		active	
	A[y]	B[x]	A	B	A	B	A	B	A	B	A	B	A	B	A	B
Gas composition																
% O_2	11.6 b	8.4 c	3.5 d	2.7 d	17.9 a	17.9 a	17.3 a	17.6 a	9.3 bc	9.2 bc	4.9 d	3.8 d	17.3 a	16.9 a	17.9 a	18.1 a
% CO_2	5.6 bcd	8.3 a	4.4 d	4.7 d	1.2 e	1.2 e	1.0 e	0.6 e	6.5 bc	7.0 ab	4.8 d	5.1 cd	2.2 e	1.7 e	0.8 e	0.6 e
Quality attributes																
OVQ[w]	5.1 bcd	4.7 d	4.9 cd	5.5 abcd	5.9 ab	5.9 ab	5.9 ab	6.1 a	4.9 cd	5.3 abcd	5.7 abc	5.7 abc	4.9 cd	5.9 ab	5.1 bcd	5.9 ab
Texture[v]	4.0	4.0	4.0	4.1	4.2	4.1	4.0	4.3	4.0	4.0	4.2	4.3	4.0	4.2	4.0	4.3 N.S.
Aroma[u]	3.0	3.0	3.2	3.0	3.0	3.2	3.4	3.0	3.2	3.0	3.4	3.0	3.4	3.2	3.4	3.2 N.S.
Off-odour[t]	1.4 c	1.5 bc	1.6 bc	1.7 abc	1.6 bc	1.7 abc	1.8 abc	2.0 abc	1.4 c	2.5 a	2.4 a	1.7 abc	2.2 ab	2.2 ab	2.2 ab	1.7 abc
Disorders																
LEB[s]	1.58 abc	1.53 abcd	1.58 ab	1.54 abcd	1.38 de	1.38 cde	1.38 cde	1.35 de	1.65 a	1.63 ab	1.49 abcde	1.45 bcde	1.62 ab	1.31 e	1.65 a	1.36 de
LSB[r]	1.39 a	1.36 a	1.18 bcd	1.30 ab	1.05 def	1.02 ef	1.06 def	1.07 def	1.16 cde	1.21 bc	1.06 def	1.05 def	1.03 ef	1.05 def	1.0 f	1.03 ef
Decay[q]	1.0 b	1.0 b	1.0 b	1.0 b	1.0 b	1.0 b	1.0 b	1.0 b	1.0 a	1.1 a	1.0 b	1.0 b	1.0 b	1.0 b	1.1 a	1.0 b

[z] Initial 4% O_2 and 0% CO_2 [y] A = tap water (1.5 ppm chlorine). [x] B = citric acid (200 ppm) + chlorine (50 ppm)

[w] OVQ (overall visual quality). Score: 9 = excellent, 7 = good, 5 = fair, 3 = poor, 1 = unusable. 5 is considered the limit of saleability

[v] Score: 5 = crisp, 3 = moderately crisp, 1 = none. [u] Score: 5 = full, 3 = moderate, 1 = none or not characteristic

[t] Score in the recently open bags: 1 = none, 3 = moderate, 5 = severe

[s] LEB (leaf edge browning), [r] LSB (leaf surface browning) and [q] Decay, calculated by multiplying the scores of severity (1 = none, 3 = moderate, 5 = severe) by the percentage of weight affected

Mean separation at P = 0.05 according to the Duncan Multiple Range Test. N.S. not significant

The observed decay was identified as bacterial soft rot (*Pseudomonas* and *Erwinia* species). Aerobic microbial count increased from about 5.6 log cfu/g at harvest to about 7.5 log cfu/g at the end of the experiment. According to French legislation[1], a maximum of 4.7 log cfu/g at production and 7.7 log cfu/g at consumption must be the acceptable range for minimally processed lettuce. Using this criterion the microbiological quality of the lettuce pieces at the beginning of the experiment was not acceptable. However, that at the end of the storage period was good. More studies are needed to know adequate range for microbial counts in order to establish microbiological quality in shredded lettuce.

References

1. G. López-Gálvez, G. Peiser, X. Nie and M. Cantwell, *Z. Lebensm. Unters. Forsch. A*, 1997, **205**, 64.

2. J. A. Magnuson, A. D. King, Jr. and T. Török, *Appl. & Environ. Microbiology*, 1990 **Dec.**, 3851.

3. J. K. Brecht, HortSci., 1995, **30**(1), 18.

4. F. Hamza, F. Castaigne, C. Willemot, G. Doyon and J. Makhlouf, *J. Food Qual.*, 1996 **19**, 177.

5. N. Aharoni and S. Ben-Yehoshua, *J. Amer. Soc. Hort. Sci.*, 1973, **98**(5), 464.

6. F. Artés and J.A. Martínez, *Lebensm. Wiss Technol.*,1996,**29**, 664.

Acknowledgements

Authors are grateful to the Consejería de Medio Ambiente, Agricultura y Agua - Fundación Séneca de la Región de Murcia for financial support, and to Agrícola Mar Menor, S.L. for providing lettuce and other facilities.

PHYSICAL METHODS FOR PREDICTION OF RIPENING OF APRICOTS

Cs. Balla[1] , A. Fekete[2] and J. Felföldi[2]

Department of Refrigeration and Livestock Products Technology[1]
Department of Physics and Control[2]
University of Horticulture and Food Industry
Ménesi út 45. Budapest,
H-1118 Hungary

1 INTRODUCTION

Horticultural products still live and respire after harvesting, their quality and appearance change during post-harvest handling. Products - including apricot - are affected by a lot of factors during the post-harvest chain, which affect their biochemical and ripening processes, so investigation of ripening is a basic condition of modelling. Apricot is a very popular fruit in summer time in Hungary. The fruits can reach their full ripening stage on the tree, but they have post-ripening ability as well. Apricot harvested at 70-75 % ripe stage can be stored up to one month, but after this long period the post-ripening is difficult and weight loss can increase up to 15-20 % . For long storage the 80-85 % ripe stage is more suitable (3).

Since there is correlation between ripening and colour, the colour development of fruits is used very often for modelling of ripeness (1), (4), (5). On the other hand, the good appearance in colour can stimulate purchase and consumption. Pedryc and Zana (6) confirmed that the a* values of consumer preferred apricot fruits are higher than 21-22.

The measurement of colour is a nondestructive method, but the analysis of fruits texture profile by non-destructive methods is very limited. During ripening the protopectin content of apricot fruits turns to water-soluble pectin content, which causes the softening of fruits. The Magness-Taylor penetration measurement test is very often used for texture analyses (2), but in the case of this method only the maximum penetration-force can be used to describe the firmness. For the quasi-nondestructive texture analysis only a little part of the penetration curve can be used, and in other cases *firmness assessment by acoustic parameters* is used as nondestructive method for the determination of the texture profile of fruits during ripening.

The aims of our experiment were to determine some physical parameters of apricots during ripening, to describe the ripening of fruits, and to find out how the temperature affects the ripening process of apricots during post-harvest storage.

2 MATERIALS AND METHODS

- *The raw material* for the experiment (variety: "Gönci Magyar Kajszi") was collected from an eight years old apricot orchard in 1997. The fruits used for the experiment were harvested from the same trees.
- *Creating visual colour scale*: The visual colour scale (from 1 to 8) suggested by Stenvers and Stock (7) for tomato was used for the experiment.
- *Storage:* change in ripeness and colour of the apricots was investigated at 5, 10, 15, and 20 °C in a temperature controlled cabinet.
- *Colour measurement:* Minolta CR 200 tristimulus colorimeter (45° illumination angle, 0° viewing angle, CIE C illumination, 8 mm diameter measuring area) was used. For the calibration white etalon was used: Y= 92.1, x= 0.3137, y= 0.3220.
- *Determination of colour characteristic (R-G):* The difference between red (R) and green (G) components was used, and it was measured by a video camera-based machine vision system.
- *Determination of rupture ratio (rr, kPa/mm):* The rupture stress (σm, kPa) and the deformation (zm, mm) which occurred with rupture stress was measured by penetrometer, and the rupture ratio was determined as follows:

$$rr = \sigma m \, / zm \tag{1}$$

- *Determination of coefficient of elasticity (ec, kPa/mm):* Compression stress (σ, kPa) occurred and z = 1.5 mm deformation was measured by quasi-nondestructive electronic penetrometer, where the coefficient of elasticity was calculated as follows:

$$ec = \sigma/z \tag{2}$$

- *Firmness assessment by acoustic parameters*: acoustic stiffness factor (s), as firmness parameter was determined by a characteristic sound frequency of the fruit's response to a mechanical input. Fast Fourier Transformations of the response were used to determine the characteristic frequency (fo, Hz), resolution is 4 Hz, and the acoustic stiffness factors were calculated, where: m = the mass of tested fruits, kg.

$$s = fo^2 * m \tag{3}$$

3 RESULTS

3.1 Correlation between visual colour estimation and instrumental colour measurement

Figure 1 shows the correlation between visual colour estimation and ground colour (A), the sunny-side colour (B), and flesh colour (C) of the same fruit. Sunny-side colour has higher standard deviation, flesh colour has a narrow range and it starts earlier than superficial colour development. As we had supposed, the visual colour estimation of ground colour has good linear correlation to CIELAB a* value (a*= 12.6+3.89 V, where V= visual colour estimation). If there is overlapping between neighbour groups, using this linear correlation it can be supposed that the a* value of ground colour will move between -12 (green) and 20 (orange) during the ripening of apricot.

3.2 Assessment of ripening stage of apricots by firmness and colour

Before instrumental investigation the fruits were separated into three different groups: immature (V=1-2), medium ripe (V=4-5), ripe (V = 7-8). Using multilinear regression analysis for the investigation of the correlation between physical parameters and ripening stage of fruits, there was good correlation found between the physical parameters and the estimated ripening stage (ERS). For the mathematical definition the following values were ordered to the different real ripening stages (RRS) of the fruits individually: immature =1, medium ripe=2, ripe=3.

$$ERS=0.608+0.00127*rr - 0.0000882*e_c + 0.0238*[R-G] \qquad (4)$$

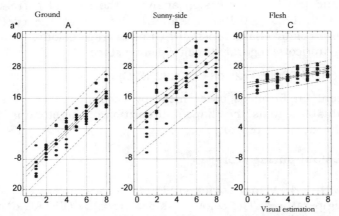

Figure 1 *The correlation between visual colour scores and CIELAB a* values of "Gönci Magyar " apricot*

There was good correlation (R^2=83.1, LSD = 0.1%) found between ERS and RRS, and the calculated ERS values could be used for classification of a batch, but classifying an individual apricot fruit is difficult, because of the overlapping of neighbour categories. To solve the problem discriminate analysis was used to determine the ratio of proper-improper fruit according their maturation. The efficiency of ranking was 85%, 63%, 82% using σ_m and e_c for selection, and there was no higher efficiency using e_c and r_r for it. Combining the colour characteristic (R-G) value and e_c for selection the efficiency of ranking increased to 91%, 76%, 100%, and there was no higher efficiency if three independent values (σ_m-e_c-[R-G]) for discriminant analysis were used (Figure 2).

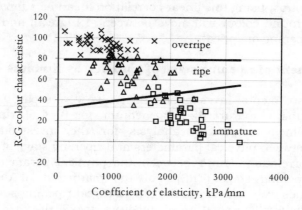

Figure 2 *Discriminant analysis for three ripening stage by the coefficient of elasticity and R-G colour characteristic*

Using e_c, and (R-G) for regression analysis, the level of correlation was same (R^2=83.1, LSD = 0.1%). It can be stated that the coefficient of elasticity and colour characteristic (R-G) as average colour, can be used for ranking apricots regarding their ripening stage.

ERS=0.682-0.00006472* e_c + 0.0232*(R-G)   (5)

Experiment on firmness assessment by acoustic parameters did not result in valuable signals, the shape and position of the stone in the fruits disturbed the acoustic response.

3.3 Fruit colour during storage

Fruits with 2-4 visual colour grade of 1-8 scale were stored at four different temperatures. The colour was described by CIELAB a* values of ground colour of the fruits. The colour development of 24 individual fruits was measured at each temperature. The colour changes of four selected apricots at four different temperatures are presented in Figure 3. The results confirmed that the change in colour of apricots during artificial ripening followed the sigmoid type model, and the storage temperatures affected the rate of ripening. Applying the sigmoid model and investigating the correlation, the parameters of the model are collected in Table 1.

$$a^*_x = C_1 + \frac{C_2 - C_1}{1 + e^{C_3(x - C_4)}} \tag{5}$$

$$R_{inf} = \frac{C_3(C_2 - C_1)}{4} \tag{6}$$

a_x^*= a* value at day X   C3= rate at inflection
C1=   a* value at the C4= time of inflection (day)
beginning
C2= a* value at the end R_{inf}= rate of changing colour and inflection (a*/day)

The data can describe the effect of temperature on ripening. If standard deviation represents the difference between fruits, both the rate and final colour are affected by storage temperature. At low temperature (below 15 °C) the final a* value cannot reach the value 21-22 described by Pedryc and Zana (6) as the lower limit of consumer preference (Figure 3).

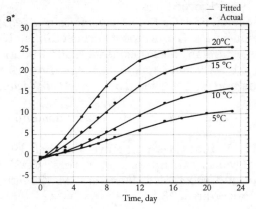

Figure 3 *Changing colour of apricot described by CIELAB a*values*

4 SUMMARY

The results confirm that there is close correlation between the visual colour estimation using eight categories for colour stage and CIELAB a* values. Using three ripening categories, the coefficient of elasticity and image analysed colour can be used for separating the immature, ripe and overripe stages of apricots. The superficial colour (CIE a*) follows sigmoid type curves during ripening, and the parameters of curves (maximum, and inflection) can describe the effects of different storage temperatures on ripening velocity.

Table 1 *The parameters of the sigmoid model used to describe colour during ripening*

Temperature	5°C		10°C		15°C		20°C	
C1 and stnd. error	- 3.10	1.80	3.38	0.67	4.92	0.88	-5.50	0.46
C2 and stnd. error	11.50	0.33	17.30	0.43	23.7 5	0.33	25.70	0.20
C3 and stnd. error	0.19	0.02	0.21	0.01	0.24	0.14	0.33	0.01
Rinf (a*/day)	0.67	0.16	1.05	0.15	1.72	0.14	2.58	0.19
Time of inflection (day)	9.70	0.31	9.39	0.29	7.41	0.25	5.55	0.11

References

1. Cs. Balla, T. Sáray, K. Horti, Á Koncz and K. Polyák, *COST94 Post-harvest treatment of fruit and vegetables, Workshop on "Quality Criteria", Bled, Slovenija* .1994, Proceedings 81.
2. A. Fekete, J. Felföldi and Cs.Balla, *Gyümölcs-zöldség postharvest és logisztikai konferencia* 1977. Budapest, Proceedings 139.
3. P. Sass, 'Gyümölcstárolás', Mezõgazdasági Kiadó Budapest 1986. 460.
4. C. N. Thay, and R.L.Shewfelt, *Americ. Soc. Agri. Eng.* 1990. **34**, 950.
5. L. M. M. Tijskens and R.G. Evelo, *Postharv. Biol. Techn.* (1994) **4**. 85-98.
6. A. Pedryc and J. Zana, *Hort.sci. - Kertészettudomány* 1995. **27**. (3-4). 55.
7. N. Stenvers and H.W. Stock, Gatrenbauwissenschaft 1976. **41**, 167
8. P.E. Zerbini, C. Liverani, G. Spada and E Tura, *COST94 Post-harvest treatment of fruit and vegetables, Workshop on "Quality Criteria", Bled, Slovenija* 1994, Proceedings 93.

EFFECT OF CALCIUM AND PRESTORAGE HEAT-TREATMENTS ON THE STORAGE BEHAVIOUR OF 'RAIKE' AND 'RED ATLAS' APPLES

R. Dris and R. Niskanen

Department of Plant Production
Horticulture, P.O. Box 27,
FIN-00014 University of Helsinki,

1 INTRODUCTION

Preharvest calcium spraying on fruitlets have been reported to reduce the incidence of physiological storage disorders in apples[1,2,3]. Prestorage heat treatment of apples has been shown to affect the postharvest quality by delaying ripening and reducing ethylene production rate[4,5]. The aim of this study was to investigate the effect of calcium chloride spraying or heat treatments on storage behaviour of two commercial apple cultivars.

2 EXPERIMENTAL

The preharvest spraying treatments on fruitlets of apple cultivars Red Atlas and Raike was done six times during the fruit development with $CaCl_2$ solution (Ca 1.3 g/l) at two week intervals. During the prestorage heat treatment, apples were kept in a warming chamber at 38°C for 4 days[5]. The relative humidity aimed at was 90-95%, but only about 50% could be reached with the equipment available.

Calcium or heat treated and control apples were stored for six months at +2°C and 90-95% RH. Nutrients in fruit flesh and other quality characteristics were analysed and physiological disorders scored after 0, 1, 3 and 6 month storage. The diameter, firmness, juice pH and soluble solids concentration (SSC) were measured and fruit flesh nutrient contents were analysed. Fruit firmness was measured by a penetrometer and SSC by refractometer. Freeze-dried apple slices were pulverized and N was determined by the Kjeldahl method. Fruit samples were ashed at 475°C and dissolved in 0.5 M HCl. P in the ash extracts was determined by the ammonium vanadate method, K by flame photometry, Ca and Mg by atomic absorption spectrophotometry (AAS).

Two replicates of two uniform sized apples of each treatment were kept in an open 1-liter jar in the storage room. The CO_2 and C_2H_4 production rates were monitored 7 times during storage. For measurements, apples were weighed and kept in jars tightly closed with a metal lid for 12 to 14 h. The jars were uncovered after taking gas samples with a syringe through the rubber septum in the lid. A 5 ml sample of gas was injected into a gas chromatograph equipped with a Porapak Q column for CO_2 and a flame ionization detector for ethylene.

Results were tested by analysis of variance. In the tables, mean values followed by the same letter are not significantly different at P=0.05.

Mean N of fruit flesh was higher in 'Raike', P and K in 'Red Atlas' and there were no significant differences in Ca, Mg and K/Ca between cultivars (Table 1). The total DM in fruit flesh was higher in 'Red Atlas' and there were no significant differences in juice pH, soluble solids concentration, diameter and firmness/diameter (Table 2).

As mean of both cultivars, the calcium treatment increased N and decreased P, heat treatment increased N, P and K, but both treatments did not affect Ca and Mg (Table 3). The calcium treatment decreased the K/Ca ratio in flesh of Red Atlas (Table 4). The heat treatment increased the pH in juice of 'Raike' apples and decreased the ethylene production rate of 'Red Atlas' apples. The SSC in juice of heat treated 'Red Atlas' apples was higher than in juice of calcium treated apples.

After three month storage, the incidence of physiological disorders in control apples (7.5%) was significantly higher than those in calcium treated (2.5%) and heat treated apples (2.5%), but after 6 month storage there were no significant differences between treatments. The disorders found were mainly senescence breakdown in 'Raike' and lenticel spot in 'Red Atlas'.

Table 1 *Mean Nutrient Contents in Apple Flesh* mg kg^{-1} *Fresh Weight and K/Ca Ratio of Cultivars*

Cultivar	N	P	K	Ca	Mg	K/Ca
Red Atlas	353b	124a	1470a	55a	49a	26.8a
Raike	466a	101b	1375b	75a	49a	19.3a

Table 2 *Mean Juice pH and Soluble Solids (SSC),Total Dry Matter of Apple Flesh, Diameter (D) and Firmness/Diameter (F/D) of Apple Cultivars*

Cultivar	pH	SSC (%)	DM (%)	D (cm)	F/D (N cm^{-1})
Red Atlas	3.26a	11.3a	14.8a	6.33a	85.5a
Raike	3.47a	10.4a	13.6b	6.24a	77.9a

Table 3 *Effect of Calcium and Heat Treatments on Mean Nutrient Concentrations* mg kg^{-1} *Fresh Weight in Flesh of Apple Cultivars During Storage (0 = Control, Ca = Treated with CaCl$_2$, Ht = Heat Treated)*

Treatment	N	P	K	Ca	Mg
0	380c	112b	1412b	69a	49a
Ca	434a	103c	1312b	64a	48a
Ht	416b	124a	1544a	63a	51a

The preharvest CaCl$_2$ treatment did not increase Ca level in apple flesh, because the small amount of Ca absorbed from foliar sprays might fail to penetrate into the fruit and

Table 4 *Effect of Calcium and Heat Treatments on K/Ca Ratio in Apple Flesh, Juice pH and Soluble Solids (SSC), Mean Respiration and Ethylene Production Rates at +2⁰C of Apple Cultivars During Storage (0 = Control, Ca = Treated with CaCl₂, Ht = Heat Treated)*

Cultivar	Treat-ment	K/Ca	pH	SSC %	CO₂	C₂H₄
					mg h⁻¹ kg⁻¹	
Red Atlas	0	29.2a	3.25 a	11.1 ab	3.21 a	0.186 a
	Ca	22.4b	3.25 a	10.7 b	2.60 a	0.167 a
	Ht	28.8a	3.28 a	12.1 a	3.13 a	0.133 b
Raike	0	16.2a	3.33 b	10.6 a	2.67	0.174
	Ca	20.0a	3.30 b	10.0 a	3.95	0.172
	Ht	21.8a	3.78 a	10.5 a	3.44	0.176

be analytically masked by background Ca levels in the whole fruit samples[6].The increase in fruit N by CaCl₂ treatment might be associated with delaying of maturation process by preharvest Ca spraying treatment[2,3].The fruit firmness per diameter was not affected by the low numbers of CaCl₂ sprays. The increase in fruit firmness is obtained by very frequent CaCl₂ spraying treatments only[1].

As in the earlier studies[5], the heat treatment increased pH and decreased C₂H₄ production rate. Both CaCl₂ and heat treatments reduced the incidence of physiological disorders until three month storage. In studies on other apple cultivars, breakdown, core browning and decay were almost eliminated by heat treatment at 38°C for four to six days[4].

References

1. F. J. Peryea, *Good Fruit Grower*, 1991, **42**,12.
2. R. Dris and R. Niskanen, Report, Department of Plant Production, Horticulture Section, University of Helsinki, 1997.
3. R. Dris and R. Niskanen, *Acta Hort.*, 1997, **448**, 323.
4. S. W. Porritt and P. D. Lidster, *J.Am. Soc. Hort. Sci.*, 1978, **103**,584.
5. J. D. Klein and S. Lurie, *J. Am. Soc. Hort. Sci.*, 1990, **115**, 265.
6. J. R. Davenport and F.J. Peryea, *J. Plant Nutr.*, 1990, **13**, 701.

NITROGEN AND CALCIUM STATUS IN LEAVES AND FRUITS OF APPLE TREES GROWN IN THE ÅLAND ISLANDS

R. Dris and R.Niskanen

Department of Plant Production
Horticulture, P.O. Box 27
FIN-00014 University of Helsinki

1 INTRODUCTION

Quality maintenance of apple fruit is influenced by mineral composition and maturity stage at picking time[1,2,3]. High N in the tree favors vegetative growth rather than fruiting and affects fruit quality[4,5,6]. Apples high in Ca have low respiration rate and longer potential storage life than low-calcium fruits and high Ca also effectively counteracts the detrimental metabolic effects of high N in the fruit[7,8]. The objective of this study was to investigate the N and Ca status in leaves and fruits of apple trees.

2 EXPERIMENTAL

Leaf and fruit samples of 12 apple cultivars (Maikki,Transparente Blanche, Melba, Red Melba, Samo, Ranger, Jaspi, Raike, Red Atlas, Åkerö, Aroma and Lobo) were collected in nine orchards during 1993-1995. A total of 80 leaves were sampled during the period 22.7.- 26.8. from two randomly chosen plots with two trees per cultivar and dried at 70°C for 24 h. Four fruits per tree and eight fruits per plot were collected one week before predicted commercial maturity.During 1993-1994, the sliced fruit samples were freeze-dried and vacuum packed. During 1995, samples were dried at 60-65°C for 48-96 h. Leaf and fruit samples were pulverised and N was determined by the Kjeldahl method. Fruit and leaf samples were ashed at 475°C, dissolved in 0.5 M HCl and Ca determined by atomic absorption spectrophotometry (AAS).

The average N and Ca contents in leaves during the sampling period (Table 1) were within the recommended limits: N 20-30, Ca 10-20 g/kg of DM^9. In some samples, the leaf N concentration was below the lower limit, but Ca concentration often exceeded the recommendations. For preventing the storage disorders in apple fruits, leaf Ca should be more than $1.1-1.2\%^{10}$. In our study, this limit was exceeded.Mean leaf nutrient ratio N/Ca was in the tentative range 1.0-1.5, calculated on the base of recommended limits[9].

According to the recommendations in the United Kingdom for apple fruit of good keeping quality, N and Ca concentrations should be 500-700 and 50 mg/kg fresh weight, respectively[5].Mean fruit N was lower than recommendation but Ca was equal to recommended value (Table 2). Fruit mean N/Ca was in the tentative range 10-14, based on the recommendations[5]. Fruit N correlated positively with leaf N ($r=0.28*$) and fruit Ca

Table 1 *Means, Standard Deviations and Ranges of Nitrogen and Calcium Concentrations g kg^{-1} Dry Matter in Apple Leaves (Sampling 22.7.-26.8.) During 1993-1995*

	Number of Samples	Mean	Standard Deviation	Range
N	171	22.1	2.6	16.0-29.7
Ca	171	17.4	4.5	7.0-33.3
N/Ca	171	1.35	0.38	0.64-3.54

Table 2 *Means, Standard Deviations and Ranges of Nitrogen and Calcium Concentrations mg kg^{-1} Fresh Weight in Apple Fruit During 1993-1995*

	Number of Samples	Mean	Standard Deviation	Range
N	49	469	128	221-867
Ca	49	50	18	24-111
N/Ca	65	11.0	5.2	4.4-32.4

negatively with leaf Ca (r=-0.36**), but the values of linear correlation coefficients were not high. It has often been found that there are no close correlations between corresponding leaf and fruit nutrients, even when fruits and leaves are sampled on the same day[11].

References

1. J. van der Boon, *J. Hort. Sci.*, 1980, **55**, 313.
2. R. D. Marcelle, *Acta Hort.*, 1989, **258**,373.
3. I. B. Ferguson, R.K. Volz, F. R. Harker, C.B. Watkins and P.L. Brookfield, *Acta Hort.*,1995, **398**, 23.
4. W. J. Bramlage, M. Drake and W. J. Lord, In: D. Atkinson, J. E. Jackson, R.O. Sharples and W. M. Waller (eds), 'Mineral nutrition of fruit trees', London,1980, p 29.
5. R.O.Sharples, In: D. Atkinson., J.E. Jackson, R.O. Sharples and W.M.Waller (eds), 'Mineral nutrition of fruit trees', Butterworths, London, 1980, p 17.
6. R. Dris, R. Niskanen and I. Voipio, Department of Plant Production, Horticulture Section Publ.32, University of Helsinki, 1997.
7. M. Faust and C.B. Shear, *J. Am. Soc. Hort. Sci.*, 1972, **97**, 437.
8. M. Faust and J.D. Klein, *J. Am.Soc. Hort. Sci.*, 1973, **99**, 93.
9. Viljavuuspalvelu Oy, 'Viljavuustutkimuksen tulkinta avomaan puutarhaviljelyssä', 1997.
10. H. Clysters, *Fruit Belge,*1967, **35**, 247.
11. R. D. Marcelle, *Acta Hort.,*1990, **274**,315.

CURRENT STATE AND FUTURE REQUIREMENTS OF THE IRISH VEGETABLE INDUSTRY

A.Greene, M. Cowley and J. Mulcahy

Dublin Institute of Technology
School of Food Science and Environmental Health
Cathal Brugha St.
Dublin 1
Ireland.

1. INTRODUCTION

The Irish food industry is a vital national industry accounting for 37% of all industrial output. The industry is largely Irish owned, with 75% of employment in indigenous Irish owned companies[1]. The agricultural sector is one of the most important and traditional sectors of Irish industry. Agriculture represents 8.3% of Ireland's gross domestic product and employs 11.4% of the workforce. The fruit and vegetable industry, as part of this sector, accounts for 6.1% of gross agricultural output[2].

Statistics from the Department of Agriculture, Forestry and Food, identify that 90% of Irish grown vegetables are sold unprocessed with just 10% being processed[3]. Consumer trends towards healthy eating, convenience and freshly prepared produce are creating a market for produce prepared using minimal processing techniques such as modified atmosphere packaging (MAP). In addition, retail and food service sectors aim to purchase vegetables in order to decrease costs and labour input, and to improve hygienic practices[4].

The market for minimally processed fruit and vegetables has grown explosively in Europe in the early 1990's, especially in the UK and France[5]. The UK and France produce 40% and 25% respectively of all modified atmosphere packaged fresh chilled produce[6].

2. RESEARCH AIMS

The aim of this research is to establish the purchasing behaviour of Irish caterers and retailers with respect to sourcing vegetables. The main objectives of this study were to:
- establish the most frequently purchased vegetables.
- determine the presentation format of these vegetables.
- establish the origin of these vegetables.
- identify the main reasons for purchasing particular products.
- determine the level of satisfaction with vegetables currently available.
- identify vegetables/vegetable mixes which would benefit the consumer if their shelf life could be extended

A sample of one hundred Irish catering outlets (hotels, restaurants, catering companies), and retail outlets (major supermarket chains) was surveyed using a detailed postal questionnaire or by interview.

3. METHODOLOGY

The survey addressed a number of issues. Respondents were asked to indicate the following:

• seasonality of purchase of each vegetable.
• presentation format of each vegetable.
• vegetable mixes currently purchased and any vegetable mixes respondents would like to see available on the market.
• vegetable quantities purchased.
• the percentage of organically grown vegetables purchased.
• vegetables for which an extended shelf-life would be desirable.
• further preparation carried out on vegetables purchased.
• uses of the purchased vegetables.

4. RESULTS

This study has established the purchasing behaviour of Irish caterers and retailers with respect to sourcing fresh vegetables.

The most popular vegetables purchased in Ireland by caterers and retailers are carrots, lettuce, onions and potatoes. A higher proportion of imported vegetables was rated by respondents as excellent in quality, e.g. of all imported carrots purchased, 50% were rated as excellent while only 29% of all home produced carrots was rated as excellent (Figure 1). In addition, a higher percentage of home produced vegetables was rated as fair/poor in quality, e.g. 20% of Irish grown lettuce was rated as fair in quality while all imported lettuce was rated as excellent/good in quality.

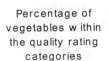

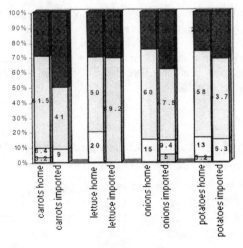

Figure 1 *Quality by origin of the four most frequently purchased vegetables in Ireland*

Two of the most popular vegetables (carrots, onions), were chosen in order to compare the origin of the vegetables purchased by caterers and retailers (Figure 2). Retailers purchased a high percentage of both home and imported carrots (66.7%), while the majority of caterers (81.5%) purchased home produced carrots only.

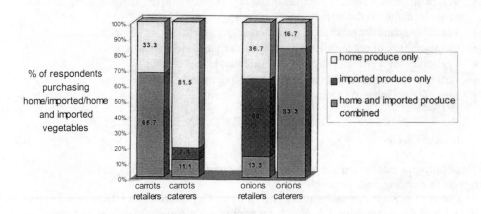

Figure 2 *Comparison of the origin of two vegetables purchased by caterers and retailers*

A large percentage of retailers (50%) purchased imported onions only while the majority of caterers (83.3%) purchased a combination of both home and imported produce.

Both caterers and retailers purchased a large proportion of imported produce. Reasons cited by respondents for purchasing imported vegetables were quality, regularity of supply and cost advantage.

Table 1 shows that it was primarily retailers (80%) who were interested in prolonged vegetable shelf life at extra cost. A smaller percentage of caterers (38%) were in favour of prolonged vegetable shelf life at extra cost.

Table 1. *Percentage of caterers/retailers in favour of prolonged vegetable shelf life at an extra cost.*

	% in favour	% not in favour
CATERER	38	62
RETAILER	80	20

5. DISCUSSION

Statistics from the Irish Central Statistics Office show that in 1996, total Irish imports of carrots and other root crops were 20,234 tonnes, accounting for 38% of total consumption[7]. This study showed the dependence of Irish caterers and retailers on imported produce and the generally higher level of satisfaction with the quality of imported vegetables. Respondents were concerned with quality, seasonal availability and cost of Irish vegetables. Reports from An Bord Glas, the Irish Horticultural Development Board, state that these factors are often due to reasons beyond our control such as climate, increasing costs and competitiveness from abroad. An analysis of imports shows that a large proportion are brought into Ireland at times when they cannot be produced from domestic sources[8]. Also, costs of production of many vegetable lines have increased without a corresponding increase in returns from the marketplace. For some vegetables, this has led to a decrease in the acreage of certain crops being grown[9].

Many Irish farmers and food service outlets are seeking to increase demand for home produced vegetables through the production of value added products e.g. stir fry mixes, sliced/diced carrots. High consumer demand for convenient and freshly prepared produce coupled with the high demand among retailers for prolonged vegetable shelf life, demonstrates the opportunities for further development and application of MAP in Ireland. Reasons for the lower percentage of caterers in favour of prolonged vegetable shelf life were the daily availability of fresh vegetables and the desire not to interfere with the natural deterioration process of the vegetables. This highlights the need for food service outlets and consumers to be informed of the benefits of MAP as a chemical, preservative and additive-free method of prolonging vegetable shelf life.

References
1. Food, Drink and Tobacco Federation, 'The Irish Food Sector, Competitiveness with the UK', Irish Business and Employers Confederation, Ireland, October 1995, Section 1, p. 1.
2. Fingleton, W. A., Cushion, M., 'Irish Agriculture in Figures', September 1997, Teagasc, Ireland, 6th Edition, p. 3.
3. 'Estimate of Gross Horticultural Output 1996,' Department of Agriculture, Forestry and Food, Ireland.
4. Ahvenhainen, R., Hurme, E., Mattila, M, Proceedings of the Sixth International Symposium of the European Concerted Action Program COST 94. 'Post Harvest treatment of fruit and vegetables'. 1-2 October, 1992, Istanbul, Turkey.
5. Day, B.P.F., *European Food and Drink Review,* Autumn 1989, p47.
6. Day, B.P.F., *Food Science and Technology Today,* 1990, **4**, 215.
7. 'Trade Statistics of Ireland 1995-1996,' Central Statistics Office, Ireland.
8. 'Hort. Report', An Bord Glas, The Horticultural Development Board, Ireland, Summer 1997, p. 2.
9. 'Hort. Report', An Bord Glas, The Horticultural Development Board, Ireland, Spring 1997, p. 2.

CHANGES IN THE CHEMICAL AND SENSORY QUALITY OF GREEN PEAS DURING DEVELOPMENT

M. Hansen[1], J. N. Sørensen[1] and L. Wienberg[2]

[1]Department of Fruit, Vegetable and Food Science, Danish Institute of Agricultural Sciences, Research Centre Aarslev, DK-5792 Aarslev and [2]Department of Dairy and Food Science, The Royal Veterinary and Agricultural University, DK-1958 Frederiksberg C, Denmark

1. INTRODUCTION

A number of chemical and physical processes take place in green peas during growth and development which influence both the flavour and texture of the product. During the first approx. 3 weeks after flowering, there is a large reduction in moisture content and a great increase in seed weight and content of soluble fructose, glucose and sucrose (1). Soluble carbohydrates decrease thereafter and are used to form starch (1) and other structural components. During the period from immature to mature seeds, weight, size and therefore yield increase considerably (2). This yield increase is, however, often followed by a severe decrease in the sensory quality of the processed product.

According to Danish consumers, pea flavour and mealiness are the most important attributes of green pea quality followed by juiciness, crispness, sweetness and skin toughness (3). The aim of the present work was to study changes in the chemical and sensory quality of two pea cultivars during seed development. Peas were analysed for alcohol insoluble solids (AIS), starch and sucrose content, and evaluated for pea flavour, sweetness, bitterness and mealiness by sensory descriptive analysis

2. MATERIALS AND METHODS

Peas *(Pisum sativum* L.) of two cultivars were commercially-grown and processed and supplied frozen in three sizes by a Danish company (Danisco Foods) in 1993. The sizes were: fine (8.2-8.75 mm), medium (8.75-10.2 mm) and large peas (> 10.2 mm). The same cultivars were grown and processed experimentally in 1996 to study the effect of successive harvesting on pea quality retention. The total heat units above 4.5°C (4) were recorded from sowing to flowering, from flowering to harvest and during harvesting. At harvest, peas were threshed, washed and graded into size using a Jel shaking equipment (Engelsmann, Ludwigshafen, Germany) with stainless steel sieves. Following grading, peas of the medium size (8.75-10.2 mm) were blanched for 2½ min to inactivate peroxidase, frozen and stored until later analysis. Samples were analysed for tenderometer value (TV), AIS and starch content according to Hansen et al. (5). Sugars were extracted with water and the sucrose content determined by HPLC (6). Sensory evaluation was carried out according to procedures described by Hansen et al. (5) and Wienberg et al (7). The evaluation was made by 11 panelists using a set of 14 descriptive terms (7). Only results on pea flavour, sweetness, bitterness and mealiness are given here. Specific pea samples were used as references for pea flavour, sweetness and bitterness. Mealiness was defined as the starch-like sensation of 2-3 pea cotyledons (5). Means with corresponding standard errors (SEM) were calculated for the physical-chemical variables using SAS (Cary, N.C.). Means with corresponding standard errors for the sensory variables were calculated using mixed model Analysis of Variance (ANOVA) taking the combined residual structure into account (7).

3. RESULTS AND DISCUSSION

Changes in the chemical and sensory quality of two pea cultivars in three sieve sizes each are shown in Figure 1. The TV, starch and AIS content increased as size increased from fine to large peas (Figure 1A, 1B, 1C). Young seeds (size fine) had a high content of sucrose and little starch (Figure 1B). As the seeds developed, the amount of starch increased and the sucrose content either increased (cultivar 1) or decreased (cultivar 2) and this resulted in a rise in the AIS content and the sensory scores for mealiness in both cultivars (Figure 1C, 1F). Others (8, 9) also report that the TV, starch and AIS content increase with size.

Peas of cultivar 1 were less mature at harvest as indicated by a lower starch and AIS content than those of cultivar 2 (Figure 1B, 1C). This result was surprising because there was only a slight difference in the TV between cultivars regardless of size (Figure 1A). The TV is used indirectly to express maturity of raw peas at harvest (10). As the TV was measured on the same machine, differences in maturity were not due to errors in the readings but probably genetic or environmental factors between cultivars (2, 4). In general, peas of

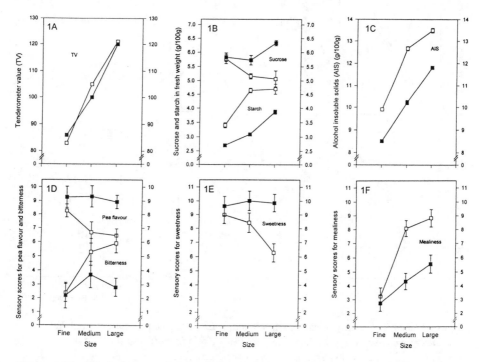

Figure 1 *The effect of size on the chemical and sensory quality of cultivar 1 (■) and 2 (□) peas. Vertical bars represent ± SEM, n=3 for all except for AIS (n=7)*

cultivar 1 had higher sensory scores for pea flavour and sweetness and lower scores for bitterness and mealiness than peas of cultivar 2 (Figure 1D, 1E, 1F). These differences were especially pronounced in the medium and large peas. The relative increase in AIS content from fine to large peas was 40 % in cultivar 1 and 35 % in cultivar 2. Despite almost similar relative increases in AIS content from fine to large peas in both cultivars, the relative decrease in sensory scores for pea flavour and sweetness was only 5 % in cultivar 1 compared with 20-30 % in cultivar 2. In addition, the relative increase in sensory scores for bitterness was also higher in cultivar 2 than in 1. The quality of large peas were tested among Danish consumers, who preferred peas of cultivar 1 for those of cultivar 2 (11). A larger pea size of cultivar 1 may therefore be used in high quality products when both crops are harvested at identical TV.

Retention of quality during harvesting is an important factor in industrial processing. Cultivars with rapid changes in quality are less suitable for industrial processing than cultivars which change slowly. The effect of successive harvesting on the quality retention of peas grown under controlled conditions is shown in Figure 2. Only the quality of the medium sized peas are shown. The accumulated heat units during harvesting are shown in Figure 2A. In total, cultivar 1 accumulated 57 heat units from harvest day 1 to 5 and cultivar 2 accumulated 52 units (Figure 2A). Cultivar 1 had accumulated a total of 676 heat units from sowing to harvest and cultivar 2 had accumulated 775 units. The vegetative period was

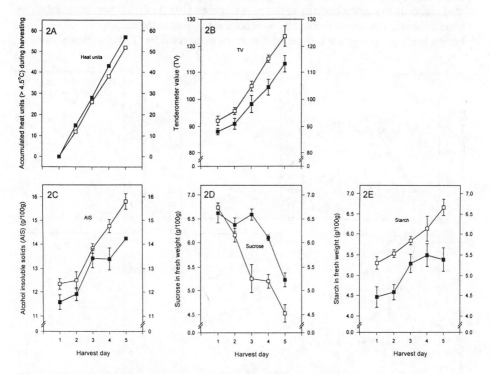

Figure 2 *Chemical quality of cultivar 1 (■) and 2 (□) peas (the medium size) during successive harvesting. The total heat units were 676 in cultivar 1 and 775 in cultivar 2 at day 1. Vertical bars represent ± SEM, n=3.*

longer for cultivar 2 (488 units) than for cultivar 1 (370 units). On the other hand, the generative period (from flowering to harvesting) was longer for cultivar 1 (306 units) than for cultivar 2 (287 units).

The changes in physical and chemical quality during harvesting are shown in Figure 2B-2E. At day 1 there was only a slight difference in the sucrose content between cultivars (Figure 2D). However, as harvesting proceeded, the sucrose content decreased faster in cultivar 2 and this was accompanied by a faster increase in starch content as compared with cultivar 1 (Figure 2D, 2E). During the 5 day period, the sucrose content decreased 20 % in cultivar 1 and at the same time the starch and AIS content increased 20 %. In cultivar 2, the sucrose content decreased 35 % and the starch and AIS content increased 25 %. These results indicate that conversion of sugars to starch may have been initiated earlier and/or that the conversion to starch was faster in cultivar 2 than in cultivar 1. In general, the relative change in physical and chemical quality per heat unit during the 5 day period was lower in cultivar 1 than in cultivar 2. For every one heat unit, TV, AIS and starch content increased 0.44 units, 0.064 % and 0.016 g/100 g, respectively, in cultivar 1. In contrast, the values were 0.62, 0.085 and 0.026 in cultivar 2.

In conclusion, TV, starch, sucrose and AIS content of green peas and sensory mealiness were dependent on size and maturity at harvest. Sensory pea flavour, sweetness and bitterness were dependent on size in cultivar 2 but size independent in cultivar 1. For these reasons, a larger pea size of cultivar 1 may be used in high quality pea products when the crops are harvested at identical TV. Cultivar 1 had the best quality retention during harvest, which makes the timing of harvest of this cultivar less critical.

References

1. R.G. Stickland and K.E. Wilson, *Ann. of Botany*, 1983, **52**, 919.
2. L. Ottosson, L, 'Growth and Maturity of Peas for Canning and Freezing', Almqvist & Wiksells Boktryckeri AB, Uppsala, 1958.
3. A.C. Bech, M. Hansen and L. Wienberg, *Food Quality and Pref.*, 1997, **8**, 329.
4. K. Grevsen and U. Kidmose, *Danish J. Plant Soil Sci.*, 1992, **96**, 279.
5. M. Hansen, A. Thybo, L. Erichsen, L. Wienberg and L. Andersen, submitted to *J. Food Sci.*
6. M. Hansen, A.C. Bech, P.M. Brockhoff and L. Wienberg, in preparation for *J. Sci. Food Agr.*
7. L. Wienberg, M. Martens and P. Brockhoff, in preparation for *J. Food Quality*.
8. G. Ros and F. Rincón, *Lebensm.-Wiss. u.-Technol.*, 1991, **24**, 549.
9. M.J.Periago, G. Ros, C. Martínez, F. Rincón, G. Lopez, J. Ortuno, and J. Rodrigo, *J. Food Quality*, 1996, **19**, 91.
10. P.J. Rutledge and P.W. Board, *J. Texture Stud.*, 1980, **11**, 379.
11. C.S. Poulsen, H.J.Juhl and A.C. Bech, Personal communication, 1996.

DO CHANGES IN CHEMICAL COMPOSITION DURING STORAGE OF RAW ONIONS (*ALLIUM CEPA*) INFLUENCE THE QUALITY OF DEEP FAT FRIED ONIONS?

Susanne Lier Hansen
Danish Institute of Agricultural Sciences
Department of Fruit, Vegetable and Food Science
Kirstinebjergvej 6, DK-5792 Aarslev

1 INTRODUCTION

Fried onion pieces are brown and crispy and a popular condiment for hot dogs and other bread and meat combinations. According to the industry, darkness and oil content of deep fat fried onion slices depends among other things on storage time of the raw onions. This makes it difficult to produce a homogeneous product throughout the year.

It is well known that dry matter content and composition change during storage of onions[1,2]. Especially the quantitative changes in carbohydrate constituents are of interest, since fructose and glucose participate in the Maillard reaction responsible for the browning during frying. It was previously shown that free amino acid content, another reagent in the non-enzymatic browning process, plays a synergistical role in the colour production, but reducing sugar content is the limiting factor in influencing fry colour of potatoes[3].

Relationship between initial dry matter content and final oil content in different varieties was confirmed in potato chips production[4]. In another study an increased initial surface moisture content resulted in increased oil content[5]. Increasing the dry matter content by pre-fry drying caused a decrease in the oil content in the fried product[6,7]. The influence of the natural changes in dry matter content and composition during storage on the quality of fried onions is still obscure.

The purpose of the study was to investigate how the quality of deep fat fried onions depended on the changes in dry matter content and composition of raw onions during storage.

2 MATERIALS AND METHODS

For two seasons (1993 and 1994) onions (*Allium cepa* L. CV Hyduro) were grown by a local grower in Denmark. Tubers were harvested by hand, when 80% of the plants had collapsed tops, dried at 24^{0}C for 5 weeks and stored for 8 months at 1^{0}C, RH 75-80%. The bulbs were not topped.

2.1 Sampling and Preparation Procedure for Analysis

Representative samples for analysis were taken immediately after harvest and at suitable intervals during storage. They were hand trimmed, bisected and sliced into 2 mm lengths using a slicer (DITO SAMA TR-21).

Onions for frying were mixed with a proprietary batter and deep fat fried at 150⁰C as previously described [8].

2.2 Analysis

2.2.1 Raw onions. Dry matter content, N-total and soluble carbohydrate[2]. Free amino acid[9].

2.2.2 Deep fat fried onions. Oil and moisture content and colorimetric evaluation of darkness[7].

3 RESULTS AND DISCUSSION

The dry matter content changed during storage (table 1) for onions harvested in 1993 and for onions harvested in 1994. A pronounced difference in dry matter content from time of harvest until the end of storage was found. A well-defined difference in the dry matter level was seen between the two years. Total-N content remains constant except the period of drying, 1-5 weeks. Onions harvested in 1994 had the highest dry matter content during the entire storage increase after one week of storage/drying. No detectable differences for the two years was found. (Data not shown.) For both years the total fructan content was reduced during storage.

Table 1 *Changes in Dry Matter Content and Composition in Onions during Long-term Storage*

Storage (weeks)	Dry matter [A] (g/100g onions)		Total-N[B] (mg/100g onions)		Fructans[C] (g/100g onions)		#Reducing Sugars[D] (g/100g onions)	
	1993	1994	1993	1994	1993	1994	1993	1994
1	12,33	11,10	143	149	5,32	4,56	2,05	1,80
5	12,36	11,28	178	202	4,82	4,15	1,70	1,87
10	12,10	11,03	179	205	2,69	1,91	2,41	3,03
16	12,06	10,78	179	206	2,59	1,66	2,91	3,30
23	11,31	10,55	182	211	2,29	-	2,76	3,52
29	11,21	10,38	185	203	1,61	1,33	2,84	3,52
35	11,16	10,02	175	199	1,58	0,97	2,92	3,41
41	11,50	10,33	177	205	1,45	0,89	2,93	3,36

- Missing value
(glucose + fructose)
The standard errors were in the range [A] 0,01-0,57, [B] 1-11, [C] 0,01-0,30 and [D] 0,01-0,24.

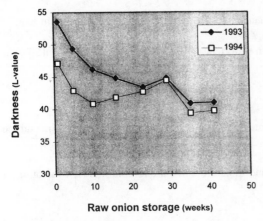

Figure 1 *Changes in Darkness (L-value) of deep fat fried onion pieces during long term storage of raw hole onions.*

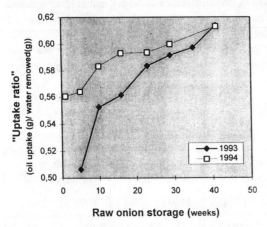

Figure 2 *Changes in "uptake ratio" (oil uptake (g) / water removed (g)) of deep fat fried onion pieces during long term storage of raw hole onions*

The fructan content in the onions harvested in 1994 was during the entire period lower than in the onions harvested in 1993. Reducing sugars, representing glucose and fructose increased until 15-20 weeks of storage, afterwards it remained almost constant, with the highest content in onions harvested in 1994. Enzymatic hydrolysis of fructan to fructose and glucose might explain the change in carbohydrate content in raw onions during storage[10].

The darkness of the fried product evaluated colorimetrically is shown in figure 1. In 1993 the darkness was increasing during storage, except a decrease in darkness at week 29. In 1994 the darkness increased only until week 9, afterwards there was a decrease until week 29. This development can partially be explained as change in content of reactants in the Maillard-reaction. The higher the content of reducing sugars the darker the fried product. This is in agreement with a previous study[3]. The content of total amino acids, which also participate in the Maillard reaction, do not change during storage, which means that they are not the limiting factor in the Maillard reaction. This agrees with conclusions drawn from a potato chip study[3]. The decrease in darkness of the fried product at week 29 might be due to sprouting. Sprouting lead to a re-allocation of metabolites for synthesis of new cells in the sprout. As a result the carbohydrate contents increase in the area around the new sprout and decrease in the remaining part. Because of the high content of reducing carbohydrates in the inner part of the onions some minor parts get very dark during frying while the remaining part developed a less dark colour.

Since one main effect of deep fat frying is water replacement by oil, the "uptake ratio" U_r brilliantly describes changes in the water holding capacity during deep fat frying[11]. The U_r express the weight ratio between oil uptake and moisture loss[11]. As can be seen in figure 2 U_r increases as a function of raw onions storage time. This means that the older the raw onions used for frying the greater is the ratio of oil entered and moisture removed. The figure also shows, that onions harvested in 1994 always obtained the highest U_r criterion. The study shows a relationship between dry matter content in raw onions and the U_r criterion. This is in agreement with a pre fry drying study, where increasing the initial dry matter content in onions resulted in a decrease in the fat content. The moisture content of the fried product was independent of the manipulated pre frying dry matter content[7].

4 CONCLUSION

Deep fat fried onions get darker the longer the period of raw onion storage. Exception is at the beginning of sprouting where reagents in the Maillard reaction are re-allocated, and the fried onions get less dark. The darkness of the fried product not only depends on the dry matter content and constituents of the raw onions it also depends on the allocation in the onion. Reducing sugars are the limiting factor in colour development during deep fat frying of onions.

The oil content of deep fat fried onions depends on the dry matter content at time of harvest and the raw onions storage time prior to frying. The lower the dry matter content the higher is the ratio oil uptake/water removed in the fried onions.

5. REFERENCES

1. B. Darbyshire, *J. Hort. Sci.*, 1978, **53**, 195.
2. S.L. Hansen, *Acta Agr. Scand* (submitted).
3. M.A. Roe et al., *J. Sci. Food Agric.*, 1990, **52**, 207.
4. M.H. Gamble and P. Rice, *Lebensm. -Wiss.u. -Technol.*, 1988, **21**, 62.
5. I. Lamberg et al., *Lebensm. -Wiss.u. -Technol.*, 1990, **23**, 295.
6. M.H. Gamble and P. Rice, *Intl. J. Food Sci. Technol.*, 1987, **22**, 535.
7. S.L. Hansen, *J. Food Quality* (submitted).
8. S.L. Hansen, *Irish J. Agr. Food Res.*, 1998, **37**, (in press).
9. S.L. Hansen, (writing paper).
10. H.G. Pontis, `Methods in Plant Biochemistry`, Academic Press Ltd., London, 1990, Vol. 2, Chapter 10, p. 353.
11. E. J. Pinthus et al., *J. Food Sci.*, 1993, **58**, 204.

POSTHARVEST QUALITY OF STRAWBERRIES

Margareta Hägg[1,3], Ulla Häkkinen[1], Mirja Mokkila[2], Kati Randell[2] and Raija Ahvenainen[2]

[1]Agricultural Research Centre of Finland (MTT), Laboratory of Food Chemistry, FIN-31600 Jokioinen.
[2]Technical Research Centre of Finland (VTT), Biotechnology and food research, P.O. BOX 1500, FIN-02044 VTT, Finland.
[3]Present address: Centre for Metrology and Accreditation, P.O.Box 239, FIN-00181 Helsinki, Finland

1 INTRODUCTION

Strawberries are extremely perishable fruits. Temperature management is crucial in prolonging the postharvest quality of strawberries. Prompt storage at low temperatures extends storage life. Critical stages influencing postharvest quality are harvesting method, cooling and transport. Improvement of the postharvest quality of strawberries is being studied in a joint research project to be carried out between VTT, MTT, the Finnish packaging and refrigeration industry and domestic producers during 1995-1997. Results of the 1995 trials on the effects of storage temperatures (+2° and +5 °C) and degree of ripeness on the nutritional and sensory quality of two strawberry cultivars will be reported in this paper.

2 MATERIALS AND METHODS

Trials were performed in Suonenjoki, Finland during the summer (4.7-21.7). The main cultivars grown in Finland, *Jonsok* and *Senga Sengana*, were obtained from a commercial farm. The *Jonsok* cultivar was studied in all three weeks and *Senga Sengana* the last week. Strawberries were harvested into 1 liter perforated cellular containers and then packed into corrugated cardboard cartons, with 8 containers in each. Ripe or 3/4 ripe strawberries were picked and stored at +2° and +5°C in 80 % (MTT) and 65 - 80 % (VTT) relative humidities. Strawberries were harvested on Mondays and determinations were performed on Tuesdays through Fridays.

Vitamin C was determined according to Speek *et al* [1] with minor modifications according to Hägg *et al.*[2]. Strawberries were homogenized with a Bamix mixer. About 5 g were then accurately weighed and 0.3 N trichloroacetic acid and octanol immediately added, homogenized, diluted and filtered. 4.5 M sodium acetic acid buffer and one spatula oxidase ascorbate were added. The sample was mixed and placed in a water bath at 37 °C, and 0.1 % 0-phenyldiamide was added. An HPLC instrument equipped with a fluorometric detector set at 365 nm:418 nm was employed. The column used was a Spherisorb 5 μm 125 mm. Vitamin C is determined as dehydroascorbic acid.

Sugars and organic acids were assayed by GLC using the method of Li and Schuhman[3] and some modifications according to Haila *et al* [4] . Samples were homogenized as above. The homogenate was immediately extracted twice with 80 % ethanol then centrifuged at 3000 rpm for 10 min and derivated. The instrumentation employed comprised a Sigma 300 duel gas chromatograph (Perkin-Elmer) equipped with a flame ionization detector, and a stainless steel column (3 m x 0.085 in i.d., 1/8 in. o.d.) packed with 3 % OV-1 on Gas Chromo Q (Applied Science).

Moisture was determined in a 3 g sample kept in a vacuum oven at 45 °C for 24 h in small, metal containers and then in a desiccator for 1 h before measurement.

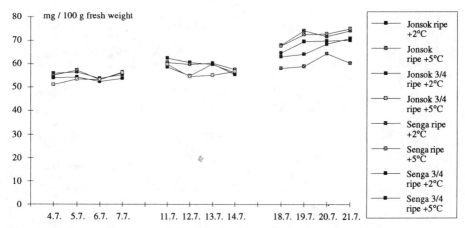

Figure 1 *Vitamin C contents of strawberry cultivars Jonsok and Senga Sengana*

Mouldy and bruised strawberries were counted. Then, the mouldy berries were removed, and sensory quality of the remaining berries was evaluated. Appearance, structure, flavour, and overall quality were rated by 8-10 trained judges according to the Karlsruhe quality scale from 1 to 9, where 9 is excellent and 1 unacceptable. Strawberries with a score of 1 to 3 were rejected as unacceptable for sale (VTT-4440-91 method).

3 RESULTS AND DISCUSSION

Vitamin C contents of strawberries are shown in Figure 1. Vitamin C was retained well during the 4 d storage. Vitamin C contents were about 55 mg/100 g fresh weight (FW) in the *Jonsok* cultivar and about 70 mg/100 g FW in *Senga Sengana*. In a previous study (Hägg *et al.* 1995) the vitamin C contents of *Jonsok* were also slightly below those of *Senga Sengana*. Vitamin C contents of *Jonsok* were lowest during the first week and increased slightly thereafter. Vitamin C contents of ripe strawberries were slightly higher than in 3/4 ripe strawberries of both cultivars.

Glucose, fructose and sucrose contents of the present strawberry cultivars were determined. Lower postharvest sugar contents were found in 3/4 ripe compared to ripe strawberries (Figure 2). Sugar contents of ripe strawberries stored at + 5°C decreased slightly during the 4 d storage and those of 3/4 ripe strawberries increased somewhat. The *Jonsok* cultivar contained about 2.1 % glucose, 2.2 % fructose and 1.8 % sucrose on a fresh weight basis. The *Senga Sengana* cultivar contained about 2.0 % glucose, t 2.3 % fructose and 3.6 % sucrose.

Only malic acid and citric acid were determined in the present strawberry cultivars. Organic acid contents differed considerably and slightly higher organic acid contents were detected in the 3/4 ripe strawberries of the *Jonsok* cultivar compared with ripe *Jonsok* berries, however, organic acid contents were notably higher in *Senga Sengana* strawberries (Figure 3). Malic acid contents varied between 0.4 and 0.5 % of the fresh weight and citric acid contents between 0.65 and 1.0 %.

When strawberries were harvested ripe the percentage of bruised berries was higher and increased further during the storage period, especially at +5 °C (Figure 4).

Percentage of mouldy strawberries increased during the 4 d storage when the strawberries were harvested ripe, yet only increased slightly in 3/4 ripe strawberries during a 7d storage period (Figure 5). The storage temperature of +5 °C caused a higher amount of mouldy berries compared to +2 °C.

The flavour of ripe strawberries was superior to that of 3/4 ripe berries, especially at the beginning of the storage period. However, the flavour of 3/4 ripe strawberries improved during the storage period, particularly in the strawberries ripened at +2 °C, whereas the

% of fresh weight

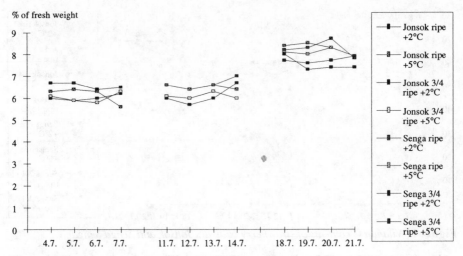

Figure 2 *Sugar contents of strawberries. Cultivars Jonsok and Senga Sengana*

% of fresh weight

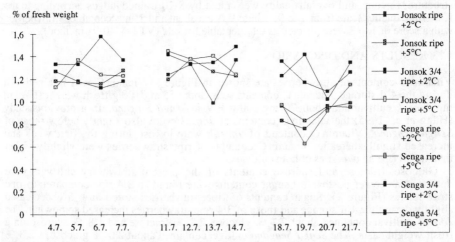

Figure 3 *Organic acid contents of the Jonsok and Senga Sengana strawberry cultivars*

flavour of ripe strawberries either deteriorated or remained unchanged. Overall quality of ripe strawberries improved after harvesting compared to that of 3/4 ripe strawberries.

The 4 d storage caused a slight loss of moisture in the strawberries. Moisture contents of the *Jonsok* cultivar were nearly similar in ripe and 3/4 ripe strawberries, while ripe strawberries of the *Senga Sengana* cultivar were slightly drier compared to 3/4 ripe berries.

4 CONCLUSIONS

Vitamin C and sugar contents in ripe and 3/4 ripe strawberries were retained very well during the 4 d storage at both temperatures. The amount of mouldy and bruised berries were higher when the strawberries were harvested ripe than at 3/4 ripeness and also at the +5 °C storage temperature. The sensory quality of 3/4 ripe strawberries improved during the storage period while that of ripe strawberries either deteriorated or remained unchanged.

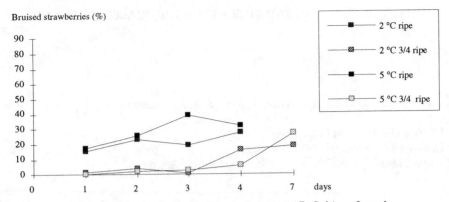

Figure 4 *Percentage of bruised strawberries during week 27. Cultivar Jonsok*

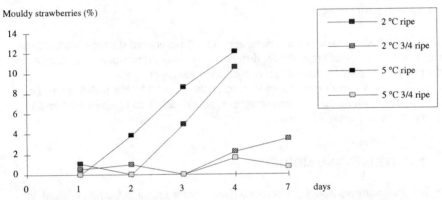

Figure 5 *Percentage of mouldy strawberries during week 27. Cultivar Jonsok*

Flavour and overall quality of both ripe and 3/4 ripe strawberries was quite acceptable after the 4 d storage at both temperatures. Extending the storage period of Finnish strawberries appears to be possible if rapid cooling and good storage conditions are ensured.

Acknowledgements
Many thanks to Ms Tuula Kurtelius, Ms Kaija-Leena Ristisuo, Ms Kirsi Norberg, Ms Anna-Leena Lamberg, Ms Ulla Österlund, and Ms Heidi Eriksson for their skillful technical assistance.

References

1. A. Speek, J. Schrijver, W. Schreurs, *J. Agr. Food Chem.*, 1984, **32**, 352-355
2. M. Hägg, S. Ylikoski, J. Kumpulainen, *J. Food Comp. Anal.*, 1995, **8**, 12-20.
3. B. Li and P. Schuhmann, *J. Food Sci.*, 1980, **45**,138-141.
4. K. Haila *et al.*, *J. Food Comp. Anal.*, 1992, **5**, 100-107.

THE EFFECT OF CULTIVAR AND STORAGE ON THE QUALITY OF POTATO

R. Kervinen, M. de Prado[1], E. Laurila, K. Autio & R. Ahvenainen

VTT Biotechnology and Food Research
P.O. Box 1500, FIN-02044 VTT, Finland
[1]Present address: Gabriel Aresti, 13 9°B, 48004 Bilbao, Spain

1 INTRODUCTION

The quality of potato varies with variety, agricultural factors and storage conditions. In order to select raw potato material for different uses, more information is needed on the effect of internal potato quality on processing and cooking quality.

This study is a small part of a wider project, the aim of which is to determine the effect of nitrogen fertilization, harvesting time, potato variety and storage time on the quality of raw and cooked potato.

2 MATERIALS AND METHODS

The potatoes (varieties Asterix, Nicola and Fambo) were grown in Lammi, Finland, in 1995. The growing time for Asterix and Fambo was 15 weeks and for Nicola 16 weeks. The potatoes were analysed four times during storage (4 °C) from November 1995 till May 1996. The following characteristics of the raw potato were analysed: dry matter, starch content, pectinesterase activity[1,2] and α-amylase activity (VTT-4404-97). The cell wall area and the average cell size of cooked potatoes were analysed using microscopy and image analysis. The sensory quality (colour, evenness of colour, flavour, sweetness, doneness, evenness of doneness, mealiness and dryness) of 8 mm steam cooked (15 min) potato cubes after different storage times (from November 1995 till May 1996) of the potato raw material was studied using ten trained panellists. Sensory evaluation was made after the cooked potato cubes had been stored for one day at 5 °C. Samples were evaluated using the scale from 1 to 9. The data from all sensory evaluations were subjected to multifactor ANOVA (STATGRAPHICS) to compare potato varieties and storage times. Fisher's least significant difference (LSD) was used to discriminate among the means.

3 RESULTS AND DISCUSSION

3.1 Composition of Potato

During storage there were only relatively minor changes in the dry matter of Asterix (22.7–23.4 %) and Fambo (23.5–24.7 %) and in the starch content of Asterix (16.9–17.9 %), Nicola (14.0–14.3 %) and Fambo (18.4–19.6 %). However, the dry matter in Nicola increased noticeably between March and May (20.1→24.5 %). The differences between varieties were more significant. The pectinesterase activity of each variety was highest in November (Asterix 0.059 µM GA/min, Nicola 0.029 µM GA/min and Fambo 0.042 µM GA/min) and lowest in May (Asterix 0.020 µM GA/min, Nicola 0.016 µM GA/min and Fambo 0.024 µM GA/min). The α-amylase activity, though very low, seemed to increase from November until March and then started to decrease again.

3.2 Cell Wall Area and the Average Cell Size

The cell wall areas of cooked potatoes were the following: Asterix 19.3 %, Nicola 21.3 % and Fambo 16.6 %. The average cell sizes were the following: Asterix 11860 μm^2, Nicola 9110 μm^2 and Fambo 10340 μm^2.

3.3 Sensory Quality

The sensory quality of the cooked Asterix and Fambo varieties remained stable during raw material storage: there were no statistically-significant differences ($P<0.05$) between cooked Asterix and Fambo over the whole of the storage period. The sensory quality of cooked Nicola, on the other hand, decreased during raw material storage. The odour and flavour of Nicola were more typical in January than in March and May and the colour was yellower in November and January than in March and May (Figure 1a).

The most common differences between the potato varieties were in colour, evenness of colour and dryness (Figure 1). The following differences in Figures 1a, 1b and 1c were statistically significant ($P<0.05$): Nov. 95 Nicola was yellower than Fambo, Jan. 96 Nicola and Asterix were yellower than Fambo, May 96 Asterix and Fambo were yellower than Nicola (Figure 1a); Jan. 96 the colour of Asterix was more even than the colour of Nicola and Fambo, Mar. 96 the colour of Asterix was more even than the colour of Nicola and Fambo, May 96 the colour of Asterix and Fambo was more even than the colour of Nicola (Figure 1b); Jan. 96 Asterix was dryer than Nicola, May 96 Asterix and Fambo were dryer than Nicola (Figure 1c).

The differences between potato varieties increased as the storage time of raw material increased. In May, in addition to the differences shown in Figure 1, the following differences were also found: the odour and flavour of Nicola was less typical than that of Asterix and Fambo, Nicola was sweeter than the other varieties, Fambo was mealier than Asterix and Nicola, and the doneness of Fambo was more even than that of Asterix and Nicola.

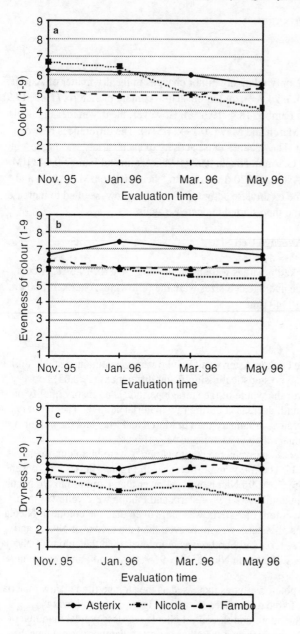

Figure 1. *The sensory characteristics of steam-cooked potato cubes after different storage times of the potato raw material. The statistical significancies are shown in the paragraph 3.3.*
(a) Colour, 1 is grey and 9 is extremely yellow
(b) Evenness of colour, 1 is extremely uneven and 9 is extremely even
(c) Dryness, 1 is extremely watery and 9 is extremely dry

4 CONCLUSIONS

Even though there were remarkable differences in raw material composition (dry matter, starch content, pectinesterase activity, α-amylase activity) between potato varieties and during storage, the most common differences found in the sensory quality of cooked potato were in colour. In the main, structural differences were found at the end of storage period.

The cooking quality of Asterix and Fambo varieties remained stable over the whole period of storage as raw material, while the cooking quality of Nicola fell, especially at the end of the storage period.

To correlate the quality of raw potato with cooking or processing quality, more investigation is needed. At VTT, potato samples from the summer of 1996 have also been analysed and a statistical treatment of the results will be completed in the near future.

References

1. C.B. Djerf, P.A. Andersson & R.E. Öste, The effect of preheating on potato pectin methylesterase. In: A. Andersson (Ph.D. thesis), Modelling of Potato Blanching, University of Lund, Lund, Sweden, 1994.
2. A.E. Hagerman & P.J. Austin, *J. Agric. Food Chem.*, 1986, **34**, 440.

NUTRITIONAL VALUE OF SOME POLISH VEGETABLE AND FRUIT-VEGETABLE MIXED JUICES

H. Kunachowicz and I. Nadolna

National Food and Nutrition Institute
Powsińska 61/63
02-903 Warsaw (Poland)

1. INTRODUCTION

The modern status of medical knowledge unambiguously points out that it is beneficial for health to consume vegetables, fruits and their products, including juices and nectars, several times every day. It is recommended in Poland to include these products in every meal.

Domestic vegetables and fruits are available in our climatic zone for only a relatively short time, and for that reason the role of vegetable and fruit juices in permanent nutrients supply should be emphasized.

Vegetable and fruit-vegetable juices are produced by the Polish industry in a wide assortment. These products include tomato, multivegetable, carrot, carrot-fruit juices and others. These juices are of pomace type which, in contrast to clear ones, are an important source of dietary fibre and many minerals and vitamins in diet, and their energy value is relatively low. The basic raw material for their production is carrot pomace, thus these juices are the main source of β-carotene in diet.

In the present work the nutritional value of the vegetable and fruit-vegetable juices produced in Poland have been assessed, especially with respect to their content of β-carotene and vitamin C.

2. METHODS

The examination covered juices produced in the years 1993-1997. Samples for analysis were prepared out of 3-5 unit packings depending on their size. A total of 16 assortments of vegetable and fruit-vegetable juices produced by various Polish producers was analysed.

Vitamin C in the juices was determined using spectrophotometric method[1] according to ISO/DIS 6557, 1984. β-carotene content was determined by column chromatography[2].

In the selected juices the values for energy, basic components, minerals and other vitamins were calculated on the basis of recipes of the juices and the present knowledge of the nutritional value of raw materials, with adequate correction made for the effectiveness of the technological process.

3. RESULTS

The energy value of the vegetable and fruit-vegetable juices is lower than that of fruit juices[3,4]. The lowest energy values were found in tomato and multivegetable juices (13 and 25 kcal/100 g, respectively). For the remaining juices this value ranged from 40 to 50 kcal per 100 g. The content of dietary fibre varied from 0.4 to 2.6 g per 100 g. Among the minerals the presence of potassium and magnesium was worth stressing.

Vitamin content in these juices varied. β-carotene in the juices ranged from about 500 µg in 100 g of tomato juice to about 5500 µg in 100 g of carrot juice with small addition of lemon juice.

The content of vitamin C in the non-enriched juices was from 4.8 mg in 100 g of carrot juice to 12.8 mg in beetroot, plum and pear juice. Many assortments of juices are enriched with synthetic vitamin C in amounts from 20 to 30 mg per 100 g. The enriched juices contained 19.0 to 34.9 mg of this vitamin in 100 g. Besides that, differences in the content of β-carotene and vitamin C were found both between the samples of juice produced by various producers and between batches with different dates of production. For example, tomato juice contained vitamin C in amount from 6.3 to 18.3 mg in 100 g, and β-carotene from 411 to 540 µg in 100 g. It is worth stressing that vegetable and fruit-vegetable juices are a worse source of vitamin C than fruit juices.

To allow assessing of the nutritional value of vegetable and fruit-vegetable juices the coverage of the Recommended Dietary Allowances[5] for β-carotene and vitamin C by one portion (200 ml) of the studied juices has been presented in diagrams. Figure 1 shows that 200 ml of the juices covers the Recommended Dietary Allowances for β-carotene calculated as vitamin A in 20 to 225%. These values stress the high nutritional value of the juices prepared with carrot pomace and thus their importance in providing β-carotene to the daily diet.

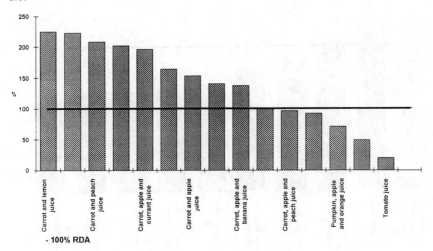

Figure 1 *Coverage of the Recommended Dietary Allowances for β-carotene by 200 ml of vegetable and fruit-vegetable mixed juices*

In the case of vitamin C the 200 ml portion of the examined vegetable and fruit-vegetable juices provided 13% to 43% Recommended Dietary Allowances for that vitamin

(Figure 2). The juices enriched with synthetic vitamin C covered these allowances from 67% to 116% (Figure 3).

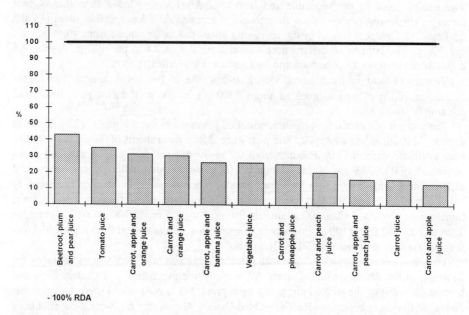

Figure 2 *Coverage of the Recommended Dietary Allowances for vitamin C by 200 ml of vegetable and fruit-vegetable mixed juices*

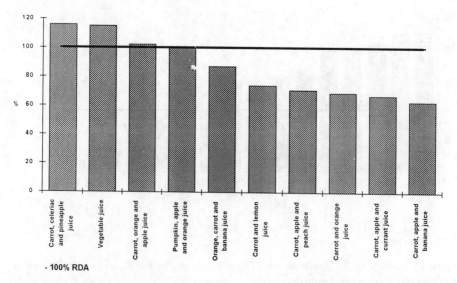

Figure 3 *Coverage of the Recommended Dietary Allowances for vitamin C by 200 ml of vegetable and fruit-vegetable mixed juices enriched with vitamin C*

4. CONCLUSIONS

Vegetable and fruit-vegetable juices usually have a high nutritional value, mainly in view of their two important nutrients that are β-carotene and vitamin C. Analyses show, however, considerable variations in the content of these vitamins in different batches of these products.

Carrot and carrot-fruit juices are important sources of β-carotene in diet. Vitamin C content in these juices is lower than in many fruit juices. Thus the composition of juices containing carrot juice with pomace or juices of fruit as a natural vitamin C source is an interesting and reasonable solution in view of raising their vitamin content and improving their taste. Another interesting method is juice enrichment with synthetic vitamin C introduced by the producers.

The 200 ml portion of vegetable or fruit-vegetable juice of low energy value covers in a high degree the Recommended Dietary Allowances for vitamin A and less, although still significantly, the allowances for vitamin C.

References

1. H. Kunachowicz (ed.), 'Wybrane metody analityczne oceny wartości odżywczej', Prace IŻŻ 83, Warszawa, 1997, Chapter 8, p.43.
2. U. Rutkowska (ed.), 'Wybrane metody badania składu i wartości odżywczej produktów spożywczych', PZWL, Warszawa, 1981, Chapter 7, p. 327.
3. H. Kunachowicz, I. Nadolna, K. Iwanow, B. Przygoda, 'Wartość odżywcza wybranych produktów spożywczych i typowych potraw', PZWL, Warszawa, 1997, Chapter 2, p. 90.
4. H. Kunachowicz, I. Nadolna, B. Przygoda, K. Iwanow, 'Tabele wartości odżywczych produktów spożywczych', Prace IŻŻ, Warszawa, 1998 (in print).
5. Ś. Ziemlański, B. Bułhak-Jachymczyk, J. Budzyńska-Topolowska, B. Panczenko-Kresowska, M. Wartanowicz, *New Medicine*, 1996, **1**,1.

EFFECT OF EXTRUSION ON THE NUTRITIONAL QUALITY OF *PHASEOLUS VULGARIS* BEAN FLOUR

M.A. Martín-Cabrejas, C. Karanja*, G. Maina*, E. Herrero, L. Jaime, R.M. Esteban, C.M-F. Mbofung[&], A.C. Smith[$] and K.W. Waldron[$].

Dpto Química Agrícola. Facultad de Ciencias. Universidad Autónoma de Madrid. 28049 Madrid. Spain
* Kenyan Agricultural Research Institute, P.O. Box 57811, Nairobi, Kenya.
[&]Dpt. of Food Science and Nutrition, E.N.S.A.I., School of Food Technology, P-O. Box 455, Ngaoundere, Cameroon.
[$]BBSRC Institute of Food Research, Norwich Research Park, Colney, Norwich, U.K.

1 INTRODUCTION

Grain legumes are rich and less expensive sources of dietary proteins and contribute substantially to the protein content of the diets of a large part of the world's population. Among legumes for human food, dry beans are widely consumed and have beneficial effects on human health, being a good source of proteins, water-soluble vitamins and certain minerals.

However, the utilisation of dry beans as human foods is below their potential. Many varieties of grain legumes, particularly those of *Phaseolus vulgaris* become progressively harder to cook when stored under the high temperature, high humidity conditions prevalent in Tropical countries[1]. Not only does this result in a reduction in organoleptic quality, excessive use of wood fuel and water, but it is also accompanied by an increase in the levels of certain antinutrients, particularly lectins, and a decrease in digestible starch and protein.

This paper presents the results of an EC-funded study on the use of single screw extrusion to improve the nutritional quality of bean flour. Extrusion is a useful process which inactivates antinutrients[2], particularly protease and amylase inhibitors, and importantly lectins; in addition, increases protein digestibility and soluble fibre fraction. It is also useful because the extruded product is a dry material which can be milled easily, and stored cheaply.

2 MATERIALS AND METHODS

Beans (*Phaseolus vulgaris* cv Canadian Wonder) are a widely used cultivar in Kenya and exhibit the hard-to-cook (HTC) defect as a consequence of tropical conditions of storage. The extrusion cooking process was carried out by IFRN group. The beans were pin-milled and conditioned to moisture content of 25%. The bean flours were extruded using a Bradender bench top extruder 20DN. The extruder barrel was heated at 140°, 160°, or 180°C and screw speed of 90 rpm was used. A circular die of 3 mm diameter was used. The extrudate was cooled, allowed to dry at room temperature and then stored in sealed plastic bags at 1°C.

2.1. Chemical determinations.

Haemaglutinating activity (lectin) and trypsin inhibitors content were estimated according to the method of Grant *et al.*[3] α-Amylase inhibitor content was determined by starch iodine procedure of Piergiovanni[4]. Protein digestibility was evaluated by the use of a modified *in vitro* method[5].

Total, insoluble and soluble dietary fibre were determined using the enzymatic-gravimetric method of Lee *et al.*[6] The principles of the method are the same as those for the AOAC dietary fibre methods 985.29 and 991.42. The neutral sugar composition of the dietary fibre fraction was determined by HPLC[7] and uronic acid by an automatic colorimetric method[8].

3 RESULTS AND DISCUSSION

The effect of extrusion cooking on Canadian Wonder bean flour (CW) shows a general inactivation of antinutrients depending on temperature intensity. When the extrusion treatments are made at 120°C, lectin activity decreases dramatically to 2.5%, however protease inhibitors (trypsin and α-amylase) are still present with important percentages (45 and 40% of residual activity, respectively). Thus, the above inhibitors need higher temperatures to be destroyed. At an extrusion set temperature of 180°C the activities of inhibitors are not detected. The use of single screw extrusion not only inactivates antinutrients but also causes important changes on digestibility of protein in HTC CW beans[9,10]. Raising the extrusion temperature to as high as 180°C equally brought about a much higher increase in digestibility of raw CW from 64 to 78.4%.

The processing of extrusion also modifies the dietary fibre fractions. The results obtained can be observed in Figure 1.

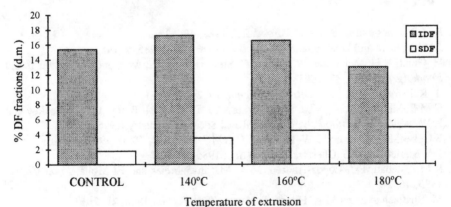

Temperature of extrusion
Figure 1. Effect of extrusion on Insoluble (IDF) and Soluble (SDF)
Dietary Fibre fractions.

Extrusion cooking of milled HTC Canadian Wonder bean flour under different temperature conditions shows decreases in insoluble dietary fibre (IDF) fractions, being more accentuated at 180°C. In this fraction, it is observed an increase of resistant protein, except in samples extruded under the most severe conditions. Regarding soluble dietary fibre (SDF), a significant increase is exhibited. Greater increases of SDF are obtained with higher temperatures. In general, total dietary fibre (TDF) increases from 17 to 21% d.m. This could be accounted by modification of enzymatically indigestible starch in this process as suggested Ralet *et al.*[11] in extruded barley products.

HPLC sugar analysis of IDF and SDF fractions confirm the solubilization observed in the gravimetric assay. Insoluble fibre is composed mainly by arabinose glucose followed by xylose and uronic acids. Extrusion cooking causes solubilization of insoluble fibre constituents: arabinose and uronic acid contents of IDF decrease 27.5 and 31%, respectively. Glucose increases at lower set temperatures (30%), while it decreases dramatically at 180°C. In general, a decrease of total sugars is observed when temperature increases.

SDF of CW shows an important amount of arabinose and galactose/rhamnose. Extrusion treatments result in significant increases of xylose, arabinose, uronic acids and galactose/rhamnose, and in minor extent glucose. This might be due the solubilization of glucoxylans and also glucans from the IDF fraction.

A good correlation is seen between DF values obtained with the gravimetric method and the method based on specific analysis of the monomeric composition of the fibre fraction. The solubilization of fibre might probably be a factor in increasing the fermentability characteristics of bean flour in the gut as previously reported for extruded wheat[9]. Likewise, the increase solubility of fibre may modify the functional properties of bean flour leading to new technological applications.

These results show that the different degrees of inactivation of antinutrients, digestibility of protein and redistribution of fibre fractions are depending on temperature conditions in this processing, however there is no doubt of the efficacy of extrusion cooking to improve the usage of HTC beans.

References

1. P.M.B. Jones and D. Boulter, *J. Food Sci.*, 1982, **48**, 623.
2. J.M. Aguilera and D.W. Stanley, *J. Food Proc. & Pres.*, 1985, **9**, 145.
3. G. Grant, N.H. McKenzie, W. Watt, J.C. Stewart, P.M. Dorward and A. Pusztai, *J. Sci Food. Agric.*, 1986, **37**, 1001.
4. A. R. Piergiovanni , *Lebensm. Wiss. Techol.*,1992, **25,** 321.
5. C.M.F. Mbofung, L.L. Niba, M.L. Parker, A.J. Downie, N. Rigby, C.L.A. Leakey and K.W. Waldron, *Agrifood Quality.* The Royal Society Chemistry. 1996, 433.
6. S.C. Lee, L. Prosky and J.W. DeVries, *J. AOAC Int.*,1992, **75(3)**, 395.
7. J.A. Marlett and J.G. Chesters, *J. Food Sci.*, 1985, **50**, 410.
8. R.M. Esteban, F-J. López-Andréu, M.A. Martín-Cabrejas and E: Mollá, *Food Chem.*, 1993, **46**, 289.
9. M. Bhattacharya and M.A. Hanna, *Lebensm. Wiss. Techol.*,1988, **21**, 20.
10. O. Fapojuwo, J.A, Maga, and G.R. Jansen, *J. Food Sci.*,1987, **52**, 218.
11. M.C. Ralet, J-F. Thibault and G. Della Valle, *Lebensm. Wiss. Techol.*, 1991, **24,** 107.
12. M. Nyman, K.E. Palsson and N.G. Asp, *Lebensm. Wiss. Techol.*, 1987, **20,** 29.

Acknowledgements

The authors wish to thank the European Communities for their financial support (STD-3 project TS3-CT92-0085) and also partly financed by the UK Office of Science and Technology (Dr. Waldron and Dr. Smith).

STORAGE CONDITIONS IN THE NORWEGIAN DISTRIBUTION OF FRESH PRODUCE

B. K. Martinsen, G. B. Bengtsson and T. Sørensen

MATFORSK - Norwegian Food Research Institute
Osloveien 1
N-1430 Ås, Norway

1 INTRODUCTION

Fruit and vegetables are living plant materials. Ideal storage temperature and relative air humidity are of great importance to achieve a long shelf life.

During three two-week periods from November 1994 to June 1995 the losses of potatoes, ten types of vegetables and three fruits were recorded by weighing rejected amounts through the distribution chain[1,2]. Thirty-nine producers, 28 wholesalers and 140 retailers all over Norway participated in the registration. This mapping was carried out as an aid to reduce the distribution losses by identification of high benefit to cost measures.

Different sources recommend different temperatures because the ideal temperature may depend on sort (e.g. apple) and ripeness (e.g. tomato). Generally it is difficult to keep the temperature in a storeroom at a constant level, due to fluctuations by the temperature control machinery and due to energy losses or gains from open doors, staff working in the room or outside weather conditions. For this reason most of the answers in the mapping of losses were given as an interval.

2 RESULTS

2.1 Temperature

Most of the producers kept the temperature very close to the ideal values in their stores. However, the maximum temperatures were too high, especially for onions, carrots and cabbage (Table 1).

The temperature of the wholesale stores was on an average 2-3°C higher than the ideal temperatures for "cold-stored" vegetables, i.e. produce with ideal temperatures of 0-1°C. In a few cases the temperatures were up to 9°C higher than the ideal temperature. Onions were kept above 6°C in 83% of the stores. On the other hand, cucumbers were stored at too low a temperature by 25% of the wholesalers. A few degrees under ideal temperature will not damage the cucumber seriously, but it will shorten the shelf life.

With the retailers the situation was different from the wholesalers. Most of the produce was placed in a storage room and only a small part of it was placed at the counters. Some of the retailers moved the products to cold storerooms in the evening, and others had no cooling at all. In this mapping we found that 92% of the retailers had cold storerooms and 75% had cold counters.

In the retail stores the temperature was on an average 2.5°C higher than in the wholesale stores for "cold-stored" vegetables. The temperature in the counters was additionally 5°C higher for the same vegetables. Onions, apples and pears were stored at even higher temperatures. Cucumbers and bananas were kept at a too low temperature in 33% and 5 % of the shops, respectively.

Table 1. *Temperature (° C) in distribution of fruit and vegetables (Nov. 1994)*

Culture	Ideal[1]	Producers[2]		Wholesale (24 distributing stores)		Retail (140 shops) Storeroom		Counters		Produce moved to store-room in the night
		Mean value	(min-max)[3]	Mean value	(min-max)[3]	Mean value	(min-max)[3]	Mean value	(min-max)[3]	%[4]
Potato	4-6	6	(4-8)	6.4	(2.5-10)	8.3	(2.0-20.0)	16.1	(3.5-22.5)	9
Carrot	0-1	1.5	(-1-4)	3.6	(1-9.5)	6.1	(2.0-19.0)	10.7	(2.0-22.5)	8
Swedish turnip	0-1	3	(0-6)	3.7	(1-9.5)	6.2	(2.0-19.0)	10.4	(2.0-22.5)	8
Onion	0	2.5	(0-5)	8.6	(3-13.5)	7.1	(2.0-19.0)	13.6	(2.0-22.5)	6
Leek	0-1	-		3.1	(1-9)	5.4	(2.0-19.0)	10.1	(0.5-22.5)	21
Cabbage	0-1	1	(1-2)	3.3	(1-9)	6.0	(2.0-19.0)	10.5	(2.0-22.5)	10
Cauliflower	0-1	-		3.0	(1-9)	5.4	(2.0-19.0)	10.6	(2.0-22.5)	31
Chinese Cabbage	0-1	1.5	(1-3)	3.0	(1-9)	5.4	(2.0-19.0)	10.8	(2.0-22.5)	29
Iceberg lettuce	0-1	-		3.1	(1-9)	5.6	(2.0-19.0)	10.3	(2.0-22.5)	25
Tomato	11-14	-		9.9	(5-13.5)	9.1	(2.0-20.0)	11.8	(2.0-22.5)	17
Cucumber	12.5	-		10.5	(6-15)	12.5	(2.0-20.0)	13.1	(2.0-22.5)	10
Apple	0-6	3		3.6	(2-9)	6.0	(1.0-20.0)	15.0	(2.0-22.5)	13
Pear	0-1	-		3.5	(1.5-9)	5.8	(1.0-19.0)	14.2	(2.0-22.5)	22
Banana	13.5	-		13.2	(9-16)	16.0	(2.5-21.0)	18.0	(5.5-22.5)	5

1) Ideal temperature is from Godt Norsk (1994)[3] except for cucumber and banana.
2) Number of producers: Potato. 14; carrot. 20; Swedish turnip. 13; onion. 7; cabbage. 11; Chinese cabbage. 9; apple. 1
3) (min-max) is the variation in temperature between the different enterprises.
4) (%) is the share of shops which moved the goods between counters and storerooms.

2.2 Relative humidity

The relative humidity should be about 80% for tomato and onion and above 95% for all the other products but it was not sufficiently high in the producers stores (Table 2). Among the wholesale stores the relative humidity was very varying and it was on an average lower than with the producers. The relative humidity was not asked for in the retail stores but we expect it was even lower there.

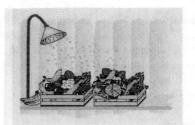

The loss of produce due to evaporation is estimated to an average of 3 - 4% during the distribution chain.

Table 2. *Relative air humidity (RH %) in distribution of fruit and vegetables (Nov. 1994)*

Culture	Ideal[1]	Producers[2]		Wholesale (14 distributing stores)	
		Mean value	(min-max)[3]	Mean value	(min-max)[3]
Potato	95-98	91	(86-96)	80	(50-100)
Carrot	95-100	90	(81-98)	83	(50-100)
Swedish turnip	95-100	91	(81-100)	84	(50-100)
Common onion	75-85	64	(58-70)	71	(50-85)
Leek	95-100	-		81	(51-93)
Cabbage	95-100	91	(85-96)	83	(51-100)
Cauliflower	95-100	-		83	(51-100)
Chinese cabbage	95-100	92	(85-98)	83	(51-100)
Iceberg lettuce	95-100	-		83	(51-100)
Tomato	75-80	-		81	(69-93)
Cucumber	95-100	-		78	(68-90)
Apple	95-100	90		82	(50-100)
Pear	95-100	-		82	(50-100)
Banana	95-100	-		79	(50-95)

1) Ideal relative humidity is from Godt Norsk (1994)[3]
2) Numbers of producers: Potato. 14; carrot. 20; Swedish turnip. 13; onion. 7; cabbage. 11; Chinese cabbage. 9; apple 1
3) (min-max) is the variation in RH % between different enterprises

3 CONCLUSION

The frequency and magnitude of deviation from ideal postharvest storage conditions increased in the distribution chain from the producer to the retailer for fourteen important products of fresh produce. The variation between different enterprises was very large. This variation in storage conditions between various enterprises depended upon the presence or the efficacy of their cooling and humidifying systems. Storage of ethylene sensitive together with ethylene producing products resulting in shortened shelf life was recorded in some wholesale stores and by a lot of the retailers. This project has resulted in several activities in the distribution chain to reduce the losses and improvements are still going on.

References

1. G. B. Bengtsson. T. Sørensen and B. K. Martinsen, Tema fra MATFORSK, 1997, 135.
2. G. B. Bengtsson, InforMat, 1997, 1, 14.
3. Godt Norsk, Konkurransestrategier for norsk mat, 1994, 10.

FREEZING SUITABILITY OF SOME SNAP BEAN (PHASEOLUS VULGARIS) AND PEPPER (CAPSICUM ANNUM) VARIETIES

Mencinicopschi Gh. and Popa M.

Institute of Food Chemistry
1D Garlei street,71576
Bucharest, Romania

1. INTRODUCTION

The effect of the variety and technological conditions influencing frozen product quality of frozen snap bean and peppers was studied.

In addition the climate conditions of the harvest was studied through making experiments for 3 harvests from 3 different years.

The selected analysed varieties were harvested from an experimental farm - Vidra Romania.

2. SELECTION CRITERIA OF SUITABLE VARIETIES FOR FREEZING

❑ a better resistance of nutritional properties factors on freezing process;
❑ a more intense shade of variety colour;
❑ a good firmness with a good resistance on freezing process;
❑ intense taste - flavour;
❑ the harvesting of the pods at an immature stage (filled section pod without formed seeds);
❑ juicy firmness and strong firmness, undetachable epitelial surface of the pods on scalding and freezing processes.

3. COMPARATIVE INDICATORS FOR FREEZING SUITABILITY COMPARISON

The following physico - chemical and sensorial indicators have been established at some particular times of experiments, namely:
➤ starting moment (raw materials);
➤ immediately after freezing (at t =-30^{0}C and forced air circulation);
➤ after 5 and 9 months storage at -20^{0}C, without light exposure.

Snap bean
■ titrable acidity (ml NaOH/100 g edible);
■ pH in minced vegetal material;
■ AIS (alcohol insoluble substances %);
■ refractometer degree (%);
■ peroxidase activity test (qualitative determination).

Sweet pepper
■ Titrable acidity (ml NaOH/100 g edible);
■ pH in minced vegetal material;
■ ascorbic acid content (mg/100 g edible);
■ refractometer degree (%);
■ thawing drip loss;
■ peroxidase activity test (qualitative determination).

For both species, the sensorial indicators have been computed according to the Karlsruhe method, using the following factors:

$F_{texture} = 4$;
$F_{appearance\ and\ colour} = 3$;
$F_{taste\ and\ flavor} = 3$.

$$N(\text{total score}) = \frac{F_{texture} \cdot N_{texture} + F_{appearance\ and\ colour} \cdot N_{appearance\ and\ colour} + F_{taste\ and\ flavor} \cdot N_{taste\ and\ flavor}}{10}$$

4. CONCLUSIONS AND RECOMMENDATIONS

Snap bean
• As Figure 1 illustrates, sensory scores are higher when the snap bean variety has a higher carbohydrate content and a pH value about 8 – 8.5.
At the same time, sensory scores are higher when AIS (alcohol insoluble substances) value is below 8.5 - 9.
In spite of a high maturometric grade (AIS ≈ 12 %), NERINA variety got the highest sensory scores value due to its very good taste and low level of cellulosic compounds.
• As Figure 2 illustrates, the most suitable varieties for freezing are classified for quality as follows: NERINA, FINA VERDE, ECHO, CREST, ATLANTIC, AURELIA, VIDRA 9, LENA, LAVINIA.
Sweet peppers
• As Table 1 illustrates, the percentage loss of ascorbic acid in green and yellow blanched mild peppers, after freezing and 9 months storage is about half of percentage loss in mild unblanched frozen peppers, after the same 9 months storage.
So, the advantages of the enzymatic inactivation (i. e. blanching) before freezing green and yellow mild peppers are obvious.
• The experimental work demonstrated that orange and red long and bell peppers can be frozen without enzymic inactivation, due to the antioxidant role of carotenoidic pigments.

- As Figure 3 illustrates, the most suitable varieties of peppers for long freezing storage may be classified, based on the quality criteria as follows:

MILD PEPPERS (Capsicum anuum grossum): L 186, L 126, L 108+127, L 104/88, OPAL, AROMA, L 190, A/95;

LONG PEPPERS (Capsicum anuum longum): ORANGE, L 46 – 1;

BELL PEPPERS (Capsicum anuum grossum):SPLENDID, MADALINA, CORNEL, GLOBUS,L 546, L 710.

REFERENCES

1.   G. Crivelli, 'Ricerche sul comportamento alla congelazione, degli ortaggi', *ANNALI IVTPA*, vol.X, 1979.

2.   F. Guttierrez and J. M. Garcia , 'Influence of postharvest handling on commercial losses and quality of strawberry' at *Contribution du froid a la preservation de la qualite des fruits, les legumes et produits halientiques-conference proceedings*, Maroc, 1993.

Table 1   *Ascorbic acid content at different processing stages for some mild peppers varieties*

Variety name	ASCORBIC ACID CONTENT mg/100 g edible						
	Color	Raw material	After2 min. blanching at 95-98°	Immediately after freezing	After 5 months storage	After 9 months storage	Loss % to raw material
AROMA	Green	110.52	*	76.22	70.79	68.38	38.13
			-	91.95	57.38	33.88	69.35
L 186	yellow	169.12	80.51	79.08	59.09	55.65	67.1
			-	79.55	60.62	29.74	82.42
L (108-127)	yellow	120.05	72.88	70.50	63.79	57.38	52.2
			-	71.46	59.27	34.67	71.12
OPAL	dark green	105.76	*	82.83	58.21	52.85	50.03
			-	86.70	51.40	24.46	76.9
L 190	dark green	102.90	*	66.69	62.44	58.43	43.22
			-	82.41	45.03	25.41	75.31
L 104/88	yellow	124.34	*	71.92	60.90	57.20	53.99
			-	79.08	47.04	17.20	86.7

Note: * *frozen samples after blanching;*
      - *frozen samples without blanching.*

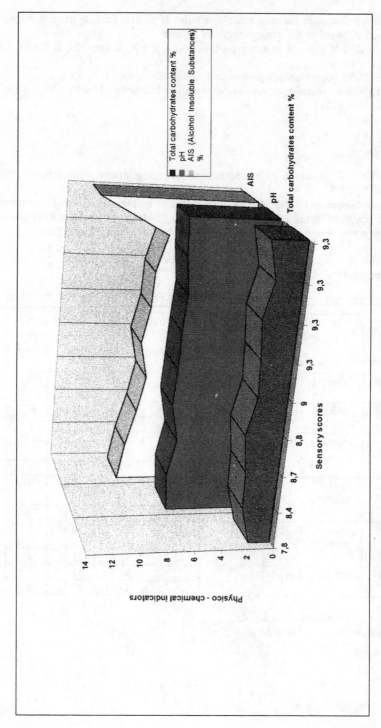

Figure. 1. *Correlation between AIS, pH, total glucidic content and sensorial value score (Karlsruhe) for snap bean varieties*

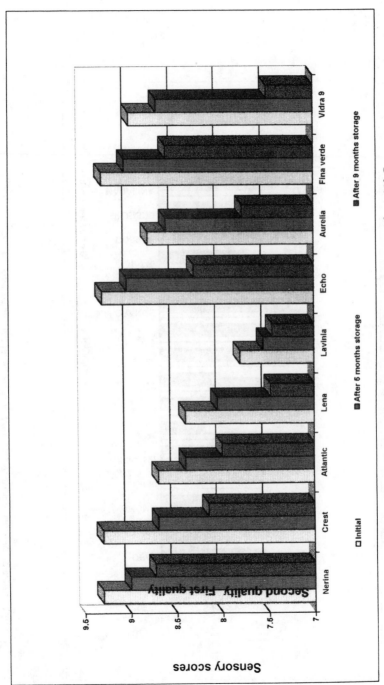

Figure 2. *Sensory scores (Karlsruhe) of snap bean varieties stored at −20° C*

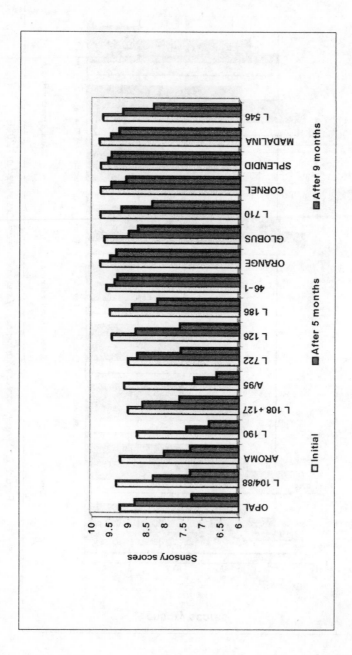

Figure 3. *Sensory scores (Karlsruhe) of some sweet pepper varieties stored at – 20 °C*

THE EFFECT OF MODIFIED ATMOSPHERE PACKAGING ON THE SHELF LIFE OF STRAWBERRIES

M. Mokkila[1], A.-L. Lamberg[1], U. Häkkinen[2], A. Kinnunen[1], K. Latva-Kala[1] and R. Ahvenainen[1]

[1] Technical Research Centre of Finland (VTT), Biotechnology and Food Research, P.O. Box 1500, FIN-02044 VTT, [2] Agricultural Research Centre of Finland (MTT), Laboratory of Food Chemistry, FIN-31600 Jokioinen

1 INTRODUCTION

Strawberries are extremely perishable fruits. The texture of Finnish strawberry varieties is soft and they normally have very short shelf lives, of only one to three days. On the other hand, Finnish strawberries are superior in flavour and exportation would be possible if their shelf life could be prolonged. Modified atmosphere packaging is a known method for prolonging the shelf life of strawberries, but unfortunately the information available from the literature is often contradictory and impossible to apply in practice.

The aim of this study was to determine the dependence of the quality of strawberries on packaging parameters. The results of the 1997 season experiments are discussed.

2 MATERIALS AND METHODS

The experiments were carried out in preliminary and primary tests. In primary tests strawberries (greenhouse-grown cultivar 'Elsanta') were packed in gas mixtures with high oxygen concentrations (20, 50 and 80%). Primary tests with the field-grown cultivar 'Jonsok' were carried out in three series:
1) Air was used as the packaging gas and the following packaging films were tested: two permeable films (oxygen transmission rate 6600 and 10000 $cm^3/m^2/day$), one commercial film for strawberry packages (film A) and two microperforated films.
2) Gas mixture was used in packaging and the experiments were carried out based on a statistical experimental design (Table 1).
3) Strawberries were packed with ethene absorbent and film having a high oxygen transmission rate (8000 $cm^3/m^2/day$).

All three primary test series were repeated twice in July. Strawberries were picked as fully-ripe, pre-cooled using forced air, packed into MA-packages and stored at 2 °C for 1, 5 or 9 days. The quality parameters used to measure the quality and the shelf life of strawberries are shown in Table 2.

3 RESULTS

According to the preliminary tests high oxygen concentrations in the packaging gas did not appear to be beneficial.

In the primary test series based on statistical experimental design good or excellent models were achieved for appearance, faultlessness of flavour, odour, intensity of strawberry flavour and sweetness. The optimum packaging conditions established for strawberries (Figure 1) were:
- An oxygen transmission rate of 6600 $cm^3/m^2/day$.

Table 1 *The factors and factor levels used in the statistical experimental design*

Factor	Low level	Intermediate level	High level
O, concentration of packaging gas (%)	10	20	30
CO, concentration of packaging gas (%)	0	10	20
N,O concentration of packaging gas (%)	0		35
Oxygen transmission rate of packaging material ($cm^3/m^2/day$)	2600	4600	6600

Table 2 *Quality parameters used to measure the quality and shelf life of strawberries*

Property analysed	Determined as	Method used
Sensory quality	Appearance and odour	Sensory evaluation by 2 trained judges according to the Karlsruhe quality scale from 1 to 9
	Texture and flavour	Sensory evaluation by 8-10 trained judges according to the Karlsruhe quality scale from 1 to 9
	Mouldy berries	Counting
Nutritional quality	Vitamin C	HPLC (Speek et al[1], Hägg et al[2])
	Sugars	GLC (Li and Schuhmann[3], Haila et al[4])
	Organic acids	GLC (Li and Schuhmann[3], Haila et al[4])
Gas concentrations in the package headspace	Oxygen and carbon dioxide	PBI Dansensor Combi Check analyser
	Ethanol and ethylacetate	Gas chromatography with flame ionisation detector (FID)

- Carbon dioxide concentration at the lowest level, 0%.
- Oxygen concentration at the lowest level, 10%.

The odour of the strawberries improved as the oxygen transmission rate increased. Both odour and faultlessness of flavour improved as the carbon dioxide concentration in the packaging gas decreased. Appearance improved as the oxygen concentration in the packaging gas decreased. An increase in the permeability of the packaging material did not affect the appearance of the strawberries. Neither the use of nitrous oxide or ethene absorbent had any effect on the shelf life of strawberries.

Good results were achieved by using film with an oxygen transmission rate of 8000 $cm^3/m^2/day$. In this case, appearance was well maintained and the odour and faultlessness of flavour of strawberries improved in comparison to situations in which films of lower permeability were used (Figure 2).

Mould growth could be limited to some extent by using packaging gas containing carbon dioxide, but the amount of off-flavours then increased.

The shelf life of strawberries could also be improved by using normal air as the packaging gas. Appearance was not maintained as well as in the optimal gas packages (Figure 3) but the strawberries retained their faultlessness of flavour better than in gas packages if the oxygen transmission rate of the film was sufficiently high or microperforated films were used. The quality of strawberries was also well maintained in air-containing packages in which commercial film for strawberries (film A) were used.

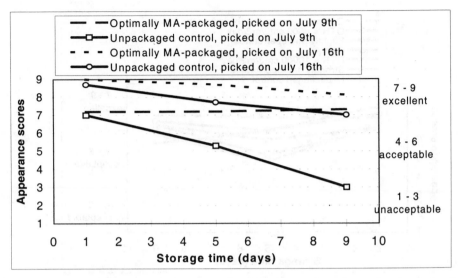

Figure 1 *Appearance of optimally MA-packaged and unpackaged strawberries during 9 days of storage*

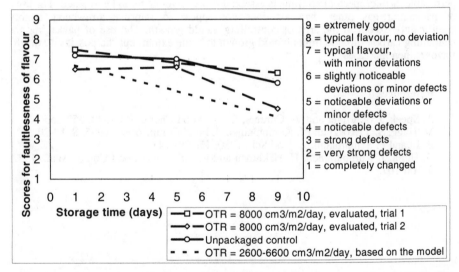

Figure 2 *Improvement in faultlessness of flavour of strawberries picked on July 16th when very permeable film (oxygen transmission rate (OTR) = 8000 cm³/m²/day) was used. The carbon dioxide concentration in the packaging gas was 20% in all cases*

4 CONCLUSIONS

The shelf life of strawberries can be improved by means of MA-packaging. MA-packaging had no effect on the nutritional quality of strawberries. The use of very

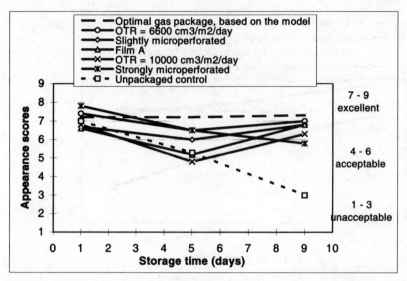

Figure 3 *Effect of packaging material on the appearance of strawberries picked on July 9th when air was used as the packaging gas*

permeable or microperforated films is essential if off-flavours are to be avoided. The use of packaging gas other than air does not appear to improve strawberry quality significantly except in the area of controlling mould growth. The use of packaging gas containing carbon dioxide limits mould growth to some extent, but the possibility of off-flavours then increases.

References

1.　A. Speek, J. Schrijver and W. Chreurs, J. Agr. Food Chem., 1984, **32**, 352-355.
2.　M. Hägg, S. Ylikoski and J. Kumpulainen, J. Food Comp. Anal., 1995, **8**, 12-20.
3.　B. Li and P. Schuhmann, J. Food Sci., 1980, **45**, 138-141.
4.　K. Haila, J. Kumpulainen, U. Häkkinen and R. Tahvonen, Food Comp. Anal., 1992, **5**, 100-107.

EFFECT OF PRESSURE-COOKING AND EXTRUSION-COOKING ON CELL WALLS OF ONION WASTE

A. Ng, S. LeCain, M.L. Parker, A.C. Smith and K.W. Waldron

Institute of Food Research,
Norwich Research Park,
Colney,
Norwich NR4 7UA,
United Kingdom

1 INTRODUCTION

Onion (*Allium cepa* L. outer fleshy scale leaf bases) is a bulb, the edible portion being the swollen leaf bases. In a well-ripened onion, the outer tissues form the skin. In the European Economic Community, over 450 000 tonnes of onion waste are produced annually. There is considerable interest in valorization of the wastes which will benefit the onion producers and processors. The cell walls of onion parenchyma are virtually devoid of lignin and phenolic components, and could provide a source of dietary fibre supplements[1,2]. Recently, studies have demonstrated considerable variation in the carbohydrate composition of cell-wall pectic polysaccharides between tissues of onions, particularly the brown outer skin and fleshy outer and inner scale leaves[3]. However, there is little information on the effect of heat treatment on the chemistry of cell-wall polysaccharides of outer fleshy scale leaves of onions.

2 MATERIALS AND METHODS

2.1 Materials

Onion waste were obtained from the British Onion Producers Association. The outer fleshy scale leaves of the onions were subjected to pressure-cooking at 120 °C for 0, 10, 20, 30 and 50 min. Cell-wall material prepared from such tissues was also subjected to extrusion-cooking over a range of temperatures, pressures and moisture contents.

2.2 Cell-wall preparation

Cell wall materials (CWMs) of fresh material (for extrusion) and pressure-cooked material (for analysis), were prepared[4,5].

2.3 Water extraction of cell wall polymers

Cell-wall material was suspended in water (pH 5.1) and stirred for 2 h at 20 °C. The supernatants were filtered and dialysed exhaustively with water prior to concentration and freeze drying.

2.4 Carbohydrate composition

Carbohydrate composition was determined by GL-separation of alditol acetates after hydrolysis and reduction using standard techniques[6].

3 RESULTS AND DISCUSSION

3.1 Effect of pressure-cooking

3.1.1 *Instrumental textural Measurement.* Pressure-cooking at 120 °C reduced the tissue firmness of both varieties (Delta; Figure 1).

3.1.2 *Water Extraction.* Pressure-cooking resulted in a substantial increase in water-soluble pectic polymers (Figure 2).

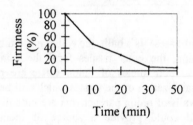

Figure 1 *Change in firmness of onions (Delta) during pressure-cooking*

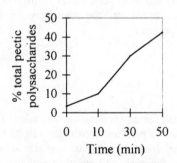

Figure 2 *Change in % pectic polysaccharides that are water soluble during pressure-cooking*

3.2 Effect of extrusion

As for pressure-cooking, heating-induced swelling of cell wall material (CWM) was also observed during extrusion-cooking. Increasing barrel temperature resulted in an increase in water solubility (Figure 3).

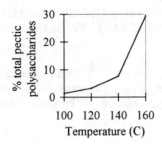

Figure 3 *Change in % pectic polysaccharides that are water soluble during extrusion-cooking*

4 CONCLUSION

Waste onion tissues contain significant quantities of cell-wall polymers which could provide a source of dietary fibre for incorporation into foodstuffs.

The cell-walls are significantly modified by heat treatments during pressure-cooking and extrusion, resulting in an increase in solubility. This is consistent with the general depolymerisation of pectic polymers through β-elimination.

References

1. P.K. Chatterjee and H.V. Nguyen, Mechanism of Liquid Flow and Structure Property Relations. In: Absorbency, ed. Chatterjee, P.K. Elsevier, Amsterdam. The Netherlands, 1985, pp 29-84.
2. G.R. Fenwick, *CRC Critical Reviews in Food Science and Nutrition*, 1985, **22**, 273.
3. A. Ng, A.C. Smith and K.W. Waldron, *Food Chem.*, 1998, in press.
4. S. LeCain, A. Ng, M.L. Parker, A.C. Smith and K.W. Waldron, *Carbohydr. Polym.*, 1998a, (submitted).
5. A. Ng, S. LeCain, M.L. Parker, A.C. Smith and K.W. Waldron, *Carbohydr. Polym.*, 1998b, (Submitted).
6. A. Ng and K. W. Waldron, *J. Sci. Food. Agric.*, 1997, **73**, 502.

EFFECT OF THREE TEMPERATURES OF STORAGE ON THE ACTIVITIES OF POLYPHENOLOXIDASE AND PEROXIDASE IN MAMEY SAPOTE FRUITS (*Pouteria sapota*).

G.O. Pérez-Tello; M.A. Martínez-Téllez,*; B.O. Briceño; T.I. Vargas-Arispuro and J.C. Díaz-Pérez. Centro de Investigación en Alimentación y Desarrollo, A.C. Apdo. Postal No. 1735; 83000, Hermosillo, Sonora, México. *Principal Author

1. INTRODUCTION

The market of exotic tropical fruits is restricted to developed countries that search for flavors, textures and new aromas.[1] Most tropical fruits are sensitive to chilling temperatures. Therefore, low storage temperatures are a serious problem in the postharvest handling and quality of tropical fruits, specially, the exotic fruits. Many enzymes are involved in the internal metabolic alterations of fruit tissue. The oxidases, peroxidase (E:C: 1.11.1.7; POD) and polyphenoloxidase (E.C. 1.14.18.1, PPO) have been implicated in many metabolic changes and reactions of fruit tissues. [2,3] POD enzyme has been related with quality control of processed fruits as a blanching indicator in the processing of vegetables, and PPO are implicated in tissue browning. It has also been associated with antocyanin degradation in various horticultural commodities. [2] This has served also like an index of global quality. The lower temperatures of storage are considered as an external factor that modify certain metabolic processes.

In Mexico, mamey is cultivated in small plantations situated on southeast and central sites of the country. The fruit is very sensitive to chilling temperatures, microbial contamination and to the physical damage. Its market is totally domestic. [1,2]

In order to understand some functions and effects of refrigerated storage on postharvest behaviour of PPO and POD activities in mamey sapote (*Pouteria sapota*) at low and high temperatures, this study was undertaken among 25 days, also evaluating some physical attributes.

2. METHODOLOGY

2.1 Samples and storage conditions

Mamey sapote fruit was grown in a native cultivar in Morelos, Mexico. Maturity index, external damage and flesh firmness were evaluated in 4 fruits[4,5]. Each fruit was weighed and its colour attributes measured. Ten fruits were allocated into each temperature (2, 10 and 20°C) by random procedure.

Each group of fruits were placed on cardboard trays and were stored in refrigerated coolers at temperatures of 2.5, 10 and 20°C, 80-85 % HR for 25 days. Weight loss, colour, fruit flesh firmness, chilling injury index, PPO and POD activities, were examined each four days.

2.2 Physical evaluation

Weight loss (g/100 g of flesh fruit) was evaluated by difference in fruit weight at the first day of the experiment and each sampling. Total colour change (DT) was calculated by square root of square difference of Luminosity (L) and (a) and (b) parameters measured by Hunter Lab colorimeter. Hue angle was calculated as tan^{-1}(b/a). Flesh firmness was measured in Newtons according to the method of Bourne[4]. Chilling injury index was evaluated by the subjective scale described by Martínez-Téllez.[5]

2.3 Enzyme assays

Extraction of enzyme was carried out from acetona powders of mamey pulp.[3] The acetona powders obtained were stored at -47°C until used. PPO (EC 1.14.18.1) activity (units of activity per gram-minute) was determined according to the method of Flukey and Jen.[6] POD (EC 1.11.1.7) activity was determined using a modification of method of Flurkey and Jen.

2.4 Statistical analysis

Statistical analysis data was from a completely randomised design by analysis of variance. Time and storage temperatures were the treatments with α= 5 %. Several lineal regressions were evaluated by coefficient of determination (r^2) carried out between enzymatic activities (PPO and POD), firmness and weight loss against storage temperatures.

3. RESULTS

Fruit weight was very variable, mean was 484 ± 50 g and average of flesh firmness of 189.4 ± 2.6 N. Size fruit had great variation. PPO activity showed a range between 3166 and 8973 units of activity/g-min for the three storage temperatures (Figure1). Enzyme activity of PPO increased as ripening and time progressed at 20°C and 2.5°C, observing a behaviour of increase followed by a decrease every 4 days of sampling for the three temperatures. This result agrees with the data informed by Siddiq et al., in plum pulp extracts.[7] The fruits stored at 2.5°C exhibited an important increase in their PPO activity after 16 days of storage, meanwhile at 10°C, the fruits showed a constant pattern and diminish subsequently. PPO activity was not different (p<0.05) at three storage temperatures.

POD activity (Figure 2), was stable and remaining constant (85.3 ± 4.6 units/g-min) at 2.5°C, however, it was growing at 10°C and more at 20°C. POD increased in a sigmoidal manner along the 25 storage days at 20°C with a high coefficient of determination (r^2=0.954) and a significant difference at 2.5°C (p<0.05). Pulp deterioration was observed in mamey storage at 20°C more than at 10 and 2.5°C. The deterioration observed in the samples became particularly notable after 12 days, but peroxidase activity could not be assigned as responsible for these changes.

Weight loss and fruit flesh firmness were both significantly affected by storage time and temperature (p<0.05). Weight loss increased at 20°C along the days (32.5 g / 100g at 25 days) showing high correlation with the time (r^2= 0.997). Meanwhile at 2.5°C weight loss was 8.37 g/ 100g (r^2= 0.899) and was slightly superior than that at 10°C. These results showed a highly significant linear relationship between storage temperatures and the increase on weight loss due to fruit respiration.

Abnormal ripening in 30 % of fruit flesh close to skin (±1.5 cm) was observed at 2.5°C. This portion had a yellow pale colour. The pulp softening was progressive at 20°C showing a quick ripening of fruits. The index of external damage was worthless. However, pulp browning was observed at 20 and 10°C as well as internal growth of grey mould situated in the poles of fruits after day 12 of the storage. The lower temperature could be affecting normal ripening and cause partial flesh browning close to skin in some fruits.

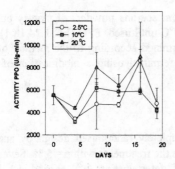

Figure 1 *Activity of PPO in mamey sapote under 3 temperatures of storage. Values are the average of 3 analysis ± SD*

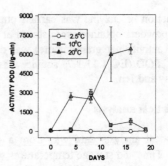

Figure 2 *Activity of POD in mamey sapote under 3 temperatures of storage. Values are the average of 3 analysis ± SD.*

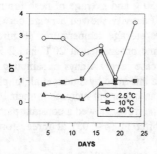

Figure 3 *Total colour change (DT) in mamey peel*

Hue angle values (50-58°) were not different (p<0.05) at 3 storage temperatures. Total colour change (DT) showed significant difference at 2.5 and 20°C (Figure 3). The results suggest that the fruit skin suffers higher desiccation at 20°C which affected the measurement of external colour (L, a and b), given clearer tones than at low temperatures with minor respiratory rate.

Results suggest that POD activity was inhibited in the pulp of mamey at low temperatures, although PPO activity was not. Weight loss was very superior at 20°C, meanwhile at low temperatures, it showed similar behaviour.

Fruits showed chilling injury as abnormal ripening in fruit flesh, close to skin at 2.5°C. Fruit flesh firmness was higher at 2.5 and 10°C than 20°C. Hue angle showed a similar tone in the stored fruits at three temperatures.

References

1. A. E. Casas, Bachelor Sci. Thesis. Instituto Politécnico Nacional, 1977.
2. C. Y. Wang, *HortScience,* 1994, **29,** 986.
3. J. R. Powers, M. J. Costello and H. K. Leung, *J. Food Sci.,* 1984, **49,** 1618.
4. F. S. Burnette, *J. Food Sci.,* 1977, **42,** 1.
5. M.A. Martínez-Téllez and M.T. Lafuente, *Acta Horticulturae,* 1993, **343,** 257.
6. W. H. Flurkey and J.I. Jen, *J. Food Sci.,* 1978, **35,** 1826.
7. M. Siddiq, N. K. Sinha and J. N. Cash, *J. Food Sci.,* 1992, **57,** 1177.

SOME ASPECTS REGARDING MODIFIED ATMOSPHERE PACKAGING OF MUSHROOMS

M. Popa, D. Stanescu, M. Herascu, A. Ilie, R. Dumitrescu and I. Vraci

Institute of Food Chemistry
1D Garlei street, 71576
Bucharest, Romania

1. INTRODUCTION

The aim of the study was to increase the shelflife of mushrooms. The shelflife and quality changes of 2 different mushroom species (Agaricus bisporus and Pleurotus) packaged in 4 different gaseous atmospheres and stored at 3 different temperatures. Also studied were washing process influence and modified atmosphere packaging with or without humidity absorbers.

2. MATERIALS AND METHODS

2.1 Harvesting
Cultivated mushrooms (Agaricus bsp. and Pleurotus) grown at a mushroom farm, were picked early in the morning of the experimental work day. Agaricus bsp. has
been harvested at the "button" stage and matched according to cap diameter with the velum differentiated but unbroken.

2.2 Washing
The mushrooms were dipped in chlorinated water(96 ppm) at a 1 / 5 rate (mushrooms / chlorinated water), for 1-2 min.

2.3 Packing
The unwashed and washed containing normal air control samples were packaged in LDPE (low density polyetylene) bags.
The samples were packaged in modified atmospheres of the following compositions:
MAP I: 3% O_2, 5% CO_2, 92 % N_2;
MAP II: 5 % O_2, 5 % CO_2, 90 % N_2;
MAP III: 1 % O_2, 5 % CO_2, 94 % N_2;
MAP IV: 3 % O_2, 7 % CO_2, 90 % N_2.
The packaging material for the MAP samples was LDPE, too.
Samples weighed approximately 150 g and a head space volume,was almost 1000 cm^3
The amount of silicagel was 7 g / sachet/packs.

The MAP machinery was a vacuum / modified atmosphere machine equipped with a gas blender KM 100 – 3M and packaging room A 300 / 16, (MULTIVAC, Germany).

2.4 Storage
3 different storage conditions, have been used namely: 4 °C – constant;
18 °C – constant, 26 – 28 °C.
Storage devices were:
- a commercial cabinet for 4°C, without light exposure;
- a climaroom from FEUTRON Germany, for 18 °C.

2.5 Quality assessment
The quality of the mushrooms during storage was assessed by the following methods:
- microbiological analysis;
- sensory evaluation;
- growth of cap / stem measurements.

Microbiological analysis were:
♦ Total plate count (CFU/g) according to ISO 4833 / 1993.
♦ Clostridium species – incidence,according to Romanian standards.
♦ Pseudomonas aeruginosa – incidence, according to Romanian standards.

Sensory evaluation consisted of color assessment during storage time.

Growth measurements
The growth measurements were made manually using a micrometric gauge.

3. RESULTS AND CONCLUSIONS

As Figures 1...8 illustrate, the most efficient MAP composition was MAP III, i.e. 1 % O_2; 5 % CO_2; 94 % N_2 modified atmosphere packaging and storage temperature of 4°C lead to the greatest microbiological inhibition expressed by total plate count.
- washing procedure reduces microbial load by one decade
- humidity absorbers presence in the MAP packs leads to   one decade reduction in total plate count
- Agaricus ssp. had a greater shelflife in the same conditions compared to Pleurotus ssp.
- lower temperatures and MAP conditions reduced the mushrooms growth.
As conclusion,the best way to process Agaricus mushrooms is:
- washing in clorinated water
- modified atmosphere packaging at low O_2 levels (1....3 %, at least 5 % CO_2 and N_2 balance and humidity absorbers.
- storage at 4°C.

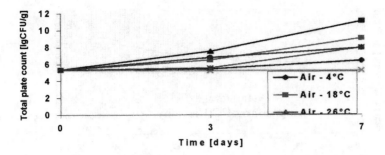

Figure 1 *The effect of temperature and storage time on total plate count of mushrooms*

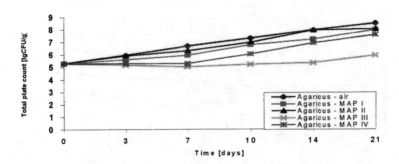

Figure 2 *The effect of MAP composition and storage time on total plate count for washed mushrooms*

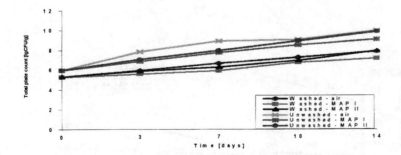

Figure **3** *The effect of washing on total plate count on mushrooms stored at 4°C*

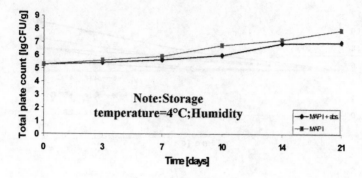

Figure **4** *The efect of humidity absorber on total plate count of mushrooms*

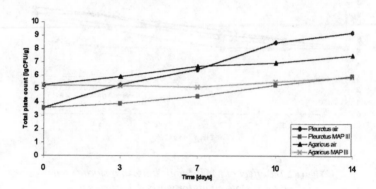

Figure **5** *The effect of MAP on mushrooms shelf-life stored 4 °C depending on species.*

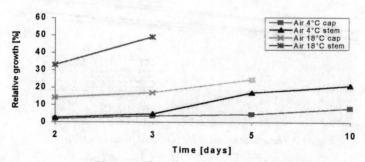

Figure 6 *The effect of temperature on cap/stem growth of mushrooms Agaricus bsp. (packaged in air)*

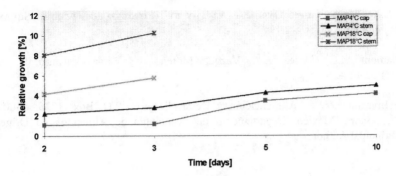

Figure 7 *The effect of temperature on cap/stem growth of
mushrooms Agaricus bsp.(packaged in MAP)*

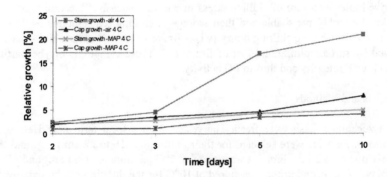

Figure 8 *The effect of MAP on cap/stem growth of
mushrooms stored at 4°C*

References

1   G. M. Sapers, R. I. Miller, F. C. Miller, P. H. Cooke and Sang-Won Choi,
    J. Food Sci., 1994, 59, No.5, 1042

2   H. W. Peppelenbos, J. van`t Leven, B.H. van Zwol and L. M. Tijskens, ' 6th
    Int. Cont. Atmos. Res. Conf. Ithaca ', NY. 15-17 June, 1993, 2, 746-758.

3. R. Nichols and J. B. Hammond, *J. Food Tech.*, 1975, 10 , 427.

4. H. Sugiyama and K. S. Rutledge, *J. Food Prot.*, 1978, 41 , No.5, 348.

Effect of Low Temperatures Over PG Activity and Firmness in Zucchini Squash (*Cucurbita pepo* L).

Ramos Clamont, M.G., Gardea, A.A, Vázquez-Moreno, L., Vargas-Arispuro, I. and Martínez-Téllez, M.A.*

Centro de Investigación en Alimentación y Desarrollo A.C. P.O. Box. 1735, 83000. Hermosillo, Sonora, México. Departamento de Tecnología de Alimentos de Origen Vegetal. *Principal Author

1. Introduction

Polygalacturonase (PG), a cell wall enzyme which cleaves non esterified linkages in the 1-4 polygalacturonic acid backbone of cell wall pectins, has been long thought to control fruit tissue softening [2,3]. Activation of transcription of the PG gene occurs after the rise in ethylene synthesis that triggers tomato ripening [10] . PG activity is absent in immature fruit, however is induced by ethylene under plant stress [6] . Stimulation of ethylene production is often observed after exposure of chilling stress in zucchini squash[7]. Zucchini squash are considered to be highly perishable and their storage at chilling temperatures (< 5° C) is limited to 1 or 2 days before chilling injury (CI) is irreversible [5]. CI in zucchini squash is characterized by surface pitting and lost of firmness [8]. These symptoms maybe partially related to cell wall integrity and thus to PG activity.

2. Materials and Methods

Raben zucchini squash were freshly harvested from a local farm near Hermosillo Sonora, México. Samples were selected for their uniformity (16 to 22 cm long) and then randomly divided in to 2 lots. First lot was stored at 2°C (chilling temperature) and 85-90% RH for 18 days. The second group was stored at 10°C for the duration of the experiment. Every 48 h eight squash were chosen at random from each group. All measurements were done in triplicate. Pulp fruit firmness was measured with a Chatillon penetrometer equipped with 10 mm plunger. Skin and pulp fruit PG activity measurements were done by Gross method [4] . Protein was determined by Bradford [1]. PG activity Unit is defined as the μmols of galacturonic acid/mg of protein / hour. Fruits were transferred from 2°C and 10°C to 20°C for measurement of ethylene production. Ethylene production was determined with a Varian gas chromatograph equipped with an alumina column and a flame ionization detector.

3. Results

Storage temperature had no significant effect ($p < 0.05$) on fruit firmness (Figure 1). Figure 2 shows that the effect of PG activity over firmness was minor ($r^2 = 0.50$). This supports the view that PG is not the primary determinant of softening. Physical factors and other cell wall hydrolases like β-galactosidase could be involved[9].

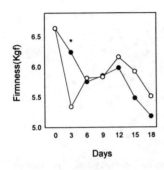

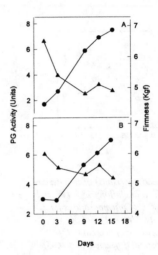

Figure 1. *Firmness of pulp tissue from zucchini squash held at chilling 2°C (O) and non chilling temperatures for 18 days. (*) represent difference (p<0.05)*

Figure 2. *PG Activity, Units (●) and firmness Kgf (▲) of pulp tissue from zucchini squash held at 2°C (A), and 10°C (B) for 18 statistical days. PG activity Unit= 1μmol of galacturonic acid / mg of protein/h.*

Figure 3 shows the changes in the skin PG activity of squash at both temperatures. At chilling temperature we observed an increase in PG activity at day nine of storage which matches with the stimulation of ethylene production by exposure at chilling temperatures (Fig4). These results suggest that PG activity may play a role in CI of zucchini squash. However, further studies are necessary.

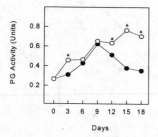

Figure 3. *PG activity (Units) of skin tissue from squash held at chilling 2° C (●) and non chilling, 10° C (○) temperatures for 18 days. PG activity Unit = 1μmol of galacturonic acid/mg of protein/h. (*) Represent statistical difference (p < 0.05).*

Figure 4. *Ethylene production, μl/Kg/h (▲) and PG activity, Units (●) of zucchini squash held at 2° C for 18 days. Fruits were warmed at 20° C for 4 h prior to ethylene analysis*

4. References

1. Bradford, M. M. 1976. Anal. Chem 72: 248-254.
2. Griegson, D. and Tucker, G.A. 1983. Planta, 157 : 174-179.
3. Gross, K.C. 1990. Postharvest News and Information. 1 : 109-112.
4. Gross, K. C.1982. Hort Sci 17: 933-934.
5. Harderburg, R.E., Watada, A.E. and Wang, Y.C.1986. USDA Hdbk. 66.
6. Hubberman, M., Pressman, E. and Jaffe, M.J. 1993. Plant Cell Physiol. 34: 795-801.
7. Martínez- Téllez, M.A. , Gardea, A. A., Mercado, J. N. y Bringas, E. 1994 .CIAD DTAOV. Reporte Anual. 3: 11- 19.
8. Mencarelli, F., Lipton, W.J. and Petterson, S. J. 1983. J. Amer. Hort. Soc. Sci. 108:884-890.
9. Parra, A. A., Saucedo-Arias, L. J., Cruz-Hernandez, A., Gutierrez-Martínez, P. y Gomez-Lim, M. A. 1996. Memorias del XXI Congreso Nacional de la Sociedad Mexicana de Bioquímica A. C. Manzanillo, Col. p 149.
10. Seymur, A ., Taylor, J. E. and G. E. Tucker. "Biochemistry of Fruit Ripening". Chapman and Hall Ed. USA.1993 pp 16-21, 65-79, 423-429.

Exogenous Polyamines Reduce Chilling Injury and Polygalacturonase Activity in Zucchini Squash (*Cucurbita pepo* L.)

RamosClamont, M.G., Gardea, A.A, Vázquez-Moreno, L., Briseño, T. O, Vargas-Arispuro I. and Martínez-Téllez, M.A.*

Centro de Investigación en Alimentación y Desarrollo A.C. A.P. 1735, 83000. Hermosillo, Sonora, México. Departamento de Tecnología de Alimentos de Origen Vegetal. *Principal Author

1. Introduction

Chilling injury (CI) is the physiological damage induced in tissues when they are exposed to low, but non freezing, temperatures. CI in zucchini squash is characterized by circular to longitudinal pits on the surface and maybe a result of a loss of cellular integrity caused by damage to the cellular wall [6].

Polyamines (PA's) such as putrescine (PUT), espermidine (SPD) and espermine (SPM) have been shown to reduce CI and thus increase the tolerance of vegetable tissues to low temperatures. Treatments with exogenous polyamines after harvest but before cold storage, increased endogenous polyamines levels and reduced CI (Kramer and Wang, 1989) also inhibit *in vitro* polygalacturonase activity (PG) in "Golden Delicious" apples inoculated with *Penicillium expansum* [3]. These results suggested the polyamine's inhibition of PG activity and that this activity may play a role in delaying CI symptoms.

2. Materials and Methods

"Raben" zucchini squash were harvested from a local farm near Hermosillo, Sonora Mexico. The fruits were pressure infiltrated with polyamines at five different concentrations. PUT was used at 0.5, 1.0, 1.5, 2.0 and three 4.0 mM. SPD and SPM were used at 0.1, 0.25, 0.5, 2.0 and 4.0 mM. Pressure infiltration involved the use of 0.9 Kg/cm^2 air pressure for 3 min [8]. After the treatment zucchinis were then stored at 2°C and 10° C for 12 days, 85-90% RH.PG Activity measurements were done by Gross method[2]. Protein was estimated by Bradford [1]. PG activity Unit is defined as the μmols of galacturonic acid/mg of protein/ h. The degree of CI as judged by the extent of surface pitting was evaluated 1 h after transfer of squash from storage chambers to room temperature (≈ 25°C) by rating on a scale of 0 to 3 with 0 = no damage, 1 = light up (10% or less), 2 = moderate (10-20%) and 3 = severe (> 20%). CI Index was determined using the following formula:Σ (pitting scale (0-3) x number corresponding fruit within each class/ total number of fruits estimated (60 per group)[5].

3. Results

Storage of the squash at 2° C resulted in develop of CI. However, the differences in the severity of this damage among different treatments were significant. Table 1 shows CI index after 12 days of storage at 2° C. Severe pitting was detected on the skin of squash from the 2° C control group. The CI symptoms in squash was reduced by PA′s infiltration. However, higher concentrations (2.0 mM and 4.0 mM) were less effective.

The SPD treatment was the most effective since its average CI value was 80% lesser than those of the control. The SPD 0.5 mM concentration was even more effective. Squash from this treatment did not develop chilling injury during storage at 2° C. This treatment thus increased time that the squash could be chilled to 12 days of storage. No CI developed during storage at 10° C .

Data presented in Figure 1A indicate that the effect of the storage temperature on PG activity was significant (p< 0.05). The activity of this enzyme in stored fruits at 10° C was higher than in fruits stored at 2° C. PA′s infiltration decreased PG activity in the storage fruits, independently of the storage temperatures.

At chilling temperature (2° C) no significant difference (p< 0.05) was found in fruits infiltrated with 2.0 mM concentrations. These treatments were the most effective since their average value of PG activity was 60% lesser than those of the control (Fig 1C). The SPD and SPM 0.1 mM and PUT 0.5 mM concentrations had no effect over enzyme activity. PG activity from 4.0 mM treatments was greater than PG activity from 2.0 mM treatments in all PA′s, these concentrations could be toxic to the fruit since its endogenic levels are much lower. Serrano et al.,[7] reported levels of PUT in squash skin from 162 to 320 nmoles/g after 12 days of conservation at 2°C, and that the levels of SPD and SPM are still minor.

Table 1   *Effect of PUT, SPD and SPM Infiltration on Chilling Index of zucchini squash Held at 2° C for 12 days.*

	Chilling Index		
Concentration	PUT	SPD	SPM
Control	1.40a	1.40a	1.40a
0.1mM	----	0.63c	0.63c
0.25mM	----	0.03h	0.30ef
0.5mM	0.79c	0.00h	0.10g
1.0mM	0.27ef	------	-------
1.5mM	0.16g	------	-------
2.0mM	0.33e	0.43e	0.56cd
4.0mM	1,30a	0.73c	1.10b

Different letters means statistical difference between treatments

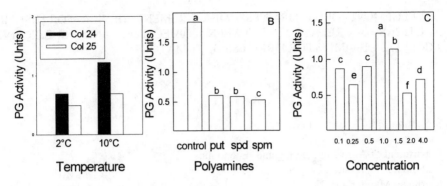

Figure 1. *PG activity (Units) of skin tissue from zucchini squash infiltrated with PA's. A). Effect of Storage Temperature. B). Effect of Polyamine type. C) Effect of Polyamine concentration. Different letters represent statistical difference (p < 0.05).*

4.References

1. Bradford, M. M. 1976. Anal. Chem 72: 248-254.
2. Gross, K.C.1982. Hort Sci 17: 933-934.
3. Kramer, G.F., Wang, C.Y. and Conway, W. 1989. J. Amer. Soc. Hort. Sci. 114:942-946.
4. Kramer, G.F. and Wang, Y.C. 1989. Hort Science. 24:995-996.
5. Martínez-Téllez, M.A. and Lafuente, M.T. 1997. J. Plant Physiol. 150 : 674-678
6. Mercer, M.D., and Smittle, A.D. 1992.. J. Amer. Hort. Sci. 117 : 930-935.
7. Serrano, M., Pretel, M. T., Martínez-Madrid, M. C., Riquelme, F., Romojaro, F.1995. IV Simposio Nacional I Iberico de Maduración y Postrecolección de Frutos y Hortalizas.273-276.
8. Wang, C. Y. and Kramer, G. F. 1990. Polyamines and Ethylene: Biochemistry, Physiology and Interactions, H E Flores, R N Arteca, J C Shannon eds. American Society of Plant Physiologists. p 411-413.

DEVELOPMENT OF A FIRMNESS MEASUREMENT DEVICE FOR USE IN AUCTIONS - RELATION BETWEEN OBJECTIVE AND SUBJECTIVE ACCEPTANCE LIMITS FOR FIRMNESS OF TOMATOES

S. Schotte & J. De Baerdemaeker

Department of Agro-Engineering and -Economics
Katholieke Universiteit Leuven
Kardinaal Mercierlaan 92
3001 Heverlee Belgium

1 INTRODUCTION

There is a growing interest world wide for non destructive methods to determine firmness of vegetables and fruits. Currently, the most frequently used method to investigate firmness of fruits and vegetables is the penetrometer. However because of the destructive character and the non homogeneous structure of tomatoes, this technique is not very suitable for these fruits [1].

A non destructive technique to measure firmness in an objective way, is the acoustic impulse response method.

2 OBJECTIVES

At the K.U.Leuven, a user friendly firmness device, based on the acoustic impulse response method, was developed for use in cooperative fruit and vegetable auctions. In Belgian auctions, firmness of tomatoes is measured with a Zwick device or judged by experts, who squeeze the fruit with their thumb to determine the acceptability of the firmness. However, the Zwick device is known to be inconsistent and user-dependent. The test is very local and it causes some damage to the point where it is applied. The classification by human experts is also subjective. Therefore the relation between subjective and objective acceptance limits for tomatoes were established.

3 FIRMNESS MEASUREMENTS

3.1 Acoustic Impulse Response Method, A User Friendly Device To Measure Firmness

In the acoustic impulse response method the fruit is gently hit with a hammer (impact). The sound pressure waves generated by the vibrating fruit are measured with a microphone (response). A Discrete Fourier transformation of the response signal yields a frequency spectrum.

In this spectrum, the first resonant frequency (RF) is selected [2,3]. Previous research did show that the resonant frequency is not significantly influenced by the place (flesh or wall of the fruit) of the measurement on the fruit. Therefore the RF was taken at 3 random locations on the equator. The averaged data are used for further processing. The firmness factor (S) is calculated from the mass and the resonant frequency according to the equation [4]:

$$S = (RF)^2 * (W)^{2/3} \qquad\qquad (1)$$

with

W : mass (g)

RF : first resonant frequency (Hz)

In the following the firmness of the fruits will be expressed in firmness units with dimensions $10^6 \, Hz^2 * g^{2/3}$.

Galily et al. [5] found that this acoustic method is more reliable and repeatable than the Zwick tester, which is used for firmness testing in Belgian auctions.

On the basis of the experiment and theoretical analyses of fruit vibration, a user friendly firmness device, with the acoustic impulse response method, was implemented for use in auctions.

The firmness device consists of a portable computer and a "measurement box". This box contains a fruit holder, connected to a small microphone, a mass balance to determine the weight and an automatic hammer to excite the tomatoes. The mass balance is used as trigger: as soon as the fruit is placed on the holder, the measurement sequence is activated.

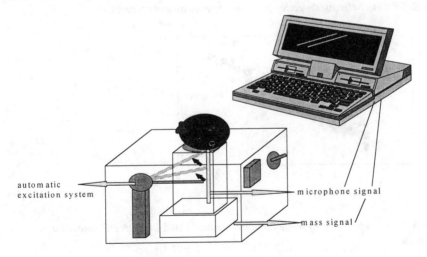

Figure 1 *Set up of the acoustic impulse response method to measure firmness*

4 RESULTS AND DISCUSSION

4.1 Relation Between The Objective And The Subjective Firmness

First the relation between objective and subjective firmness measurements was established. Therefore 100 tomatoes of different varieties and of different firmness were measured by both the experts and the automatic firmness device. The experts gave a score from 1 to 10 according to their opinion of the firmness; a good firm tomato should give a firmness of at least 6/10. The objective firmness reading was based on three measurements around the equator (Figure 2). The repeatability of the experts was tested in (another) blind experiment: the same tomatoes were tested 6 times by the same experts (Figures 3 and 4).

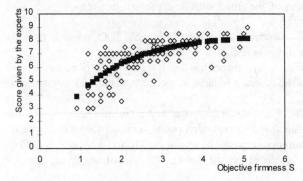

Figure 2 *Relation between objective and subjective firmness of tomatoes*

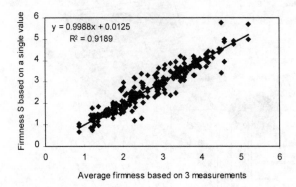

Figure 3 *Relation between average objective firmness and objective firmness based on a single value*

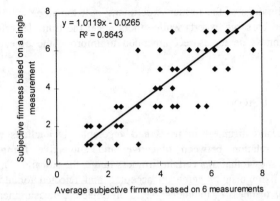

Figure 4 *Relation between average subjective firmness and subjective firmness based on a single value*

Figure 2 shows that for hard tomatoes there is a quite good correlation between the subjective and the objective firmness "scores" while for the softer tomatoes the discrepancy is much larger.

Figures 3 and 4 show that the correlation between the average firmness of the tomato and the firmness based on a single measurement, is better for the objective than for the subjective measurements.

4.2 Determination of the Acceptance Limits for Firmness of Tomatoes

Two hundred tomatoes of the market segments of Flandria 'Baron' and 'Prince' were examined by the experts, after measurement of the firmness with the acoustic impulse response method. The experts were asked to divide the tomatoes into two groups according to the firmness: accepted and rejected. The data showed that for some tomatoes there was a discrepancy between the different persons. When only the data of tomatoes on which all the experts agree are used, the results show that between a firmness value of 1.7 and 2.5 the acceptance is unclear. Below a firmness of 1.7 the experts agreed and reject the tomatoes during quality checks at the auctions.

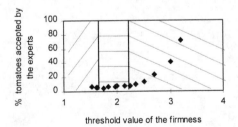

Figure 5 *Acceptance by the experts as a function of the threshold firmness (⟋ rejected ⋯⋯ acceptance not clear, ⟍ accepted)*

These results confirm the findings of the former experiment: soft tomatoes with the same firmness will be accepted by some experts while others reject the fruit. Therefore the use of the automatic firmness device is recommended to improve the objective classification.

5 SUMMARY AND CONCLUSION

A user friendly device to measure firmness of fruits and vegetables in auctions was developed. Investigating the relation between objective and subjective firmness measurements of tomatoes has shown that in a certain firmness range, the so called "gray zone", experts can not easily distinguish or agree on accepted and rejected tomatoes. Therefore acceptance limits for firmness of tomatoes, based on objective measurements, were proposed.

References

1. Thorne, S. and Alvarez, J.S.S., *J.Sci.Food Agric*, 1982, **33**, 671.
2. De Baerdemaeker, J., *Acta Hort.*, 1988, 258.
3. Chen, H., De Baerdemaeker, J. and Vervaeke, F., Proceedings of the international conference on agricultural engineering, Beijing, China, 1992.
4. Cooke, J.R., *Trans. of the ASAE*, 1972, **16 (6)**, 1075.
5. Galili, N. and De Baerdemaeker, J., ISMA conference, Leuven, Belgium, 1996.
6. Schotte, S. and De Baerdemaeker, presented at the "5th international symposium on fruit, nut, and vegetables production engineering", Davis California, 1997

EFFECT OF ETHYLENE ON SENSORY QUALITY OF CARROTS

R. Seljåsen[1], G. B. Bengtsson[1] and H. Hoftun[2]

[1]MATFORSK - Norwegian Food Research Institute, Osloveien 1, N-1430 Ås, Norway
[2]Agricultural University of Norway, Department of Horticulture and Crop Sciences, P.O. Box 5022, N-1432 Ås, Norway

1 INTRODUCTION

Bitter and turpentine-like tastes are the main reasons for off-flavour in carrots. An isocoumarin compound, 3-methyl-6-methoxy-8-hydroxy-3,4-dihydroisocoumarin (6-methoxymellein), is known to be involved in the development of bitter taste in carrots[1]. The burning, turpentine-like aftertaste in carrots seems to be a result of a high content of certain terpenes and a reduction in sugars[2]. It has been found that ethylene can induce the biosynthesis of 6-methoxymellein in carrots[3-5]. Also, it seems that bitter carrots stored in ethylene could be judged to taste more turpentine-like than non-bitter carrots stored in air[6]. In the course of investigation of various postharvest conditions, which can give an undesirable taste of carrots, we here present results from the effect of ethylene.

2 MATERIALS AND METHODS

Washed 'Yukon' carrots were stored at 15°C and RH=95-97 % in 10-litre buckets with continuous flow (50 ml/min) of ethylene (1 µl/l in air) or air (control) for 3 weeks. There were three replicates for each treatment with 10 carrots (1 kg) in each. A paper bag with 10 g of calciumhydroxide was enclosed in each bucket to avoid accumulation of CO_2. Samples for chemical and sensory analysis were taken at the beginning of the experiment and after one, two and three weeks. Data were processed by one-way analysis of variance for chemical parameters and two-way analysis of variance for the sensory parameters with panellist as random effect and treatment as fixed effect (P≤ 0.05). Tukey's pairwise comparison tests were used for differences between individual treatments (P≤ 0.05).

2.1 Preparation of Samples

The carrots were washed and approximately 2 cm pieces of the root tip and from below a green zone on the leaf-end were removed. The remaining part was cut into 10 mm cubes, which were blended thoroughly. For chemical analysis 50 g of the cubes were pulverised in liquid N$_2$ (IKA-Universalmülle M20, Janke & Kunkel GmbH & Co KG) and stored below -80°C.

2.2 Sensory Analysis

The sensory analyses were performed according to flavour profile methods (ISO 6564:1985-E) using a continuous scale for rating the samples. The left side of the scale, corresponds to the lowest intensity of each variable, was given value 1.0 and the right side, corresponding to the highest intensity, was given value 9.0 (CSA Compusense, version 4.2, Canada). The sensory panel consisted of 10 trained panellists, each receiving 25 g of carrot cubes in two replicates. The following characteristics were evaluated: Terpene flavour: turpentine-like flavour or flavour of terpenes compared to the taste of a mixture of p-cymene (10μl/l), terpinolene (50μl/l) and trans-caryophyllene (100μl/l) in soy oil; Sweet taste: taste related to the taste of sucrose; Bitter taste: sharp taste related to the taste of chinin or coffein; Aftertaste: related to sharp, burning taste in the middle of the mouth after 60 sec.

2.3 Chemical Analysis

The 6-methoxymellein was extracted as described[8] with slight modifications. The extract was filtered and purified with a silica gel (C18) Sep-Pak cartridge (Waters) and analysed by HPLC on a C18-silica gel column (150 X 3.9 mm, Nova-Pak, Waters, Milford, MA)[9]. 6-methoxymellein standard was prepared from bitter carrots[7].

Extraction of sucrose, fructose and glucose from carrots was made as described[10] and analysed by HPLC on Chrompack Carbohydrat Pb column (300 x 7,8 mm).

3 RESULTS AND DISCUSSION

3.1 Bitter Taste

The ethylene treated carrots had significantly higher sensory scores on bitter taste after 2 weeks than the carrots stored in air (Figure 1). Analysis of 6-methoxymellein show significantly higher amounts of this compound in the ethylene treated carrots compared to carrots stored in air both after 1, 2 and 3 weeks of storage (Figure 2). After 3 weeks there were no significant differences in bitter taste in spite of significantly higher amount of 6-methoxymellein in the ethylene treated carrots than in the control (Figure 1 and Figure 2). The increased bitter taste therefore may be due to other bitter compounds. Some carrots in both treatments were attacked by fungus after 3 weeks. These could be responsible for unfavourable taste components of the carrots.

The increase in 6-methoxymellein during ethylene storage agrees with several previous studies[3-5]. An increase in respiration caused by ethylene has been shown previously to give a more rapid formation of 6-methoxymellein[11].

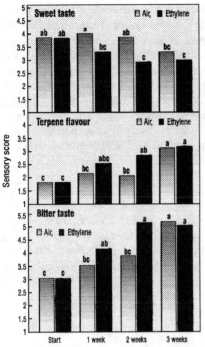

Figure 1 *The sensory score for sweet taste, terpene flavour and bitter taste of carrots stored 3 weeks in air or ethylene (1μl/l in air).Value 1.0 corresponding to the lowest intensity and value 9.0 corresponding to the highest intensity of each variable .*

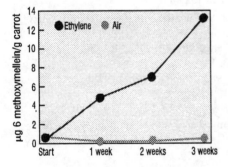

Figure 2 *Changes in content of 6-methoxymellein in carrots during 3 weeks storage in air or ethylene (1μl/l in air).*

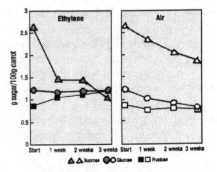

Figure 3 *Changes in the contents of sucrose, fructose and glucose during the 3 weeks storage in ethylene (1μl/l in air) or air.*

3.2 Sweet taste

Sensory analysis of sweetness showed a significantly higher score for carrots stored in air than those stored in ethylene after 1 and 2 weeks of storage (Figure 1). The total content of sugars was significantly higher at the beginning of the experiment than after three weeks of storage both in air and ethylene. Total sugar content remains mostly unchanged from one to three weeks storage, but the amounts of fructose, glucose and sucrose changed significantly during this period (Figure 3). Sucrose content was significantly lower while glucose and fructose content was significantly higher in carrots stored in ethylene compared to those stored in air. This was the result after one, two and three weeks of storage. One of the reasons for this could be that ethylene may cause an increase in respiration[11] which would give another ratio between the three sugars. The decreased sweetness in spite of an increased content of fructose, which is known to have a higher sweetness than sucrose, may be explained by the relatively large drop in the sucrose content, but also by a masking or competing action by the bitter compound 6-methoxymellein.

3.3 Turpentine-like flavour and aftertaste

No significant difference in terpene flavour (Figure 1), terpene odour or aftertaste was found between the two treatments after 1, 2 or 3 weeks. The increased terpene flavour in both the ethylene treated and the control carrots could have been an effect of the high storage temperature (15°C). Further research is needed to see if ethylene can have a significant effect on these taste characteristics in carrots.

References

1. E. Sondheimer, *J. Amer. Chem. Soc.,* 1957, **79**, 5036.
2. P. W. Simon, C. E. Peterson and R. C. Lindsay, *J. of Agric.Food Chem.,* 1980, **28**, 559.
3. Z. -J. Guo and Y. Ohta. *J. Plant Physiol.,* 1994, **144**, 700.
4. M. T. Lafuente, M. Cantwell, S. F. Yang and V. Rubatzky, *Acta Horticulturae,* 1989, **258**, 523.
5. B. C. Carlton, C. E. Peterson and N. E. Tolbert, *Plant Physiol.,* 1961, **36**, 550.
6. P. W. Simon, *Acta Horticulturae,* 1984, **163**, 137.
7. F. Kurosaki and A. Nishi, *Phytochemistry,* 1983, **22**, 669.
8. J. Mercier and J. Arul, *J. Sci. Food Agric.,* 1993, **61**, 375.
9. L. R. Howard, L. E. Griffin and Y. Lee, *Journal of Food Science,* 1994, **59**, 356.
10. Boehringer-Mannheim, 'Methods of biochemical analysis and food analysis using test-combinations', Boehringer Mannheim GmbH, Biochemica, Mannheim 31, Germany, 1989, p130.
11. M. T. Lafuente, G. Lopez-Galvarez, M. Cantwell and S. F. Yang, *J. Amer. Soc. Hort. Sci.,* 1996, **121**, 537.

ALTERNATIVE INHIBITION OF ENZYMATIC BROWNING OF APPLE SLICES

R. Vidrih and J. Hribar

Biotehnical Faculty, Department of Food Science and Technology,
Jamnikarjeva 101, 1000 Ljubljana,
Slovenia

1 INTRODUCTION

Various types of stress (low temperature, injuries, physiological disorders) significantly alter fruits metabolic activity. Wounded tissue has higher activity of enzymes involved in ethylene biosynthesis and in browning. Injuries cause the liberation of enzyme polyphenol oxidase (PPO) which is present in plastids [8], tightly bound to some membrane structure. Cellular disruption leads to mixing the PPO and endogenous phenolics which oxidise to quinons, responsible for browning. The maximum activity of apple PPO occurs before full ripeness than starts to decline [8]. Enzymatic browning is detrimental to appearance, sensory properties and nutritional quality of the food products. Inhibition of PPO activity depends on the nature and concentration of the inhibitor, source of PPO, substrate availability (O_2, phenolics), pH and temperature. Some reducing agents cause the inhibition of PPO, sulphites and ascorbic acid are used on a commercial scale. Enzymes with strong proteinase activity (papain, bromelin) have been tested against PPO [5], both enzymes inhibited browning. Honey appeared to have an inhibitory effect on PPO and consequently on browning of apple slices [6]. Authors attributed inhibitory effect to a peptide with a molecular weight of about 600. Most probably honey peptide inhibits the browning by chelating the essential copper at the active site of PPO [4].

2 MATERIAL AND METHODS

Peeled apples (cv. Idared) were cut in to slices and dipped for 5 minutes in a water solution of papain (2%), sucrose (10%), chestnut honey (10% w/v), acacia honey (10% w/v) and $Na_2S_2O_5$ (3.5 %). Slices were then left to dry on filter paper and packed in polyamid/polyethylene (PA/PE) film in normal, CO_2 (99.9%) and N_2 (99.9%) atmosphere at +1°C or at 20°C. Browning was measured daily by means of Minolta chromameter (L, a, b) on 10 slices. The activity of PPO was measured according to the method of Boyer [1] which was partially modified. 5 grams of apple slices were ground in a mortar under liquid N_2 and mixed with 5ml of phosphate buffer (0.1M pH 6.8 + NaF 0.1M) and 1 g of PVPP. Calculated amount of honey, sucrose , papain or metabisulphite to match concentration of each substance in real experiment. Mixture was left at +1°C for 10 minutes, centrifuged and measured spectrophotometrically at 475 nm. Cuvette was filled first with 0.3 ml of supernatant, phosphate buffer 0.1 M pH 6.8 and 1.25 ml of DOPA temperated at 30°C. Absorbance was measured against blind solution where 0.3 ml of water was added instead of supernatant. Ethylene accumulation of packed slices was calculated from gas chromatographic analyses of 1 ml head space gas samples taken from the PA/PE film. The tissue concentration of acetaldehyde and ethanol was analysed on frozen samples (stored at -20°C) employing GC equipped with FID detector, FFAP column by means of head space technique.

3 RESULTS

Table 1. *Values dL/L_0 *100 of apple slices stored in air at 20°C*

days after treatment	sucrose	chestnut honey	acacia honey	papain	disul- phite	control
0	0	0	0	0	0	0
1	5.51	4.61	4.78	4.46	2.05	12.1
2	6.58	5.44	4.17	4.88	3.64	18.23
3	6.99	4.35	5.78	5.5	5.76	18.29
4	6.58	6.03	8.27	5.47	4.5	20.6
5	6.73	5.38	7.79	7.02	4.75	24.8
6	7.22	5.83	8.03	3.91	4.72	23.1
7	7.54	5.93	7.97	5.9	4.66	22.7
8	6.99	5.66	7.96	6.81	4.78	24.7
9	8.4	6.05	9.12	7.66	5.77	26.7

Severe browning occurred with control fruit (air), less browning occurred when slices were treated with acacia honey, sucrose solution, papain, chestnut honey and disulphite. Disulphite and chestnut honey proved to be the most successful and nearly equal in their effectiveness (Tab. 1).

Table 2. *Polyphenol oxidase activity in apple slices (U/ml)*

	Type of		treatment		
	control	chestnut honey	honeydew	papain	disulphite
PPO activity (U/ml)	0.118	0.075	0.167	0.0026	0.0012

The lowest PPO activity was recorded in slices treated with disulphite followed in ascending order by papain, chestnut honey, control and honeydew (Table 2).

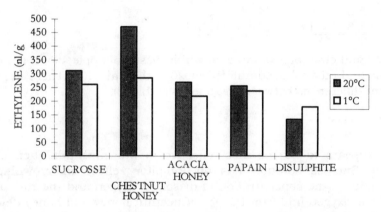

Figure 1. *Ethylene content after 10 days storage of apple slices in PA/PE film*

Ethylene concentration increased during storage, more ethylene accumulated at 20°C than at +1°C. Apple slices treated with chestnut honey produced more ethylene than other treatments, but disulphite treatment resulted in lower ethylene production (Figure 1).

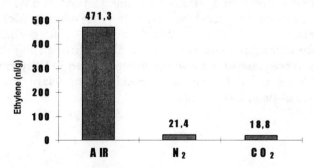

Figure 2. *Ethylene content after 10 days storage of apple slices treated with chestnut honey and stored in PA/PE film in air, N2 and CO2 at 20°C*

Air stored apple slices produced 20 fold more ethylene than N_2 or CO_2 stored (Figure 2).

Table 3. *Accumulation of acetaldehyde and ethanol (mg/100g) in apple slices stored in N_2 at 20°C and at 1°C*

Treatment	acetaldehyde mg/100g		ethanol mg/100g	
	20°C	+1°C	20°C	+1°C
sucrose	1.0	0.5	181.8	104.0
chestnut	3.2	1.4	169.4	119.0
acacia	2.8	2.3	262.4	83.4
papain	5.6	0.5	246.6	11.1
disulphite	1.3	1.6	229.5	150.5

Acetaldehyde and ethanol accumulated within tissue of apple slices, more acetaldehyde and ethanol accumulated at 20°C than at +1°C. There were no significant differences between treatments (Table. 3).

4 CONCLUSIONS

Disulphite appears to be the most effective in browning inhibition of apple slices. Sucrose solution has an inhibitory effect on browning because it reduces the concentration of dissolved oxygen and the rate of diffusion of the oxygen into fruit tissue [3]. Chestnut honey and honey dew inhibited browning more efficiently than papain and just slightly less than disulphite. This suggested that in addition to sugars honey contains proteins that act as PPO inhibitors. Dark coloured honey proved to be more efficient in browning prevention. This can be attributed to higher content of proteins compared to light coloured honey which usually contain less proteins. CO_2 storage atmosphere inhibits browning of apple slices while N_2 atmosphere has no inhibitory effect on browning. Apple slices accumulated acetaldehyde and ethanol during storage in PA/PE bags due to low O_2 level or high CO_2 level [7] and due to yeast activity. More acetaldehyde and ethanol accumulated at 20°C than at +1°C, suggesting that higher temperature stimulates anaerobic metabolism through higher activity of pyruvate decarboxylase and alcohol dehydrogenase [2]. Ethylene accumulated in the PA/PE packed slices as a result of incessant fruit metabolism and because of tissue wounding.

5 REFERENCES

1. F. B. Boyer, *Hope College*, 1977, **54**, 585.
2. K. Dangyang, Z. Lili, A.A. Kader, *J.Amer.Hort. Sci.*, 1994, **119**, 971.
3. M. A. Joslyn and J. P. Ponting, *Adv. Food Res.*, 1951, **3**, 1.
4. V. Kahn, *J. Food Sci.*, 1985, **50**, 111.
5. T. P. Labuza, *Flussiges Obst*, 1992, **59**, 15.
6. J. Oszmianski and C. Y. Lee, *J. Agricult. Food Chem.*, 1990, **38**, 1892.
7. M. Zavrtanik, R. Vidrih, M. Simčič, J. Hribar, 'Biology and biotechnology of plant hormone ethylene', Book of abstracts, June 9-13 Chania, Greece, 1996, 100.
8. J. Zawistowski, C. G. Biliaderis, N. A. M. Eskin, in: D. S. Robinson, N. A. M. Eskin, 'Oxidative enzymes in foods', Elsevier Science Publishing, New York, 1991, Chapter 6, p. 217.

Session V Effects of Pre-harvest Practice on Quality

THE EFFECT OF PRE-HARVEST PRACTICE ON THE MICROBIOLOGICAL QUALITY OF BERRIES AND VEGETABLES

R. Tahvonen

Agricultural Research Centre of Finland
SF-21500 Piikkiö
Finland

1 INTRODUCTION

High-quality vegetables and berries are free from damage or symptoms caused by diseases and insects. Plant diseases normally cause problems during the period of cultivation, but there may be problems also found during the time of marketing. Especially those microbes which do not cause severe diseases during the growing season can appear in the products during storage and marketing. In most cases the pathogen can be identified immediately, but sometimes it may be manifest only as poor quality, e. g. small size, abnormal colour or bitter taste.

In cold northern countries like Finland berries and vegetables are being stored for months with the pathogens most commonly fungi. The connection between fungi and poor quality of berries and vegetables is discussed in this paper with special reference to cultivation techniques used in Finland and storing vegetables.

2 FUNGI CAUSING QUALITY LOSSES

Those fungi which cause heavy losses to crops are normally not problematic for the quality of the yield because the plants are so badly damaged that there is no yield to market. Fungi infecting harvested yields without visible symptoms are most harmful because they continue to growth in plant tissue during storage and marketing. Another group of harmful microbes are plant root- and stem-damaging fungi which cause abnormally low growth of plants and poor yield quality. It is very difficult to find what causes the poor quality of yield without seeing or analysing the whole plant in the field. The most common fungi causing quality losses are presented in Table 1.

3 FUNGI CAUSING DAMAGE TO ROOTS AND STEMS

Many fungal diseases cause damage to roots or stems of plants, which is the reason for lack of water and nutrition in the plant. This may result to small size, abnormal colour and shape of berries and fruits (strawberry, raspberry, cucumber). The main fungi on strawberries are *Fusarium* spp., *Pythium* spp., *Phytopthora* spp. in addition to many other root fungi[3]. The primary reason for fungal root infection is normally winter damage and monoculture of crop. The stem disease of raspberries caused by *Didymella applanata* results in small size of berries. Cucumber root diseases (*Pythium* spp. and *Phomopsis sclerotioides*) kill young plants and prevent vigorous growth of older crops[1], which is the reason for small size, abnormal shape and dark green colour of fruits. As the diseases are spread by plant debris and old soils[1], the most important methods to control diseases are good disinfection of greenhouses, clean water and use of fresh growth substrates and healthy seedlings.

Table 1. *Quality losses of vegetables and berries caused by fungi* [1-7]

Plant	Fungus	Type of damage
Strawberry		
	Botrytis cinerea	Mouldy decay in field and marketing
	Rhizopus sp., *Mucor* sp.	Mouldy soft rot in marketing
	Phytopthora cactorum	Leather rot and bitter taste
	Gnomonia sp. and other fungi	Dry rot and small berries
Raspberry		
	Botrytis cinerea	Mouldy decay in field and marketing
	Didymella applanata	Small berries
Cabbage		
	Botrytis cinerea	Mouldy decay in store and marketing
	Sclerotinia sclerotiorum	Mouldy decay in store and marketing
	Phytopthora porri	Unmouldy dark decay in store and marketing
	Phoma lingam and other fungi	Dry or soft rot in store and marketing
Cucumber		
	Botrytis cinerea	Mouldy soft rot
	Didymella bryoniae	Dry rot
	Pythium spp., *Phomopsis* sp. and other root decaying fungi	Small, dark and soft fruit, bitter taste
Carrot		
	Mycocentrospora acerina	Black soft rot in store and marketing
	Botrytis cinerea	Mouldy soft rot in store and marketing
	Sclerotinia sclerotiorum	Mouldy and wet rot
	Fusarium spp. and other fungi	Dry rot in store and marketing
Lettuce		
	Botrytis cinerea	Mouldy soft rot
	Rhizoctonia solani	Bottom rot
	Pythium spp. in roots	Small and dark leaves, bitter taste
Onion		
	Botrytis alli	Neck rot in store and marketing
	Fusarium oxysporum	Dry rot in store and marketing
	Aspergillus niger	Black and green mould on the skin
	and *Penicillium spp.*	of bulb in store and marketing
Leek		
	Botrytis porri	Mouldy rot in store and marketing
Tomato		
	Botrytis cinerea	Dark ring spot on the skin of fruit
	Didymella lycopersici	Dark rot

4 FUNGI IN HARVESTED PLANT PARTS

Those fungi, which cause serious damage to yields in field conditions, are normally not harmful to the quality of berries and vegetables because the decayed plant are removed at harvesting. Latent and small invisible infections cause losses after harvesting and storing and therefore affect the quality.

Botrytis is the most harmful species for horticultural crops. They cause damage to plants both in the field and while long stored and marketed. *B. cinerea* decays berries of strawberry and raspberry in field conditions[1-2], but the latent disease infection shortens the marketing time. The fungus infects berries during the time of flowering, so the main method to controlling the disease is spraying of the crops at florescence. The same fungus is responsible also for the decay of lettuce[1], stored cabbage[7] and carrot[8]. *B. alli* is the main storage disease of onion. It spreads in seeds or in set onions. Therefore the disease can be effectively controlled by seed or set onion treatment with a fungicide[6]. *B. porri* causes in late summer latent infection of leeks which can be detected in decayed plants after storing and in shops[5].

Several minor pathogens are harmful for the quality of berries and vegetables during storing and marketing. These pathogens cause no problems during the growing period, but they start to cause decay in the store or on the market. *Mycocentrospora acerina* causes licorice rot of stored carrot and is the main reason for the poor quality of carrot in late winter[4]. Sometimes the infection rate can be over 50 %. The disease survives many years in soil and in weeds. Crop rotation is the only method to control the disease. Similar types of diseases, but not so harmful as *M. acerina*, occur also on cabbage (*Phoma lingam*, *Phytopthora porri* and other fungi), carrot (*Sclerotinia sclerotiorum*, *Fusarium spp.* and other fungi), onion (*Aspergillus niger* and *Penicillium spp)*[4], tomato (*Didymella lycopersici*)[1], strawberry (*Gnomonia* sp. and other fungi)[3] and cucumber (*Didymella bryoniae*)[1].

References

1. J. T. Fletcher, 'Diseases of greenhouse plants', Longman, London and New York, 1984.
2. M. A. Ellis, R. H. Cinverse, R. N. Williams and B. Williamson, 'Compendium of Raspberry and Blackberry Diseases and Insects', The American Phytopathological Society, Minnesota, 1991.
3. J. l. Maas, 'Compendium of Strawberry Diseases', The American Phytopathological Society, Minnesota, 1984.
4. A. L. Snowdon, 'Color Atlas of Post-Harvest Diseases & Disorders of Fruits and Vegetables Volume 2: Vegetables', CRC, Boston, 1992.
5. R. Tahvonen, J. Scient. Agric. Soc. Finl., 1980, **52**, 331.
6. R. Tahvonen, J. Scient. Agric. Soc. Finl., 1981, **53**, 27.
7. R. Tahvonen, J. Scient. Agric. Soc. Finl., 1981, **53**, 211.
8. R. Tahvonen, Ann. Agric. Fenn., 1985, **24**, 86.

THE EFFECT OF ROOTSTOCK ON YIELD AND QUALITY OF APPLES, CV. MUTSU

H. Daugaard, J. Grauslund and O. Callesen

Danish Institute of Agricultural Sciences
Department of Fruit, Vegetable and Food Science
DK-5792 Aarslev, Denmark.

1 INTRODUCTION

'Mutsu' is the main green-coloured apple cultivar grown in Denmark. It is normally stored before marketing, and during storage the fruit colour gradually changes from green to yellow. The changes in colour are attended by changes in fruit firmness and contents of acid, starch and sugar. Previously, the consumers demanded 'Mutsu' fruits of a clear green colour, but nowadays they demand 'Mutsu' fruits of a medium, green-yellow or yellow colour, mainly because of their better taste[3,4]. In 1996 a research project was initiated with the aim of finding methods and production systems resulting in an increase of the percentage of medium- or yellow coloured post-storage 'Mutsu' fruits. The choice of rootstock is a factor of importance, indicated by previous research[1,2,3,4,5], and this factor was investigated in further detail as part of the project.

2 MATERIALS AND METHODS

In an existing rootstock trial, planted in the autumn of 1989, four rootstocks were selected to be used in this research project. The planting distance in the row was variable with a fixed row distance of 3.30 m. Each plot consisted of six trees at variable distance, giving 1683, 1894, 2164, 2525 and 3030 trees per hectare. This experimental design gave information on the differences between rootstocks, but it was also possible to extract information on the relevance of rootstock choice at different orchard densities. The experimental layout comprised three replicates, giving 18 trees per rootstock.

The rootstocks selected for this experiment were M9, M26, J9 and B9, all obtained from virus-tested sources. M9, clone NIC29, was provided by Nicolai Nurseries, Belgium. M26 was the official dutch clone used in 1987, provided by NAKB, The Netherlands. J9 was supplied by the Fruit Research Institute in Jork, Germany, and B9 from Oregon Rootstocks, Inc., USA. The rootstocks were planted in the nursery in April 1988 and budded 20 cm above ground level in August the same year. One-year old maiden trees were planted in the field in November 1989.

The fruit quality experiments were conducted in 1996 and 1997. Each year fruits were picked from each tree separately and graded into 6 categories: 1) fruits < 70 mm, 2) fruits 70-80 mm and green, 3) fruits 70-80 mm and medium-coloured, 4) fruits 70-80 mm and yellow, 5) fruits > 80 mm and medium-coloured and 6) fruits > 80 mm and yellow. The fruit was graded with electronic equipment identically adjusted in both years. For a sample of 10 fruits per category, the content of starch was analysed using the Iode-Iode-Potassium test (rating 1-10, 1 highest content), the content of titratable acid and the content of sugar using refractometric index. Fruit firmness was measured using a penetrometer with 11 mm probe.

The fruit was stored in a Controlled Atmosphere (3° C, 2% CO_2 and 4-5% O_2), in 1996 for 1.5 months and in 1997 for 3 months. Following storage, the fruit was graded into the same categories as earlier, but due to the development of fruit colour during storage the grading equipment was adjusted differently compared to pre-storage. Each category of fruit was analysed in the same way as before storage, but in addition to this, an evaluation of fruit taste was carried out according to a 5 point scale (1 - bad taste, 5 - very good taste) by a panel of three persons.

All data were subject to statistical analysis using the General Model of SAS (SAS Institute, Inc., 1989-95, Cary, NC). The least significant differences between means were determined at P < 0.05 using Duncan's test.

3 RESULTS AND DISCUSSION

Total yield per rootstock and year are listed in table 1. The yields are given in kg per tree and tonnes per hectare. There are significant differences between rootstocks, but they appear to be more obvious in 1996, where the yields generally were low. In both years, M9 and J9 provided the highest total yields, corresponding to earlier results[1]. Differences in fruit size were of minor importance.

In table 2 is shown the distribution of green, medium-coloured and yellow fruit in 1996 and 1997. Presumably due to climatic variations there are considerable differences between the two years, but there are significant differences between rootstocks as well, corresponding to earlier results[1,2]. Pre-storage figures calculated in kg per tree indicate that M9 and J9 provided the highest quantity of medium-coloured fruit, but as 'Mutsu' fruits normally are stored before marketing this is only of theoretical importance. Post-storage figures are calculated in kg per tree as well, and it appears that M9 in both years provided the highest quantity of medium- and yellow-coloured fruits, followed by J9 in 1996 and M26 in 1997. This seems to be in contradiction to an earlier study, based on the same trial[1], in which the highest percentage of yellow fruits was produced by M26, but the results may not be comparable as in our study the quantity in kg per tree is considered rather than percentages. Furthermore, the grading equipment was adjusted differently in the two studies. If the results of our study are calculated as percentages instead of kg per tree , the highest percentage of medium- and yellow coloured fruit in both years was obtained by M26. In a similar study, where the effect of rootstock on fruit colour of the cultivar 'Golden Delicious' was investigated, M26 was found to cause fruits of a more green colour than several other rootstocks[3]. Thus there seems to be some inconsistency of research results, possibly caused by methodical differences.

In table 3 is shown results of fruit analyses after storage in 1996. Analyses of 'Mutsu' fruits from four rootstocks and of three colour categories are compared. When starch content is compared, significant differences between rootstocks can be seen, with M9 generally showing the lowest rating (i.e. highest content) and M26 and B9 the highest rating. For fruit firmness, differences between rootstocks are not very clear, although B9 appears to differ significantly from the other rootstocks. The amounts of titratable acid and sugar (refractometric index) are also generally showing the highest values for B9 and J9. As far as taste is concerned, no significant differences were recorded.

The corresponding results of fruit analyses in 1997 are shown in table 4. It appears that there are several differences from the preceding year. With regard to content of starch, there are generally less differences between rootstocks, and the highest ratings are obtained by J9, which

Table 1 *Yield and fruit size of 'Mutsu' apples in 1996 and 1997 as influenced by rootstock*

Root-stock	Yield Kg/tree		Yield Tonnes/ha		Fruit size Gram/fruit	
	1996	1997	1996	1997	1996	1997
M9	21.4a	31.0a	46.3a	67.1a	223b	209ab
J9	15.6ab	31.3a	33.8ab	67.8a	241a	202b
M26	10.4b	24.7b	22.5b	53.5b	229ab	214a
B9	12.1b	29.5ab	26.2b	63.9ab	233ab	203b

Table 2 *Quantities of each fruit colour of 'Mutsu' apples in 1996 and 1997 as influenced by rootstock*

Root-stock	Fruit colour 1996					Fruit colour 1997				
	Before storage Kg/tree medium-colour	After storage Kg/tree				Before storage Kg/tree medium-colour	After storage Kg/tree			
		Green	Med-ium	Yel-low	Med/Yel-ow		Green	Med-ium	Yel-low	Med/Yel-ow
M9	5.7a	4.6a	8.0a	5.4a	13.3a	12.0a	6.6a	17.7a	3.8ab	21.4a
J9	4.1ab	2.0b	6.0ab	5.2a	11.2ab	13.4a	9.8a	16.4a	1.2b	17.6bc
M26	2.4b	1.2b	4.2b	4.5a	7.0b	7.4b	2.2a	14.3a	5.6a	19.9ab
B9	3.2b	1.2b	3.4b	3.6a	8.7ab	13.6a	9.7a	15.0a	1.5b	16.5c

Table 3 *Quality of 'Mutsu' apple fruit after storage in 1996 as influenced by rootstock*

Root-stock	Starch rating 1-10			Fruit firmness kg			Titratable acid content %			Sugar content Refractometric index			Taste rating 1-5		
	Green	Med-ium	Yellow	Green	Med-ium	Yellow	Green	Med-ium	Yellow	Green	Med-ium	Yellow	Green	Med-ium	Yellow
M9	8.0c	7.7b	7.9c	6.2a	6.3ab	6.5b	0.45b	0.44b	0.45b	12.2b	13.2b	14.0c	1.7a	1.7b	1.7a
J9	7.7c	8.2b	8.7b	6.4a	6.1b	6.3b	0.51a	0.48a	0.47ab	13.1a	13.7a	14.4b	2.3a	3.0ab	3.0a
M26	9.8a	9.8a	9.7a	6.4a	6.4ab	6.6b	0.47ab	0.47a	0.46b	12.8ab	13.3b	14.1bc	3.0a	3.7a	2.3a
B9	8.8b	9.4a	9.8a	6.4a	6.5a	7.0a	0.47ab	0.47a	0.49a	12.9a	13.7a	14.9a	4.0a	4.3a	3.7a

Table 4 *Quality of 'Mutsu' apple fruit after storage in 1997 as influenced by rootstock*

Root-stock	Starch rating 1-10			Fruit firmness kg			Titratable acid content %			Sugar content Refractometric index			Taste rating 1-5		
	Green	Med-ium	Yellow	Green	Med-ium	Yellow	Green	Med-ium	Yellow	Green	Med-ium	Yellow	Green	Med-ium	Yellow
M9	9.2b	9.1ab	9.2ab	6.8a	6.7a	6.7a	0.52b	0.50ab	0.49ab	12.3ab	12.7b	13.1a	2.3a	2.3a	1.0a
J9	9.8a	9.5a	9.6a	6.7a	6.7a	6.5a	0.51b	0.50ab	0.48bc	12.6a	12.7b	13.1a	1.7a	3.0a	1.7a
M26	9.0b	8.4c	8.9bc	6.5b	6.3b	6.6a	0.55a	0.51a	0.50a	12.1b	12.4c	13.0a	1.7a	2.3a	1.0a
B9	9.0b	9.0b	8.7c	6.4b	6.6a	6.6a	0.50b	0.49b	0.47c	12.5ab	13.0a	13.1a	1.7a	2.3a	1.0a

obtained the lowest ratings in 1996. Such differences between the years may be caused by the considerable yield differences. For all rootstocks, the content of titratable acid is generally higher in 1997 and the sugar content is lower, possibly caused by climatic factors. In 1997, the highest content of titratable acid was obtained by M26. As to sugar content (refractometric index), the highest values were obtained by B9 and J9 in both years. As in 1996, no significant differences could be recorded as far as taste is concerned.

With regard to differences between fruit colours, in both years there is a general increase of sugar content and a general decrease of acid content from green to medium-coloured and yellow fruits. As to starch content and fruit firmness there are no clear tendencies, but there is a clear taste preference for medium-coloured fruit. It can be concluded that the choice of rootstock is an important factor in obtaining high yield and good fruit quality of the green-coloured apple cultivar 'Mutsu'.

References

1. O. Callesen, *Acta Hort.*, 1997, **451**, 137-145.
2. S.R. Drake, F.E. Larsen, J.K. Fellman and S.S.Higgins, *J. Amer. Soc. Hort. Sci.*, 1988, **113**, 949-952.
3. E. Fallahi, D.G. Richardson and M.N. Westwood, *J. Amer. Soc. Hort. Sci.*, 1985, **110**, 71-74.
4. J. Grauslund, and O. Callesen, Danmarks Jordbrugsforskning, *Grøn Viden* nr. **106**, 1997.
5. P. Hansen, Frugt og Bær, 1996, **25**, 160-161.

THE EFFECT OF HARVESTING TIME AND NITROGEN FERTILISATION ON ENZYMATIC BROWNING IN POTATO DURING STORAGE

M. Kari-Kärki

Potato Research Institute
FIN - 16900 Lammi

1 INTRODUCTION

Every day, approximately one million people eat warm meals during the daytime in schools canteens. Many large scale kitchens nowadays prefer pre-peeled potatoes to reduce work and dirt in kitchens. Because of this, the problem of enzymatic browning as a visual disorder of pre-peeled potatoes,   receives growing attention in the Finnish potato industry. The general negative attitude towards chemical additives is a limiting factor in treating pre-peeled potatoes. Enzymatic browning can also be a problem during traditional processing of potatoes for salads and ready-to-eat meals.

An increased level of nitrogen applied during growth has been reported to enhance enzymatic browning in potato tubers[1]. Sufficient potassium fertilisation and content in tubers is essential in controlling enzymatic browning[3]. This potassium content is in its turn reduced by an increased dose of nitrogen during growth[4]. Tyrosine content, affecting enzymatic discoloration, varies in concentration with tuber maturity[2].

This study is part of a larger study considering processing quality of potatoes. This study is focussed on the effects of harvesting time, nitrogen fertilisation and storage on enzymatic discoloration. The varieties chosen are commonly used for the production of potato salads, ready-to-eat meals, french fries or chips. The applied doses of nitrogen fertilisation are in the range of normal practice.

2 MATERIALS AND METHODS

The varieties were grown applying two levels of nitrogen fertilisation according to table 1. The nitrogen doses were chosen according to former knowledge from varieties. Tubers were sampled at three different harvesting times and harvested in intervals of two weeks. In

1996 varieties were grouped into early (Fambo, Timate) and moderately late ones (Asterix, Lady Rosetta, Nicola, Vento). Harvest with early ones was started two weeks beforehand.

Dry matter, starch and potassium content and enzymatic browning were measured.

Table 1 *Nitrogen doses applied*

Variety	1995		1996	
Bintje	60	100	-	-
Fambo	60	100	60	120
Timate	60	100	60	120
Asterix	60	100	60	10
Lady Rosetta	60	100	60	120
Nicola	40	80	40	100
Van Gogh	60	100	-	-
Vento	40	80	40	100

3 RESULTS AND DISCUSSION

3.1 Growing season

Both summers were exceptionally dry and hot. Especially in 1995 there was severe deficiency of water. After first harvest, however, potatoes were exposed to rainfall of 45 mm within two weeks. Night frost of -4°C occurred between second and third harvest.

Rainfall was more moderate in 1996. No rains occurred during or between harvest. Four later varieties were exposed to night frost after second harvest, however.

3.2 Dry matter and starch content

Dry matter and starch content fell dramatically after rainfall between first and second harvest in 1995. The fall was enhanced, but less dramatically after night frost. Very late variety Vento, was least affected by rainfall and night frost.

In 1996 two earlier varieties matured fully before last harvest. Four other varieties were exposed to night frost and they lost green canopy almost totally before third harvest. Late variety Vento was least affected by frost. Dry matter and starch content increased with maturity. Night frost, however, caused decreased dry matter and starch content.

Dry matter and starch content were lower with higher nitrogen dose. Lady Rosetta made an exception. That was not a surprise, since it is known to utilise higher doses of nitrogen than on average.

3.3 Potassium content

Content of potassium, as expected, was lower with higher amount of nitrogen, regarding to all varieties (Figure 1). Differences between varieties were remarkable. Potassium content was higher in varieties with high dry matter content.

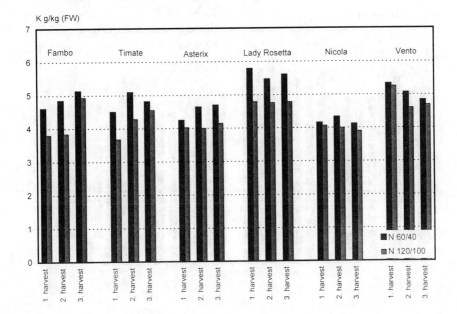

Figure 1 *Potassium content (g/kg FW) at harvest in 1996*

3.4 Enzymatic browning

Enzymatic browning was enhanced by maturity in case if canopy was not exposed to night frost. Rainfall in 1995 enhanced discoloration more dramatically than maturity. Night frost stopped or even turned the developing of discoloration (Figure 2).

In 1996 enhanced discoloration was more obviously connected to maturity. Crop was not exposed to heavy rainfall between harvesting times and only four out of six varieties were exposed to night frost (Figure 3).

Discoloration was enhanced by higher nitrogen dose. The influence was significant and it was similar irrespective of variety or production year.

Differences in enzymatic browning between treatments diminished during storage.

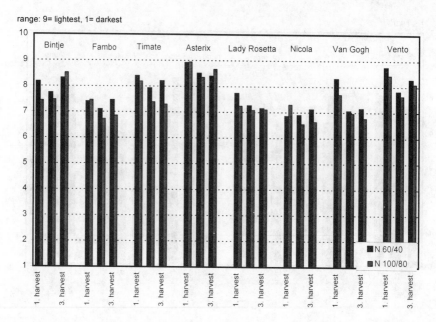

Figure 2 *Enzymatic browning at harvest in 1995*

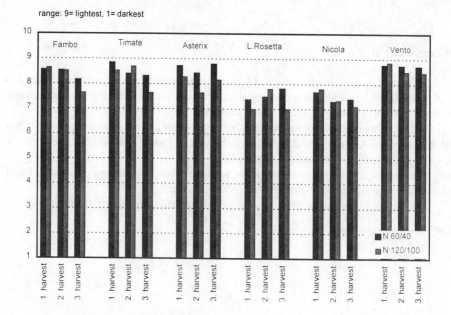

Figure 3 *Enzymatic browning at harvest in 1996*

4 CONCLUSIONS

Special attention should be paid on fertilisation and nitrogen dose if enzymatic browning tends to cause problems in certain varieties. Also rainfall before harvesting should be taken into consideration when sorting crops for different purposes. Night frost is not harmful considering enzymatic browning at the beginning of storage.

References

1.   N.I. Mondy and R.L. Koch, *J. Agric. Food Chem.,* 1978, **26**, 666.
2.   N.I Mondy and C.B. Munshi, *J. Agric. Food Chem.,*1993, **41**, 1868.
3.   R. Eltun, 1987. *Norsk landbruksforsking,* 1987, **1**, 175.
4.   W.G. Burton, 'The Potato', Longman Scientific & Technical, England 1989.

EFFECTS OF PHYSICAL WEED CONTROL ON CARROT AND ONION QUALITY

P. Vanhala

Agricultural Research Centre of Finland (MTT)
Plant Production Research
FIN-31600 Jokioinen

1 INTRODUCTION

Due to the poor competitive ability of vegetable plants, especially carrots and onions[1], weed control is essential during their production. In recent years, there has been a shift from chemical towards physical weed control. However, the question has been raised, as to whether mechanical and thermal weed control have adverse effects on yield quality. In the study of Ascard and Mattsson[2] mechanical weed control increased the proportion of branched carrots during two years, but decreased it in another. Ascard[3] found that thermal weed control (selective flaming) caused yield reduction in machine-set onion, but not in hand-set onion.

The effects of physical weed control on yield quality were studied in five field trials with carrot and six with set onion, at two locations of the Agricultural Research Centre of Finland (MTT) during 1992-94.

2 MATERIALS AND METHODS

Carrot was sown on level soil in single rows, at 80 seeds per meter. Onion was hand-set in single rows, at 20 sets per meter. For both crops 10 m rows were 40 cm apart and there were four rows in a plot. Carrot (cv. "Fontana") received 100 kg N/ha and onion (cv. "Sturon") 80 kg N/ha. The trial design was randomised blocks with four replicates.

For carrot, the weed control measures were: i) weedy = no weed control; ii) flaming whole surface before carrot emergence, hoeing between the rows, no hand-weeding; iii) three flamings (whole surface [propane 60 kg ha^{-1}] before carrot emergence, later twice (propane 45 kg ha^{-1}) between the rows, hand-weeding once or twice; iv) flaming whole surface before carrot emergence, hoeing between the rows, hand-weeding once or twice; and v) weed free (herbicide prometryn 1 kg a.i. ha^{-1} + intensive hand-weeding). In hoeing and flaming between the rows, a 10 cm wide unmanaged buffer strip was left in the row.

For onion, the weed control measurements were: i) weedy = no weed control; ii) flaming whole surface when onion leaves were 5 cm long, later two selective flamings (strong treatment, propane 48 kg ha^{-1}) in the row; iii) flaming whole surface when leaves

were 5 cm long, later two selective flamings (moderate treatment, propane 24 kg ha^{-1}) in
the row; iv) flaming (propane 60 kg ha^{-1}) whole surface when leaves were 5 cm long,
hand-weeding as necessary; and v) weed free (herbicide prometryn 1 kg a.i. ha^{-1} +
intensive hand-weeding). In the flamed plots, weeds between the rows were controlled by
hoeing.

Weeds were assessed in the rows in two quadrats of 0.125 m x 2 m, and between the
rows in two 0.25 m x 1 m quadrats per plot. Carrot and onion yields were assessed from
six meters from the two center rows of the plot. The yield was sorted into quality classes
by size and damage.

For statistical analyses the data for weed and crop biomass were square root
transformed to normalize distributions and achieve homogeneity of variances before the
data were subjected to analysis of variance. Significantly different means were separated
with Tukey's studentized range test (HSD). Statistical analyses were performed with the
GLM (General Linear Models) procedure of the SAS statistical package, version 6.09[4].

3 RESULTS

In carrot, neither hoeing nor inter-row flaming caused more branching or deformity than
intensive chemical-manual weed control in the weed-free treatment (Fig. 1). However,
competition by the surviving weeds (Table 1) had a greater negative effect on yield quality
than damage through weed control measures. The higher the weed biomass, the greater the
proportion of the yield in the 'small' category. Keeping the crop weed free resulted in
more carrots of marketable quality than allowing some weeds, although physical weed
control complemented with hand-weeding once or twice gave nearly as high quality.
Without hand-weeding in the row the yield quality and quantity was inferior, and without
weed control it was not possible to achieve marketable quality.

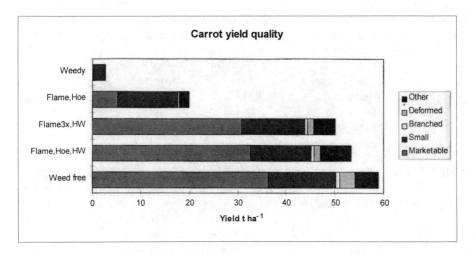

Figure 1 *Carrot yield quality as affected by weed control methods.*

Set onion was more competitive than carrot and produced marketable yield (Fig. 2) even under heavy weed competition. As for carrot, weed competition (Table 2) in onion plots decreased the total amount and proportion of marketable quality yield; 'weedy' plots (no weed control) resulting in lowest marketable yield and clearly the highest amount of small onions. However, the repeated strong flaming, although providing the best weed control, hampered onion development, thus resulting in more small onions than one flaming + hand-weeding.

Figure 2 *Onion yield quality as affected by weed control methods.*

Table 1 *Weeds in carrot in early August (untransformed data). HW = hand-weeding. Means followed by the same letter within a column do not differ significantly (Tukey's HSD, p=0.05).*

	Weeds in row, $g\ m^{-2}$	Weeds between rows, $g\ m^{-2}$
Weedy	812 [a]	545 [a]
Flame, Hoe	804 [a]	83 [b]
Flame 3x, HW	25 [c]	22 [c]
Flame, Hoe, HW	78 [b]	43 [c]

Table 2 *Weeds in set onion in early August (untransformed data). HW = hand-weeding. Means followed by the same letter within a column do not differ significantly (Tukey's HSD, p=0.05).*

	Weeds in row, $g\ m^{-2}$	Weeds between rows, $g\ m^{-2}$
Weedy	468 [a]	463 [a]
Flame 3x (strong)	27 [c]	24 [c]
Flame 3x (moderate)	38 [c]	28 [bc]
Flame, HW	66 [b]	50 [b]

4 DISCUSSION AND CONCLUSIONS

The primary factor determining crop yield quantity and quality was weed biomass, not the physical weed control methods applied. Mechanical or thermal weed control did not affect carrot quality negatively, as a 10 cm wide buffer strip was left in the row. Ascard and Mattsson[2] stated that the influence of inter-row cultivation on the yield is the sum of positive and negative effects, which depend on, for example, weed infestation, cultivation method, soil type and soil conditions. In this study the influence of control methods on weeds was the most important factor.

In onion, increasing intensity of selective flaming gave only a slight increase in weed control. Thus, the negative effect of strong flaming on yield was greater than the positive effect of better weed control. Rahkonen *et al.*[5] found in cabbage and red beet that crop tolerance to flaming is determined by size and development stage of the crop, direction of the flame and flaming intensity.

It is concluded that hoeing or flaming can be used to control weeds between carrot rows without decreasing crop yield quality, but complementary hand-weeding in the crop rows is needed to reduce competition from weeds. For set onion, very intensive thermal weed control may decrease onion quality. Thus, one should take care not to exceed crop tolerances when using thermal weed control.

References

1. H. D. J. van Heemst, *Agricultural Systems*, 1985, **18**, 81.
2. J. Ascard and B. Mattsson, *Biological Agriculture and Horticulture*, 1994, **10**, 161.
3. J. Ascard, *30th Swedish Crop Protection Conference. Weeds and Weed Control.* Swedish University of Agricultural Sciences, Uppsala, 1989, **2**, 35.
4. SAS Institute Inc., 'SAS/STAT® User's Guide', Version 6, Fourth Edition, Cary, NC: SAS Institute Inc., 1989, Volume 2, Chapter 24, p. 891.
5. J. Rahkonen, P. Vanhala and E. Kaila, *Maatalousteknologian julkaisuja* 22, University of Helsinki, 1998, Chapter 4, p. 44.

BIOACTIVE SUBSTANCES IN CRUCIFEROUS PRODUCTS

I. Schonhof, A. Krumbein , M. Schreiner and B. Gutezeit

Institute for Vegetable and Ornamental Crops
Grossbeeren/Erfurt e. V.
Theodor-Echtermeyer-Weg 1
14979 Grossbeeren
Germany

1 INTRODUCTION

In the past few years research results of food nutrition have demonstrated that not only essential compounds like vitamins and minerals are necessary to maintain human metabolic processes, but also a number of further compounds - bioactive substances. Frequently their combinations have a surplus additional health promoting effects. They cause e. g. anticancerogenic, antioxidative, immune stimulated and anti-thrombotic effects.[1]

Secondary plant products and dietary fibres, which are a part of bioactive substances, orginate from plant material. Consequently, with regard to the health promoting influence of bioactive substances, vegetable quality has to be defined not only by essential nutritive compounds, sensory attributes and undesirable attributes, but also by bioactive substances.

At present little knowledge exists about preharvest influences, e. g. effect of harvest time or crop management practices, on biosynthesis or degradation of bioactive substances. In addition the effect of different postharvest conditions on the content of bioactive substances is often unknown.

The vegetable family of the brassicaceae especially the species broccoli (*Brassica oleracea var. italica* Plenck) is a rich source of bioactive substances characterized by high contents of glucosinolates, carotenoids, chlorophylls, phenolic acids, flavonoids and sulfides. Due to the health promoting effect of carotenoids and chlorophylls and the influence of climatic factors during the production period (air temperature, irradiation), broccoli cultivar and type and the development stage at harvest on the contents of the main carotenoids -β-carotene and lutein- and chlorophylls have been investigated. Our studies of glucosinolates have been focused on genetic effects, on the influence of climatic factors and on management practices - spacing and sulphur supply were chosen as examples.

2 MATERIAL AND METHODS

2.1 Material

The genetic effect on the content of bioactive substances has been investigated using twelve broccoli cultivars grown in field. The cultivars were characterized by three types - spear broccoli, crown broccoli and violet type.

The effect of growing season has been tested in spring, summer and autumn (three years) by the cultivars 'Emperor', 'Marathon' and 'Viola'. The investigations of spacing were carried out in spring and autumn on two different soil types (sand and loess). Furthermore broccoli 'Emperor' were cultivated in a glasshouse in pots and sand culture with five sulphur treatments.

The plants were harvested, when the head diameter reached its maximum. For the investigation of head development the plants were harvested, when 30, 50, 70, and 100% of the expected head diameter were reached.

In this publication the investigations were focused on broccoli florets.

2.2 HPLC Analysis of Bioactive Substances

The carotenoids and chlorophylls were determined by high - performance liquid chromatography (HPLC) on the C - 18 reversed - phase column Lichosphere 100 using an isocratic tenary eluent of acetonitrile, methanol, and methylene chloride[2]. For the determination of glucosinolates a modified HPLC method of Lange et al.[3] was used. The desulfoglucosinolates were analysed on a Spherisorb ODS2- column using a gradient of water and acetonitrile.

3 RESULTS AND DISCUSSION

3.1 Genetic effects

According to our studies the concentration of bioactive substances depends on the choice of the broccoli type.

- Cultivars of the spear type show significantly higher contents of lutein and β-carotene than the crown type and violet cultivars.
- This type effect was also manifested by the concentration of chlorophyll a and b.
- Moreover, the broccoli types differ significantly in their glucosinolate content. In contrast to the carotenoids and chlorophyll contents the crown type shows the highest glucosinolate concentration. (Figure 1).

The glucosinolates in broccoli can be split into three classes: indole glucosinolates with glucobrassicin as the main glucosinolate, alkyl glucosinolates with glucoraphanin as the main glucosinolate and alkenyl glucosinolates, which have been especially determined by progoitrin. Due to the health promoting effect the indole glucosinolates and glucoraphanin are very important. Glucoraphanin is a precursor of sulphoraphane, a very potent inducer, which protects against carcinogenesis[4].

The choice of broccoli type determines the level of bioactive substances. These genetic relations are constant during the entire growing period of the year.

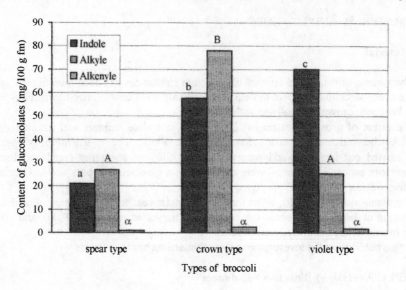

Figure 1 *Content of indole, alkyl and alkenyl glucosinolates in various broccoli types*

3.2 Effect of climatic conditions (temperature and irradiation) during the production period

Due to the differing dates of the production period in spring, summer and autumn over several years a variation of the daily mean temperature and photosynthetic photon flux density (PPFD) levels could be realized. Strong dependencies between production conditions and the contents of bioactive substances could be demonstrated.

- The content of β-carotene is mainly influenced by the daily mean temperature, shown exemplarily by spear broccoli, cultivar 'Emperor'. Daily mean temperatures below 16.5 °C lead to an obviously higher β-carotene concentration than daily mean temperatures above this temperature level (more than 30 % to 60 % in comparison with the treatment above 16.5 °C).
- The content of lutein also depends on the daily mean temperature. The lutein concentration in spear broccoli reached a maximum at a daily mean temperature of nearly 15 °C. The PPFD had no effect on the β-carotene content and no effect on the content of lutein in spear broccoli either.
- In contrast to the carotenoids the biosynthesis of glucosinolates strongly depends on sum of PPFD. A high sum of PPFD leads to an increased content of glucoraphanin. Additionally, the content of glucoraphanin in spear broccoli increased with decreasing daily mean temperature.

It should also be pointed out that the production of glucosinolates and of carotenoids is strongly temperature dependent. An increase in temperature results in an inhibited biosynthesis of carotenoids and glucosinolates. Furthermore, the glucosinolates are dependent on the daily mean sum of PPFD, because the enzymes for the glucosinolate biosynthesis are light dependent.

3.3 Development stage of the head at harvest time

Moreover, our investigations illustrate close correlation between the development stage of the broccoli head at harvest time on the one hand and the contents of bioactive substances on the other hand.

- The contents of β-carotene and lutein increased during the development of the broccoli head. This effect could also be determined with chlorophyll a and b. Fully developed heads of broccoli showed the highest contents of carotenoids and chlorophylls.
- Additionally, during the entire development of broccoli plants a strong dependence between the carotenoids - β-carotene and lutein - and both chlorophylls (a and b) has been found. Generally high carotenoid contents correlate with high chlorophyll contents.

In contrast to the carotenoids and chlorophylls the entire content of glucosinolates decreased with advanced development, due to the decreasing contents of indole glucosinolates, mainly glucobrassicin and neoglucobrassicin. It is assumed that the involvement of indole glucosinolates within the auxin metabolism may be responsible for this effect. The main glucosinolate of broccoli, glucoraphanin shows no significant changes during the head development.

3.4 Effect of sulphur supply on the glucosinolate content

The biosynthesis of glucosinolates is affected by the sulphur supply of the growing plant. According to Josefson [5] this sulphur effect is described by rape. Our investigations confirm these results for broccoli.

- A rising level of sulphur supply of up to 600 mg sulphur per plant leads to an increasing content of glucosinolates. This effect was mainly caused by the alkyl glucosinolate glucoraphanin and less so by the indole glucosinolate glucobrassicin (figure 2).

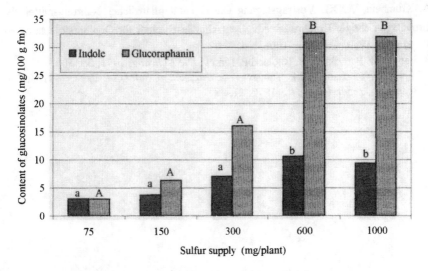

Figure 2 *Effect of sulphur supply on contens of glucoraphanin and indolglucosinolate in broccoli 'Emperor'.*

- The cause can be found in different biosynthetic pathways of alkyl glucosinolates compared with indole glucosinolates. Alkyl glucosinolates are derived from methionine and need free inorganic sulphur for their biosynthesis. In contrast, indole glucosinolates are derived from the amino acid tryptophan.
- Other glucosinolates showed no obvious changes.

3.5 Effect of spacing on the glucosinolate content

The biosynthesis of glucosinolates could also be induced by mechanical stress, e.g. a friction of leaf surfaces. Correlations between spacing and the glucosinolate content could be demonstrated by an investigation with two different dates of production period (spring and autumn) and two different soil types (sand and loess)

- A very high density of spacing (97.500 plants per ha) lead to an increasing glucoraphanin content of up to 37 % in comparison with lower spacing density in both soil types. The spacings of about 45.000 and 70.000 plants per ha show no difference in glucoraphanin.
- Indole glucosinolates are not affected by spacing, they vary according to production date and soil types.

Summarizing, high contents of the investigated bioactive substances in broccoli can be reached by a daily mean temperature below 16,5° C, a high sum of daily mean PPFD, fully developed heads, adequate sulphur supply and a high density of spacing.

References

1. B. Watzel and C. Leitzmann, 'Bioactive Substanzen in Lebensmitteln', Hippokrates, Stuttgart, 1995, p. 171
2. A. Krumbein, 'XXXI. Vortragstagung Die Qualität pflanzlicher Nahrungsmittel als Grundlage richtiger Ernährung', eds. Deutsche Gesellschaft für Qualitätsforschung e. V., Freising-Weihenstephan, 1996, p 41
3. R. Lange, M. Petrzikika, B. Rab and F. Linow, *Die Nahrung*, 1991, **35**, 385
4. J. W. Fahey, Y. Zhang and P. Talalay, *Proc. Natl. Acod. Sci. USA.* 1997, **94**, 10367
5. E. Josefsson, *J. Sci. Fd Agric.*, 1970, **21**, 98

OPTIMAL HARVEST TIME OF CARROT AND WHITE CABBAGE FOR STORAGE

T. Suojala and R. Pessala

Agricultural Research Centre of Finland (MTT), Plant Production Research, Horticulture
Toivonlinnantie 518, FIN-21500 Piikkiö, Finland

1 INTRODUCTION

Optimizing the harvest time of vegetables for storage requires knowledge about the development of the quantity, quality and storability of the yield. Cessation of yield increase is dependent on cultivar and growing conditions, especially temperature.[1-3] The effects of harvest time on the storage performance have often been unclear.[4-6]

The objective of this study was to quantify the yield development of carrot and white cabbage at the end of the growing season and to study the effect of harvest time on the storability of the carrot and white cabbage yield.

2 MATERIALS AND METHODS

Carrot experiments were performed on 9 farms in 1995 (cv. Fontana F_1, Bejo Zaden, The Netherlands) and on 15 farms in 1996 (cvs. Fontana and Panther F_1, Sluis & Groot, The Netherlands). The yield was harvested in September and October at two-week intervals. There were three and four harvests in 1995 and 1996, respectively. In 1995, farm was treated as a block factor in statistical analysis. In 1996, each farm had its own experimental design (randomized complete blocks with three replicates). In this paper, the results are presented as means of all farms.

Cabbage experiments were conducted at two locations in 1995 and at three locations in 1996. There were three to five harvests at intervals of one or two weeks in September and October. The plots for harvests were arranged in a randomized complete block design with three or four replicates. The cultivar used was Lennox F_1 (Bejo Zaden, The Netherlands).

Storage losses were analysed three times during the storage. Storage results were analysed with repeated measures of analysis of variance.

3 RESULTS

3.1 Carrot

In 1995, the carrot mean weight increased by 19 % on average between the first and the second harvest and by 13 % between the second and the third harvest (Table 1). On most farms, growth continued still in October.

In 1996, night frosts with minimum temperatures of -4--6 °C on 11-12 September and -8--10 °C on 21 September injured the plants. Therefore the yield increase was variable on different farms. On one-third of the farms, no statistically significant growth was observed after the first harvest on 10-11 September. On the contrary, the yield increased till the last harvest (21-22 October) on every third farm. In cultivar Panther, the average increase in total yield was 21 % during the six-week harvest period (Table 1). Most of the growth occurred between the first two harvests. In cultivar Fontana, the average increase was 15 % in six weeks.

The harvest time had a strong effect on the storage performace of carrot. In 1995, the storage losses, which were mainly due to storage diseases, decreased significantly with later harvest (Table 1). On average, the difference in the storage losses between the first and the third harvest was 15-20 %. The harvest time effect was greatest on farms where the yield had a poor storability.

The same trend was observed in 1996 (Table 1). In cultivar Panther, the yield of the two harvests in September had a weaker storability than the yield harvested in October. In cultivar Fontana, the first harvest differed significantly from the rest: storability neither improved nor weakened after 23-24 September.

3.2 Cabbage

Yield increase continued in October in most experiments. Usually a fast growth in September was followed by a period of slower growth which usually started at the end of September (Figure 1). At Piikkiö in 1996, the growth continued till the last harvest on 24 October.

Harvest time had only a small effect on the storage performance of cabbage (Table 2). Weight loss during storage decreased with later harvest, but the occurrence of storage diseases was not affected by harvest time. Thus the total storage loss (weight loss + trimming loss) was slightly higher in the earliest harvest.

4 DISCUSSION

The results indicate that carrot and white cabbage still grow in October if the temperature is not extremely low and the plants are in good condition. However, the yield increase slows down with decreasing temperature and radiation.

The storage performance of carrot was greatly affected by the timing of harvest. An optimal harvest time diminished the storage losses by 20 % on average. The harvest time effect was pronounced on farms where the yield was severely infected by storage diseases (mainly *Mycocentrospora acerina* and *Botrytis cinerea*). Contrary to the earlier results,[7-8] the occurrence of storage diseases was highest in the youngest carrots. The trend was uniform on all farms and in both years. What is remarkable is that the storage losses did not increase clearly with later harvest on any farm in spite of frost injuries in roots and heavy rains before and during the last harvest in 1996.

In cabbage, the harvest time had only a minor effect on the storage performance. Weight loss was highest in the early harvest, which caused a significantly higher total storage loss in the yield of the first harvest date. The occurrence of *Botrytis cinerea*, the main storage pathogen of white cabbage, was not affected by the timing of harvest.

On the basis of these results, the optimal harvest time of carrot and white cabbage intended for long-term storage begins at the end of September after which the yield increase is slow and storability is good. The sensory quality of white cabbage also improved with later harvest, which further supports the optimal timing of harvest in October. As all the yield cannot be harvested at the end of the season, knowledge about the effect of harvest time on the storage losses can be utilized in timing the marketing of the stored yield.

Table 1 *Effect of harvest time on carrot yield and storage losses at three dates in 1995 and 1996 (means of 6-9 farms).*

1995				
	Mean weight	Storage losses		
Harvest	g	% of initial weight		
time		19 Jan	28 Febr	17 Apr
'Fontana'				
12 Sept	143	42	59	67
26 Sept	166	29	43	56
10 Oct	183	21	39	52
1996				
	Total yield	Storage losses		
Harvest	1000 kg/ha	% of initial weight		
time		7-8 Jan	25-26 Febr	15-16 Apr
'Panther'				
10-11 Sept	46.7	32	42	54
23-24 Sept	53.4	22	33	48
7-9 Oct	55.9	12	23	27
21-22 Oct	57.5	11	20	32
'Fontana'				
10-11 Sept	40.6	38	53	70
23-24 Sept	42.6	17	33	39
7-9 Oct	44.4	14	23	39
21-22 Oct	47.0	17	27	38

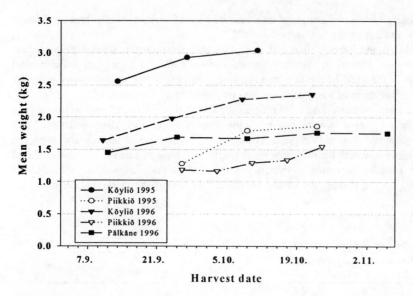

Figure 1 *Development of white cabbage yield (cv. Lennox F₁) at different locations in 1995 and 1996.*

Table 2 *Effect of harvest time on storage losses of white cabbage (cv. Lennox F₁, mean of three storage times). Experimental sites: K = Köyliö, P = Piikkiö, PÄ = Pälkäne. Year: 95 = yield of 1995, 96 = yield of 1996.*

Harvest time	Weight loss % of initial weight					Total storage loss % of initial weight			
	K95	P95	K96	P96	PÄ96	K95	K96	P96	PÄ96
10-13 Sept	6		8		9	20	15		21
24-27 Sept	5	10	7	4	8	11	13	22	14
8-11 Oct	.	9	7	4	7		12	11	14
22-24 Oct		7	6	4	9		12	17	15
p(Harvest)	0.286	0.028	0.004	0.018	<0.001	0.063	0.018	0.016	0.012

Harvest time	Infected by grey mould % of weight of cabbages after storage				
	K95	P95	K96	P96	PÄ96
10-13 Sept	31		15		16
24-27 Sept	20	2	5	9	11
8-11 Oct	23	4	5	14	10
22-24 Oct		3	15	19	9
p(Harvest)	0.563	*		0.158	0.234	0.171

* Not analysed statistically due to the low occurrence of diseases.

References

1 J. Apeland and S. Dragland, *Forskning og försök i landbruket*, 1975, **26**, 363.
2 D. Fritz and J. Habben, *Gartenbauwiss.*, 1977, **42**,185.
3 T. Nilsson, *J. Agr. Sci.*, 1987, **108**, 459.
4 D. Fritz and J. Weichmann, *Acta Hort.*, 1979, **93**, 91.
5 T. Nilsson, *J. Hort. Sci.*, 1987, **62**, 191.
6 T. Nilsson, *J. Hort. Sci*, 1993, **68**, 71.
7 W.P. Davies and B.G. Lewis, *Ann. Appl. Biol.*, 1980, **95**, 11.
8 F. Villeneuve, J.-P. Bosc and C. Luneau, *Acta Hort.*, 1993, **354**, 221.

COMPARISON OF HEAVY METAL UPTAKE IN DIFFERENT VEGETABLES AND OTHER CROPS

B. Machelett, R. Metz[1] and H. Bergmann

Friedrich-Schiller-Universität Jena, Institute for Nutrition and Environment,
Naumburger Str. 98, 07743 Jena
[1] Humboldt Universität zu Berlin

1 INTRODUCTION

Plant foods are a major source of heavy metal input into the food chain. Mainly heavy metal contents in soil are responsible for heavy metal contents in harvested crops. But plant species differ widely in their capacity for absorption of heavy metals from soil and accumulation of heavy metals varies significantly between different parts of one plant. To cultivate and utilize plants with low accumulation rates in contaminated areas can contribute to lower the exposure of man to heavy metals.

To be able to make use of this it is necessary to investigate experimentally heavy metal accumulation in different parts of plants.

It is therefore the aim of this paper to compare heavy metal uptake and accumulation in different parts of important crops by means of a field test. This shall lead to a general understanding of heavy metal accumulation in various plant parts and families. To achieve this aim a field trial with small plots containing anthropogenically contaminated soil was established growing a large number of crops next to each other until maturity.

2 MATERIAL AND METHODS

Field trial
Size of plots (containers):   60 cm in diameter, 80 cm depth
Experimental soil: soil of a sewage field, loamy sand, humus content 2,5 %,
                                pH 5,9, content of clay and fine silt 10 %,
                                content of heavy metals (mg/ kg):
                                Cd: 1.1, Ni: 13, Pb: 98, Cu: 48, Zn: 170
Analytical methods:
Soil: wet ashing in a 1:3 nitric - hydrochloric acid mix; plants: dry ashing,
flame AAS with a deuterium-arc background corrector [2]

3 HEAVY METAL CONTENT

As expected [1,3,5] it could be found that heavy metal contents differ widely between different parts of plants. Cd-contents show the widest range with a difference of nearly twice the tenth power between the lowest and the highest content in the plant under the same conditions.

To compare more easily the transfer of heavy metals from soil to plant, transfer-coefficients are calculated.

4 TRANSFER-COEFFICIENTS (TCs)

To figure out TCs plant contents are put in relation to soil contents according to the following equation [4,5,6] :

$$TC = \frac{content\ of\ the\ plant}{content\ of\ the\ soil}$$

TCs describe the heavy metal uptake independently of the concentration and thus allow a comparison even between various elements.

Coefficients calculated from this experiment (Table 1) correspond very well with values found elsewhere in the literature and reflect the different mobility of the heavy metals investigated.

Lead and copper are fixed very strongly in the soil and are not readily available for the plant. This is why contents of the plant are always lower than contents of the soil they are grown on. Maximum concentrations of Pb and Cu in plants are one third of those in corresponding soils.

In contrast to this Cd and Zn are transferred easily to plants. Accumulation of these elements in plants can be observed. Plant tissues may contain 6 times as much cadmium or zinc as the corresponding soil.

Nickel is taking an intermediate position between these two extremes

References

1. A.J.M. Baker, Journal of Plant Nutrition, 1981, **3** (no 1-4), 643.
2. M. Grün, H.D. Eschke, B. Machelett, I. Kulick, W. Podlesak, "Schwermetall-bestimmung in Klärschlamm, Boden und Pflanze mittels Atomab-sorptionsspektrometrie," Schwermetalle in der Umwelt (Kolloquien des Institutes für Pflanzenernährung Jena 1987), 92.
3. B. Machelett, R. Metz, M. Grün, "Cadmium uptake in different plants," in 6th International Trace Element Symposium, eds. M. Anke et al. (Universität Leipzig) Vol. 5, 1403.
4. B. Machelett, R. Metz, H. Bergmann,VDLUFA-Schriftenreihe, 1993, **37,** 597.
5. D. Sauerbeck, " Der Transfer von Schwermetallen in die Pflanze," in Beurteilung von Schwermetallkontaminationen im Boden, eds. D. Behrens and J. Wiesner (DECHEMA - Fachgespräche Umweltschutz Frankfurt 1989): 281-316.
6. D. Sauerbeck, S. Lübben, "Auswirkungen von Siedlungsabfällen auf Böden, Bodenorganismen und Pflanzen," Berichte aus der ökologischen Forschung 1991, 416 p.

Table 1   Transfer-coefficients of some heavy metals in different plants and plant organs

No.	plant	plant organ	Cd	Cu	Ni	Pb	Zn
1	sugar beet	leaf	5.55	0.25	0.52	0.07	6.04
2	lettuce	leaf	5.00	0.22	0.25	0.13	1.27
3	spinach	leaf	3.45	0.28	0.23	0.09	1.55
4	potato	leaf	2.82	0.14	0.30	0.07	0.61
5	celery	leaf	2.82	0.15	0.15	0.04	1.22
6	radish	leaf	2.36	0.16	0.21	0.05	1.50
7	carrot	leaf	2.27	0.13	0.32	0.06	1.97
8	sunflower	whole shoot	2.09	0.33	0.75	0.05	2.75
9	german celery	root-shoot-ME	2.09	0.32	0.18	0.04	0.74
10	scorzonera	root-ME	2.00	0.18	0.53	0.01	1.14
11	lucerne	whole shoot	1.73	0.18	0.60	0.02	1.66
12	lupin	whole shoot	1.55	0.24	0.52	0.06	2.19
13	oats	whole shoot	1.55	0.09	0.34	0.03	1.31
14	cauliflower	leaf	1.45	0.07	0.19	0.02	1.45
15	cabbage	leaf	1.36	0.08	0.34	0.05	1.53
16	leek	leaf	1.36	0.10	0.21	0.02	0.61
17	kohlrabi	leaf	1.27	0.09	0.14	0.01	0.94
18	oats	whole shoot	1.18	0.12	0.40	0.02	0.76
19	oats	fruit	1.09	0.20	1.18	0.01	0.85
20	maize	shoot (straw)	1.09	0.10	0.06	0.09	1.53
21	carrot	root-ME	1.09	0.11	0.39	0.07	0.52
22	leek	leaf	1.09	0.13	0.16	0.05	0.81
23	winter rye	shoot (straw)	1.00	0.06	0.05	0.02	1.96
24	parsley	whole shoot	0.88	0.18	0.15	0.08	0.58
25	fodder beet	root-shoot-ME	0.84	0.23	0.28	0.02	1.18
26	red clover	whole shoot	0.57	0.23	0.37	0.04	1.21
27	cock's foot	whole shoot	0.56	0.21	0.28	0.02	0.68
28	radish	shoot-ME	0.50	0.10	0.15	0.02	0.72
29	onion	leaf-ME (bulb)	0.47	0.06	0.09	0.01	0.54
30	tomato	fruit	0.38	0.18	0.15	0.03	0.21
31	strawberry	fruit	0.36	0.12	0.17	0.02	0.16
32	cauliflower	shoot-ME	0.34	0.06	0.35	0.01	0.57
33	onion	leaf	0.34	0.09	0.10	0.01	0.20
34	potato	shoot-ME (tuber)	0.33	0.18	0.14	0.05	0.21
35	maize	fruit	0.30	0.11	0.15	0.01	0.68
36	kohlrabi	shoot-ME	0.25	0.11	0.22	0.01	0.44
37	white cabbage	shoot-ME	0.20	0.09	0.55	0.01	0.44
38	winter rye	fruit	0.18	0.12	0.11	0.01	0.61
39	bush bean	fruit	0.08	0.14	0.28	0.04	0.25

ME = metamorphosis

CADMIUM IN SWEDISH CARROTS

Kerstin Olsson, Rita Svensson and Annette Hägnefelt

Svalöf Weibull AB
SE-268 81 Svalöv
Sweden

1 INTRODUCTION

Cadmium is a potentially toxic trace element which is present in steadily increasing concentration in the environment. There are great variations between parent soil materials. Atmospheric deposition, phosphor fertilizers and sewage sludge add Cd to the top soils. In Swedish farmland the average level in the plough layer is 0.26 mg/kg dry weight (dw) with great differences between areas (0.04-2.93 mg/kg dw)[1]. Low pH is favourable for the uptake of Cd in crops.

An increase of Cd in food crops leads to an increased burden on the human body by this heavy metal. The biological half-time of the absorbed Cd is up to 30 years. Due to harmful effects on the kidneys and skeleton along with possible effects on the cardiovascular system, reproduction and development, all the efforts to minimize the spread and intake of Cd are important[2].

Cereals, potatoes and vegetables are the main sources of Cd in the Swedish diet. This contributes to ca. 75% of the total intake for non-smokers. Carrot is a popular vegetable and has a relatively high concentration of Cd[3].

In this investigation Cd accumulation was demonstrated in a carrot variety grown at various distances from a heavy Cd polluting factory. The variation in Cd amounts was examined in market varieties and breeding material grown in a field with average soil-Cd levels for Sweden. Furthermore, the distribution of Cd from the carrot root suface and inwards was examined.

2 MATERIAL AND METHODS

2.1 Carrots from a polluted area

A commercial variety was grown in home-gardens at various distances from a former factory of Ni-Cd accumulators at Fliseryd, southeast Sweden. Between 1910 and 1974 the total Cd emission to air was 8 tonnes, most of which was deposited in the region close to the factory[4]. The Cd concentrations in the soils were analysed at Kalmar University[5]. Twelve carrot samples were analysed at Svalöf Weibull AB. The Cd concentrations in the outer 1 mm, in the peeled root and in the whole root were registered.

2.2 Market varieties and breeding material

Twelve commercial carrot varieties and 76 breeding lines were grown according to local practice in a field near Hammenhög, south Sweden, where the Cd level of the top soil was 0.23 mg/kg dw. The Cd concentration was analysed in 10 whole roots/sample.

2.3 Distribution of Cd in the root

Three carrot varieties from a field with high Cd levels were analysed for Cd distribution in the root. Ten roots of the same size were chosen. A plug was removed with a cork borer (19 mm in diam.) perpendicular to the longitudinal axis in the middle of each root. The plugs were cut into 7 slices from the surface to the centre. Slices representing corresponding positions were pooled and analysed.

2.4 Cadmium determination

All carrots were thoroughly washed in double de-ionized water, freeze dried and finely ground. The samples were digested in hot concentrated HNO_3 and Cd was determined by a graphite furnace atomic absorption spectrophotometer. Matrix modifier was 1% $(NH_4)H_2PO_4$. The method of additions was used for quantification.

3 RESULTS

3.1 Cd in carrots from a polluted area

Cadmium easily accumulated in the carrot roots when soil levels were high (Figure 1). Close to the accumulator factory, with a soil concentration of 9.6 mg/kg dw, the Cd in the carrots exceeded the upper recommended limit of 0.9 mg/kg dw by a factor of 4. With one exception, the concentrations gradually decreased with the distance from the Cd source as expected. Already 450 m from the factory the carrot concentrations were below the maximum limit. However, 800 m from the factory the amounts in the soil reached 5.9 and in the carrot 3.0 mg/kg dw. At this location the sub soil with a higher Cd level than the top soil had presumably been mixed into the upper layer.

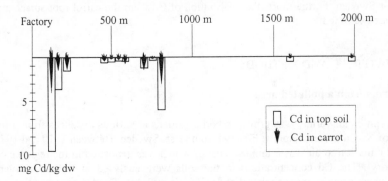

Figure 1 *Cadmium concentration in top soils and carrots at various distances from a Ni-Cd accumulator factory*

3.2 Cd in market varieties and breeding material

The Cd concentration ranged from 0.15 to 0.23 in the 12 commercial carrot varieties and from 0.09 to 0.54 mg/kg dw in the 76 breeding lines in this trial. The differences between genotypes were highly significant. During the actual growing conditions with a soil Cd level about the average for Sweden, no sample exceeded the proposed upper limit of 0.90 mg/kg. There are good possibilities to choose low-accumulating lines also for soils with a higher than average Cd level.

3.3 Distribution of Cd in the root

From the carrot root surface and ca. 7-10 mm inwards the Cd concentration showed a decreasing profile in all three varieties. In the phloem the profile increased again. In two varieties the concentration in the centre reached the same amount as in the 'peel'. In the third variety the Cd concentration was much higher in the centre than in the periphery. This material and other material from the polluted area showed that peeling reduced the Cd concentration by less than 10%.

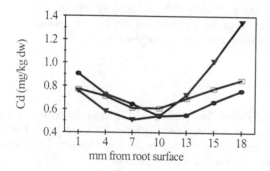

Figure 2 *Cadmium concentrations from the carrot root surface and inwards in three varieties*

4 CONCLUSION

The choice of low Cd soils for growing as well as low accumulating varieties are two ways of reducing the Cd concentration in carrot. Peeling brings down the amount by less than 10%.

Acknowledgement
This work was supported by the Swedish Council for Forestry and Agricultural Research.

References
1. J. Eriksson, M. Söderström and M. Andersson, Swedish Environmental Protection Agency, Report 4450, Solna, 1995.
2. European scientific co-operation on questions relating to food, Task 3.2.4, Cadmium exposure data, Draft report, SCOOP/CAD/REPORT/1, 1997.
3. L. Jorheim , P. Mattsson and S. Slorach, *Vår Föda,* 1984, **36** (suppl, 3), 135.
4. B. Bergbäck and M. Carlsson, *Sci. Total Environ.,* 1995, **166**, 35.
5. D. Carlsson, I. Öbom, K. Petersson-Grawé, S. Larsson and B. Bergbäck, Kadmium i trädgårdsprodukter och brunnsvatten i Fliseryd, Mönsterås kommun. Appendix to Report Project "OSKAR", Mönsterås kommun, Miljökontoret, Box 54, SE-383 22 Mönsterås, Sweden, 1998.

EFFECT OF HARVEST TIME ON THE SENSORY QUALITY OF WHITE CABBAGE

T. Suojala, R. Pessala and R.-L. Heiniö[1]

Agricultural Research Centre of Finland (MTT), Plant Production Research, Horticulture
Toivonlinnantie 518, FIN-21500 Piikkiö, Finland
[1]VTT Biotechnology and Food Research, P.O. Box 1500,
FIN-02044 VTT, Espoo, Finland

1 INTRODUCTION

When searching the optimal harvest time for vegetables for storage, development of the quantity, quality and storability of the yield need to be determined. The objective of this study was to investigate the influence of harvest time on the sensory quality of white cabbage, including changes in the sensory attributes during storage.

2 MATERIALS AND METHODS

The cabbages (cv. Lennox F_1, Bejo Zaden, The Netherlands) used in the sensory analysis were cultivated in two field experiments at Piikkiö and Köyliö in 1995 and 1996. The yield was harvested 3-4 times at two-week intervals in September and October. The plots for harvest dates were arranged in three randomized complete blocks.

The sensory quality was evaluated after harvest at the end of October and after storage in January and May. In the profiling method used, a trained laboratory panel (10-12 panellists) evaluated the following sensory attributes: juiciness, crispness, chewiness, sweetness, lack of bitterness, intensity of flavour and lack of off-flavour. The attributes were assessed from cabbage segments (4-6 per panellist) taken from different cabbages. Scores given on a graphical 10 cm scale were transformed to numerical ones (0-10) in which a high score signifies the highest intensity and a low score the lowest intensity of the attribute.

In 1995, the results were analysed by the analysis of variance with panellist as the block factor. In 1996, experimental block was the block factor and response variable was the mean of the scores of four panellists who evaluated the samples taken from the same block. In 1995, all evaluation times were analysed separately. In 1996, storage time was included in the analysis of variance as a repeated factor. SAS MIXED and GLM procedures were used.

3 RESULTS

3.1 Year 1995

Delaying the harvest improved the sensory quality of cabbages. At Köyliö in October, the cabbages from the latest harvest (11 October) were crispier than cabbages from the first harvest (13 September) (Table 1A). In January, more differences were found: cabbages from the latest harvest were evaluated as juicier, sweeter and less bitter than earlier harvested ones. In May, late harvested samples were juicier and sweeter than the other samples.

At Piikkiö, there were only slight differences in the sensory quality due to harvest time right after harvest (Table 1B). The samples from the last harvest (23 October) were less bitter than cabbages harvested four weeks earlier. On the contrary, the effect of harvest time was more pronounced after storage. In January, differences in juiciness, crispness, chewiness, sweetness, lack of bitterness and lack of off-flavour were observed between harvest dates: the intensities of these attributes were evaluated lower in cabbages from the first harvest than in the other samples. No statistically significant differences were found between the later harvests (9 and 23 October). Again in May, cabbages from the earliest harvest were evaluated as less juicy, less crispy and more difficult to chew, and they had more off-flavour than the other samples.

The effect of the storage time was not analysed statistically. Still in May, the sensory quality of cabbages was high.

3.2 Year 1996

At Köyliö, the storage time had no effect on any sensory attribute, except for lack of off-flavour (p=0.063) (Table 2A). The cabbages harvested on 9 and 23 October were juicier, crispier and sweeter than those of 11 and 25 September. As opposed to other samples, the samples of the first harvest were more bitter and more difficult to chew. Their taste was not so intensive as the taste of the other samples and it had more off-flavour.

At Piikkiö, the storage time affected the juiciness and the crispness of the cabbages (Table 2B). Scores given for crispness decreased from 6.7 - 7.8 in October to 5.3 - 5.7 in May and for juiciness from 5.9 - 7.4 to 4.5 - 5.8, respectively. The early harvested (26 September) cabbages were evaluated as less juicy, less sweet and more bitter than the cabbages harvested two or four weeks later. They had also more off-flavour than the later harvested yield. Other sensory attributes were not affected by the harvest time.

4 DISCUSSION

The results from two field experiments in two years indicate the importance of harvest time on the sensory quality of cabbage. In both years, harvesting in October improved the sensory quality of white cabbage compared to the harvest in September. Harvest time mainly influenced the juiciness, the crispness, the sweetness and the bitterness.

Another remarkable point is that the sensory quality did not deteriorate with late harvesting. Therefore, night frosts that occurred during the harvest period in September 1996 did not decrease the quality of the cabbages.

Storage time had no major effect on the attributes evaluated, and the sensory quality remained high until the end of the storage. Only juiciness and crispness decreased slightly. Differences between the harvest times found right after harvest also existed or were even pronounced after storage. Thus the correct timing of harvest is essential for maintaining the high sensory quality throughout the storage.

Table 1 *Effect of harvest time on the sensory quality of white cabbage (cv. Lennox) in 1995 and 1996.*

A. Köyliö 1995

Harvest date	Juiciness	Crispness	Chewiness	Sweetness	Lack of bitterness	Intensity of flavour	Lack of off-flavour
Analysis: 18 October 1995							
13 Sep	6.7	7.4	7.8	5.7	6.2	6.0	8.6
27 Sep	6.5	8.2	7.6	5.5	5.9	6.4	8.8
11 Oct	7.4	8.8	8.4	6.2	5.8	6.6	9.2
p value	ns	0.011	ns	ns	ns	ns	ns
Analysis: 11 January 1996							
13 Sep	7.1	6.9	7.6	5.8	6.1	6.3	7.9
27 Sep	7.1	7.4	7.1	5.8	6.1	6.2	7.1
11 Oct	8.1	7.9	7.5	7.2	7.6	6.0	8.2
p value	0.053	ns	ns	0.010	0.032	ns	ns
Analysis: 23 May 1996							
13 Sep	5.3	5.8	6.6	3.8	7.2	4.5	7.3
27 Sep	6.5	6.0	6.8	5.2	7.4	5.2	7.8
11 Oct	7.3	6.7	7.3	5.8	7.8	6.3	8.5
p value	0.076	ns	ns	0.043	ns	ns	ns

B. Piikkiö 1995

Harvest date	Juiciness	Crispness	Chewiness	Sweetness	Lack of bitterness	Intensity of flavour	Lack of off-flavour
Analysis: 25 October 1995							
25 Sept	6.7	7.7	7.6	5.1	4.7	7.0	7.3
9 Oct	7.1	7.6	7.4	5.7	6.5	6.4	8.4
23 Oct	6.7	7.7	7.6	6.4	7.1	6.3	8.8
p value	ns	ns	ns	ns	0.014	ns	ns
Analysis: 11 January 1996							
25 Sept	4.1	3.6	4.9	3.3	4.0	6.6	6.2
9 Oct	5.6	5.8	6.5	5.3	5.8	6.3	7.9
23 Oct	6.7	6.5	7.0	6.7	6.0	6.3	7.7
p value	<0.001	<0.001	0.005	<0.001	0.021	ns	0.022
Analysis: 9 May 1996							
25 Sept	5.3	5.7	6.5	4.2	5.8	5.8	7.5
9 Oct	6.8	7.0	7.9	5.3	7.0	5.3	9.1
23 Oct	6.4	7.0	7.2	5.2	6.5	5.4	8.7
p value	0.068	0.041	0.034	ns	ns	ns	0.001

Table 2 *Effect of harvest time on the sensory quality of white cabbage (cv. Lennox) in 1996 (averages of analyses in October, January and May).*

A. Köyliö

Harvest date	Juiciness	Crispness	Chewiness	Sweetness	Lack of bitterness	Intensity of flavour	Lack of off-flavour
11 Sept	4.5	5.0	5.5	3.6	5.3	4.9	7.1
25 Sept	5.0	5.6	6.0	4.8	6.2	5.4	7.9
9 Oct	6.4	6.5	6.6	6.0	7.0	5.8	8.2
23 Oct	6.5	6.9	6.8	5.6	6.9	6.0	7.9
p (Harvest)	<0.001	<0.001	0.004	<0.001	0.008	0.011	0.060
p (Storage)	ns	ns	ns	ns	ns	ns	0.063
p (H*S)	ns	ns	ns	ns	ns	ns	ns

B. Piikkiö

Harvest date	Juiciness	Crispness	Chewiness	Sweetness	Lack of bitterness	Intensity of flavour	Lack of off-flavour
26 Sept	5.5	6.2	6.7	4.7	5.7	5.7	7.3
10 Oct	6.4	6.5	7.0	5.2	6.7	5.4	8.1
24 Oct	6.3	6.6	7.2	5.7	7.3	5.3	8.1
p (Harvest)	0.030	ns	ns	0.027	0.007	ns	0.071
p (Storage)	0.046	0.005	ns	ns	ns	ns	ns
p (H*S)	0.080	0.088	ns	ns	ns	ns	ns

Session VI Quality Assessment

NONDESTRUCTIVE TECHNIQUES FOR QUALITY EVALUATION OF FRUITS

Pictiaw Chen
Department of Biological and Agricultural Engineering
University of California
Davis, California, U.S.A.

1 INTRODUCTION

Quality evaluation of agricultural products has been a subject of interest to many researchers for many years. As a result, several nondestructive techniques for quality evaluation of agricultural products have been developed. These methods are based on the detection of various physical properties that correlate well with certain quality factors of the products. This paper presents an overview of these methods.

2 REVIEW OF TECHNIQUES

2.1 Density

The density of many fruits and vegetables increases with maturity. On the other hand, certain types of damage and defects tend to reduce the density of the product. Zaltzman et al.[1] presented a comprehensive literature review of previous studies related to quality evaluation of agricultural products based on density differences. Common methods used include removing fruits that float in water or in solutions of known density, and releasing fruits from the bottom of a flowing stream of water and skimming off fruits of different density ranges from the top of the water channel at different horizontal distances from the point of release[2]. Zaltzman and his co-workers[3] developed a pilot plant unit that used fluidized bed medium to separate potatoes from clods and stones at a rate of 5 t/h with better than 99% potato recovery and 100% clod and stone removal.

2.2 Firmness

Firmness is often used for evaluating the quality of fruits and vegetables. In general firmness of fruits decreases gradually as they become more mature and decreases rapidly as they ripen. Overripe and damaged fruits become relatively soft. Thus firmness can be used as a criterion for sorting agricultural products into different maturity groups or for separating overripe and damaged fruits from good ones. Several methods for measuring fruit firmness have been developed.

2.2.1 Force-deformation. Mehlschau et al.[4] developed a "deformeter" for non-destructive maturity detection for pears based on the measurement of deformation resulting from pressing two steel balls against the opposite sides of the fruit with a fixed force. Takao[5] developed a force-deformation type firmness tester that can measure firmness of fruit nondestructively. Bellon et al.[6] built a micro-deformeter that was able to classify peaches into three texture classes with a 92% accuracy. Armstrong et al.[7] developed an automatic instrument with automatic data collection and analysis and can measure the firmness of a batch of 25 small fruits within one minute.

2.2.2   Impact. Two common impact techniques have been used to determine fruit firmness. One technique is by impacting the fruit on a force sensor, and the other is by impacting a sensor on the fruit. Nahir et al.[8] developed an experimental tomato grading machine which, by measuring and analysing the impact force response of the fruit, can separate tomatoes on the basis of weight and color.   Other sensors using similar techniques of dropping the fruit on a sensor were developed by Younce and Davis[9] and McGlone et al.[10]. A problem inherent to the technique of dropping the fruit on a force sensor is that the impact force is also a function of the mass and radius of curvature of the fruit.

A different approach is to impact the fruit with a small spherical impactor of known mass and radius of curvature and measuring the acceleration of the impactor.   The advantage of this method is that the impact-force response is independent of the fruit mass and is less sensitive to the variation of the radius of curvature of the fruit. This technique was first described by Chen et al.[11] and was used by researchers in Spain for sensing fruit firmness[12]. Results of a study by Chen et al.[13] indicated that using a low-mass impactor can result in the following additional desirable features: Higher signal strength, higher sensitivity to fruit firmness, less error due to movement of the fruit during the impact, less fruit damage, and higher sensing speed. Based on these findings, a low-mass high-speed impact sensor was designed and tested[14] with good results on kiwifruits and peaches.

2.2.3   Sonic vibration. The study of sonic vibration of fruits was pioneered by Abbott[15] and Finney[16]. In the early 1980's Yamamoto et al.[17] developed a nondestructive technique for measuring textural quality of apples and watermelons based on the acoustic response of the fruit.

The availability of high-speed data acquisition and processing technology in recent years has renewed researchers' interests in the development of sonic vibration and acoustic response techniques. Research teams currently conducting research in this area include researchers in the U.S. in Michigan[18], Maryland[19], and California[20, 21]; in Belgium[22], Israel[23], and Japan[24].

In general, the researchers detected a series of resonant frequencies. Theoretically, for the free vibration of an elastic sphere, the elastic modulus of the sphere is related to other physical properties as follows: $E \propto (1+\mu)\ f^2 m^{2/3} \rho^{1/3}$ where E is elastic modulus; μ is Poisson's ratio; f is resonant frequency of free vibration; m is mass, and ρ is density. Since μ and ρ are relatively constant, researchers have found that the value of $f^2 m^{2/3}$ is a good criterion for predicting the firmness of the fruit.

2.2.4   Ultrasonic methods. Ultrasonic techniques have been used quite successfully for evaluating subcutaneous fat, total fat, lean, and other internal properties of live animals[25]. However, because of the porous nature of fruit tissues, it is difficult to use high-frequency ultrasound to evaluate internal quality of fruits and vegetables. Mizrach et al.[26] reported some success in using low frequency (50 kHz) ultrasonic excitation to determine some basic acoustic properties of certain fruits and vegetables. They later reported encouraging findings on potential use of ultrasonic techniques for internal quality evaluation of whole avocado fruit and other fresh products[27]. Haugh[28] used a dry-coupling broad-band transducer with a frequency of 250 kHz to detect hollow heart in potatoes and found that the defective potatoes could be separated on the basis of power spectral moment of the transmitted ultrasonic signals.

2.3 Electrical Properties

The study of electrical properties of agricultural products has interested many researchers in the past two decades. A comprehensive review of the electrical properties of agricultural products was published by Nelson[29]. The electrical properties of many agricultural products, especially hygroscopic materials, are highly dependent on moisture content. This relationship between moisture content and electrical properties was used as a basis for developing commercial instruments for measuring moisture content in grains and seeds. Nelson and Lawrence developed a rapid and nondestructive technique, based on capacitance measurements at 1 and 5 MHz, for estimating moisture content in individual dates[30], and pecans[31]. They reported that this technique had potential for application in the automatic sorting of dates.

The sensitivity of electrical measurements to moisture content tends to mask changes in electrical properties associated with other variables. Zachariah[32] reported that a number of researchers have investigated the electrical properties of fruits and vegetables, but the results were not conclusive enough to permit development of a practical method for quality sorting of fruits and vegetables.

2.4 Optical Properties

In the past three decades, researchers have studied the optical properties of various agricultural products and have established correlations between optical characteristics and other quality-related properties of the products[33, 34].

2.4.1 Near-infrared analysis. Norris and his co-workers studied NIR reflectance and transmittance characteristics of many agricultural products and have found that radiation in the near-infrared region of the spectrum can provide information related to many quality factors of agricultural products. A very important contribution made by Norris and his co-workers was the development of data treatment techniques which make it possible to extract information from spectrophotometric data (curves). They found that diffuse reflectance, R, and diffuse transmittance, T, do not vary linearly with the concentration of an absorbing component in the material. Therefore, if a linear correlation between NIR measurements and the concentration of an absorber is desired, certain mathematical treatments of the reflectance or transmittance data are required.

Norris[35] gave several examples using derivative data treatments to predict the fat, moisture, and protein contents of meat from transmittance data; the oil and moisture content of individual intact sunflower and soybean seeds from transmittance data; and the composition of ground samples of wheat from reflectance data.

The research in NIR techniques has led to the development of various commercial NIR analysers for multiple constituent analysis of grains, oil seeds, meats, dairy products, feed, forages, etc. A very comprehensive book on basic fundamentals of near-infrared technology and NIR applications in the agricultural and food industries was compiled by Williams and Norris[36].

A number of researchers have conducted research to determine internal compositions of different types of fruits and vegetables. Kawano used the NIR technique to analyse sugar content of intact peaches[37] and Satsuma mandarin[38]. Similarly, Slaughter[39] successfully used the absorption characteristics of near infrared light in peaches and nectarines to predict their soluble solids content (r = 0.92). Bellon and Sevila[40] developed an NIR system, which combined a CCD spectrophotometric camera and bifurcated fiber optics, for determining soluble solids in apples.

2.5 X-rays and Gamma Rays

X-rays and gamma rays are suitable for nondestructive evaluation of quality factors that are associated with mass density variation. Lenker and Adrian[41] developed a lettuce harvester that uses X-rays for selecting mature lettuce heads. Garrett and Talley[42] also developed a lettuce maturity evaluating unit based on gamma ray transmission. Researchers have found that X-ray techniques can be used to detect bruises in apples[43, 44], hollow heart in potatoes[45], split pit in peaches[46], and granulation in oranges[47].

A number of high-speed X-ray sensors capable of detecting hollow heart in potatoes, pits in cherries or olives, and foreign objects have been developed recently for commercial applications. Tollner et al.[48] gave a comprehensive overview of ongoing research and commercial development of X-ray sensors for nondestructive detection of interior voids and foreign inclusions in fruits and vegetables.

2.6 Nuclear Magnetic Resonance

Nuclear magnetic resonance (NMR) is a technique that can detect variations in the concentration and mobility of water, oil, and sugar. Researchers have developed various techniques for evaluating a number of internal quality factors of selected fruits and vegetables. Chen et al.[49] used magnetic resonance imaging (MRI) to evaluate various quality factors of fruits and vegetables. They found that MRI can provide high-resolution images of internal structures of intact fruits and vegetables and can be used for nondestructive evaluation of various internal quality factors, such as bruises, dry regions, worm damage, internal breakdown, stage of maturity, and presence of voids, seeds, and pits. Results of recent studies by Chen and co-workers show that it is possible to use high-speed NMR techniques to evaluate maturity of avocados[50], sugar content of prunes[51], presence of pits in cherries[52], and tissue breakdown in melons. The oil content of whole avocados and sugar content of whole prunes can be determined from a single-pulse NMR spectrum using a surface coil sensor. The technique of using a surface coil to acquire single-pulse FID spectra not only facilitates rapid testing of whole fruits, but also reduces the effects of other factors such as the size of the fruit and the presence of the seed. Using a specially designed conveyor belt, Chen et al.[53] successfully acquired FID spectra of avocados while they were moving at speeds up to 250 mm/s. The oil/water resonance peak ratio, obtained from the spectrum, correlates very well ($r^2 = 0.98$) with the dry weight of the fruit.

Researchers at Purdue University have designed and built a low-resolution (5.35 MHz) proton magnetic resonance sensor capable of accommodating samples with diameters up to 30 mm[54]. Various tests were conducted using this device to measure sugar content of small fruits (with diameter less than 30 mm) and specimens of larger fruits. Ray et al.[55] used this device, after implementation of design improvements, to measure sugar content of Bing cherries and reported good correlation between the spin echo ratio (SER) and total soluble solids of the fruit ($r^2 = 0.91$).

Continuing research in NMR for quality evaluation of agricultural products should lead to the development of viable NMR sensors for agricultural applications.

2.7 Machine vision

One of the major requirements in developing machine vision systems for sorting fruits and vegetables is the ability to analyse an image accurately and quickly. Recently, new algorithms and hardware architectures have been developed for high-speed extraction of features that are related to specific quality factors of fruits and vegetables. As a result, a number of sorting systems that use machine-vision techniques for evaluating external quality factors are commercially available. It is anticipated that high-speed image analysis techniques will play a key role in expanding the capability of X-ray and NMR techniques for internal quality evaluation of agricultural products.

The machine-vision technique has been further refined to include multispectral imaging, where each point (pixel) of the image contains spectral information at several wavelengths. Alchanatis and Searcy[56] described three techniques for obtaining multispectral images. Such imaging techniques enable researchers to extract additional useful information for quality evaluation. McClure[57] predicted that just as NIR spectroscopy is the rapid nondestructive analytical technique of the 20th century, NIR imaging spectroscopy will be the analytical tool of the 21st century.

2.8 Aroma

Numerous researchers have tried for many years to develop electronic sniffers or electronic noses with limited success. A brief history of the development of electronic noses was reported by Gardner and Bartlett[58]. Benady et al.[59] developed a sniffer for determining fruit ripeness nondestructively. The sniffer used a semiconductor gas sensor located within a small cup to collect and sense gases emitted by the ripening fruit. They reported that the sensor performed successfully on three muskmelon cultivars under field and laboratory conditions over two growing seasons.

3 OTHER SOURCES OF INFORMATION

The above are highlights of only a few selected applications of nondestructive techniques for quality evaluation and sorting of agricultural products. Numerous other applications are not covered in this article. Readers who are interested in other applications are referred to works by Bellon[60], Chen[33], Chen and Sun[61], Finney[62], Finney and Abbott[63], Gaffney[64], Gunasekaran et al.[34], Mohsenin[65], Nelson[29], Williams and Norris[36], Zaltzman et al[1], and reports by Kawano and Iwamoto[66] and various other authors in the Proceedings of the International Workshop on Nondestructive Technologies for Quality Evaluation of Fruits and Vegetables, ASAE Publication 05-94.

References

1. A. Zaltzman, B. P. Verma, and Z. Schmilovitch, *Trans. ASAE*, 1987, **30**, 823.
2. R. P. Gutterman, In: Quality detection in foods. *ASAE Publication 1-76*. 1976, 211.
3. A. Zaltzman, R. Feller, A. Mizrach, and Z. Schmilovitch, *Trans. ASAE,* 1983, **26**, 987.
4. J.J. Mehlschau, P. Chen, L.L. Claypool, and R.B. Fridley, Trans. ASAE, 1981, **24**, 1368.
5. H. Takao, and S. Ohmori. JARQ (Japan Agri Res Quarterly), 1994, 28, 36.
6. V. Bellon, J.L. Vigneau, and M. Crochon, Proc. IV Int. Symp on Fruit, Nut, and Veg Production Engrng, Spain, 1994, **2**, 291.
7. P. R. Armstrong, G. K. Brown, and E. J. Timm. ASAS Paper No. 95-6172, 1995.
8. D. Nahir, Z. Schmilovitch, and B. Ronen. ASAE Paper No. 86-3028, 1986.
9. F.L. Younce, D.C. Davis. *Trans. ASAE*, 1995, **38**, 1467.
10. V.A. McGlone, R.B. Jordan, and P.N. Schaare,*Trans. ASAE*, 1997, **40**, 1421.
11. P. Chen, S. Tang, and S. Chen, ASAE Paper No. 75-3537, 1985.
12 C. Jarén, C., M. Ruiz-Altisent, R. Pérez de Rueda, AGENG 92, Paper No. 9211-113, 1992.
13. P. Chen, M. Ruiz-Altisent, and P. Barreiro, *Trans. ASAE*, 1996, **39**, 1019.
14. P. Chen and M. Ruiz-Altisent, AgEng96, Paper No. 96F-003, 1996.
15. J.A. Abbott, G. S. Bachman, N. F. Childers, J. V. Fitzgerald, and F. J. Matuski, *Food Technology*, 1968, **22**, 101.
16. E.E. Finney, 1970, *Trans. ASAE*, 1970, **13**, 177.
17. H. Yamamoto, M. Iwamoto, and S. Haginuma, J. Texture Studies, 1980, 11, 117.
18. P .Armstrong, H. R. Zapp, and G. K. Brown, *Trans. ASAE*, 1990, **33**, 1353.
19. J.A. Abbott, H.A. Affeldt, and L.A. Liljedahl, *J Amer. Soc. Hort. Sci.*, 1992, **117**, 590.
20. P. Chen, Z. Sun, and L. Huarng,*Trans. ASAE*, 1992, **35**, 1915.
21. L. Huarng, P. Chen, and S. K. Upadhyaya, *Trans. ASAE*, 1992, **36**, 1421.
22. H. Chen, J. De Baerdemaeker, and F. Vervaeke, Proc. Int. Conf. Agric. Engineering, China, 1992.
23. I. Shmulevich, N. Galili, and D. Rosenfeld, *Trans. ASAE*, 1996, **39**, 1047.
24. J. Sugiyama, T. Katsurai, J. Hong, H. Koyama, K. Mikuriya, *Trans. ASAE*, 1998, **41**, 105.
25. J.C. Alliston, Anim. Prod., 1982, 35, 361.
26. A. Mizrach, N. Galili, and G. Rosenhouse, *Trans. ASAE*, 1989, **32**, 2053.
27. A. Mizrach, N. Galili, and G. Rosenhouse, *Food Technology*, 1994, **48**, 68.
28. C.G. Haugh, Int. *Agrophysics*, 1994, **8**, 509.
29. S.O. Nelson, *Trans. ASAE*, 1973, **16**, 384.

30. S.O. Nelson, and K. C. Lawrence, *Trans. ASAE*, 1994, **37**, 887.
31. S.O. Nelson, and K. C. Lawrence, *Trans. ASAE*, 1995, **38**, 1147.
32. G. Zachariah, ASAE Publication 1-76, 1976, p. 98.
33. P. Chen, *J. Food Process Engng.*, 1978, **2**, 307.
34. S. Gunasekaran, M. R. Paulsen, G. C. Shove, *J. Agric. Engng Res.*, 1985, **32**, 209.
35. K.H. Norris, In. Proc. IUFoST Symp. on Food Research and Data Analysis, Oslo, Norway, Applied Science Pub., Ltd., England, 1983, p. 95.
36. P. C. Williams and K. H. Norris (Editors), 'Near-infrared Technology in the Agricultural and Food Industries'. Am Assoc of Cereal Chemists, St. Paul, Minnesota, 1987, 330 pp.
37. S. Kawano, H. Watanabe and M. Iwamoto, *J. Japan. Soc. Hort. Sci.*, 1992, **61**, 445.
38. S. Kawano, T. Fugiwara, and M. Iwamoto, *J. Japan. Soc. Hort. Sci.*, 1993, **62**, 465.
39. D.C. Slaughter, *Trans. ASAE*, 1995, **38**, 617.
40. V. Bellon, and F. Sevila, Proc. 4th Int. Symp. on Fruit, Nut, and Vegetable Production Engineering, Spain, 1993, p. 317.
41. D.H. Lenker and P. A. Adrian, *Trans. ASAE*, 1971, **14**, 894.
42. R.E. Garrett and W. K. Talley, *Trans. ASAE*, 1970, **13**, 820.
43. R.G. Diener, J.P. Mitchell, and M.L. Rhoten, *Agricultural Engineering*, 1970, **51**, 356.
44. T.F. Schatzki, R.P. Haff, R. Young, I. Can, L-C. Le, N. Toyofuku, *Trans. ASAE*, 1997, **40**, 1407.
45. E.E. Finney and K. H. Norris, *Amer. Potato J.*, 1973, **50**, 1.
46. S.V. Bowers, R. B. Dodd, and Y. J. Han, ASAE Paper No. 88-6569, 1988.
47. M. Johnson, Proc. Agri-Mation 1 Conf. and Expo., Illinois, 1985. p. 63.
48. E. W. Tollner, H.A. Affeldt, G.K. Brown, P. Chen, N. Galili, C.G. Haugh, A. Notea, Y. Sarig, T. Schatzki, I. Shmulevich, and B. Zion, ASAE Publication 05-94, 1994, 86.
49. P. Chen, M. J. McCarthy, and R. Kauten,*Trans. ASAE*, 1989 **32**, 1747.
50. P. Chen, M. J. McCarthy, R. Kauten, Y. Sarig, and S. Han, *J. of Ag. Engng. Res.*, 1993, **55**, 177.
51. B. Zion, P. Chen, and M. J. McCarthy, *J. Sci Food Agric*, 1995, **67**, 423.
52. B. Zion, M.J. McCarthy, and P. Chen, Lebensm.-Wiss.-u.-Technol., 1994, 27, 457.
53. P. Chen, M.J. McCarthy, S.-M. Kim, and B. Zion, *Trans. ASAE*, 1996, **39**, 2205.
54. S.I. Cho, G. W. Krutz, H. G. Gibson, and K. Haghighi, *Trans. ASAE*, 1990, **33**, 1043.
55. J.A. Ray, R.L. Stroshine, G.W. Krutz, and W.K. Wai, ASAE Paper No. 93-6071, 1993.
56. V. Alchanatis and S.W. Searcy, ASAE Paper No. 95-3210, 1995.
57. W.F. McClure, *NIR news*, 1991, 2, 8.
58. J.W. Gardner and P. N. Bartlett, Sensors and Actuators B, 1994, **18-19**, 211.
59. M. Benady, J.E. Simon, D.J. Charles, and G.E. Miles,*Trans. ASAE*, 1995, **38**, 251.
60. V. Bellon, Proc. IV Int. Symp on Fruit, Nut, and Veg Production Engng, Spain, 1994, **2**, 1.
61. P. Chen and Z. Sun, *J. Agric. Engng Res.*, 1991, **49**, 85.
62. E.E. Finney, ' Measurement techniques for quality control of agricultural products'. Amer. Soc. Agric. Eng., Michigan, 1973, 53 p.
63. E.E. Finney and J. A. Abbott, *J. Food Quality,* 1978, **2**, 55.
64. J.J. Gaffney (Compiler), 'Quality detection in foods', ASAE Pub. 1-76, Michigan, 1996.
65. N.N. Mohsenin, 'Electromagnetic radiation properties of foods and agricultural products', Gordon and Breach Science Pub., New York, 1984.
66. S. Kawano and M. Iwamoto, ASAE Publication 05-94, Michigan, 1994, p. 1.

METABOLIC CONTROL OF WALL PROPERTIES AND ITS IMPLICATION TO THE TEXTURE OF PLANT-DERIVED FOOD

G. Wende[1,2], K.W. Waldron[2], A.C. Smith[2] and C.T. Brett[1]

[1]Plant Molecular Science Group, Institute of Biochemistry and Life Sciences, Bower Building, Glasgow University, Glasgow G12 8QQ, UK and [2]Institute of Food Research, Norwich Research Park, Colney Lane, Norwich NR4 7UA, UK

1 INTRODUCTION

Controlling the textural qualities of plant-derived foods assumes an understanding of the complexity of processes involved. The texture of edible fruits and vegetables has been attributed to the physical and chemical properties of cell walls [1].

Beet root and sugarbeet (*Beta vulgaris*) are highly relevant to the food industry. They contain substantial amounts of ferulic acid which is esterified to arabinoxylans in grasses [2] and to pectin in beet [3]. The *O*-feruloyl groups are of particular interest because of their high reactivity *in vivo* to form diferuloyl cross-linkages via a hydrogen peroxide mediated radical mechanism. As a result of this mechanism, dehydrodiferulates are coupled by different linkages [4]. The 'Diferulic acid' known for more than two decades is the one with a 5–5-linkage. Dehydrodiferulates have been implicated in thermal stability of cell-adhesion and texture in Chinese Water Chestnuts [5], due to the fact that they fail to soften after heat treatment [6]. In general, tissue softening occurs due to cell separation.

The objective of the present work has been towards the elucidation of the metabolic control of texture in plant-derived foods, in order to increase our understanding of the relationships between structural and functional features.

2 MATERIAL AND METHODS

2.1 Plant Material

Suspension cultures of red beet (*Beta vulgaris*) were grown at Glasgow University under constant shaking at 110 rpm in the present of white fluorescent light (25 $\mu E/M^2/s$). Seedlings of sugarbeet were grown in the presence of white fluorescent light and in the dark (16 and 8 hours, respectively) at 25 °C. Prior to that, seeds were soaked in vigorously aerated water at 25 °C overnight. Seeds were then grown in vermiculite and transferred after 3 weeks into 5x strength Hoaglands solution with addition of 0.5% ferrous sulfate and 0.4% tartaric acid [7].

2.2 Uptake of radiolabel into plant tissues

In a time-course over several months, 14C-cinnamate was applied onto leaves of three-week old sugarbeet seedlings (3 per time-point) and its uptake monitored every two weeks. One seedling was taken for an autoradiographic image analysis, and from two seedlings the alcohol insoluble residue (AIR) was prepared separately from leaves, storage roots (divided into inner and outer tissue) and roots.

2.3 Preparation of cell wall fragments and endomembranes from suspension cultures

Suspension cultures of beet (200 ml) were pulse-labelled with 14C-cinnamate and/or with 14C-sucrose for 9 days, sub-cultured (40 ml into 160 ml) and incubated for a further 13 days. Cells were separated from medium by centrifugation at 1000g for 15 min. Cells were subsequently homogenized in buffer (5 ml of 1 M Tris-HCl, 1 M KCl, 1 M Dithiothreitol and 0.1 M $MgCl_2$) at 4 °C in a glass homogenizer to break open the cells. The rate of cell breakage was observed by light microscopy. Cell wall fragments were separated from endomembranes by centrifugation at 13000g and 4 °C for 30 min in an 8% sucrose solution, onto 8 ml of 40% sucrose solution. Endomembranes were collected from the 8%:40% interface.

2.4 Analysis

Phenolic material was extracted after alkaline treatment with 2 M NaOH either from plant material or suspension cultures according to reference [8] and separated by reverse phase HPLC according to reference [9]. Fractions were collected and monitored for radioactivity.

3 RESULTS AND DISCUSSION

HPLC analysis revealed transformation of 14C-cinnamate and also 14C-sucrose into radio-labelled ferulic, diferulic acids and other phenolic compounds. In an initial investigation, 14C -sucrose was applied onto a leaf of a 3-week old sugar beet seedling and harvested after 2 weeks. AIR was prepared from the whole plant and the phenolic material extracted as mentioned above. Figure 1 shows transformation of 14C-sucrose mainly into *trans*- and *cis*-ferulic acid and clearly also into 8–5 benzofuran form, 8–*O*–4, 5–5 diferulic acids; most likely also into 8–5, 8–8 DFA and also into the 8–8 aryltetraline form. However, the latter radio-labelled diferulates can not be clearly identified since they elute 'between' *trans* and *cis* ferulic acid (8–5, 8–8 DFA) and 'between' *p*-coumaric acid and *trans* ferulic acid (8-8 aryltetralin form). However, all three can be clearly separated on HPLC as judged by UV absorbance. In future experiments more samples will be collected especially between 13 and 18 min of elution time.

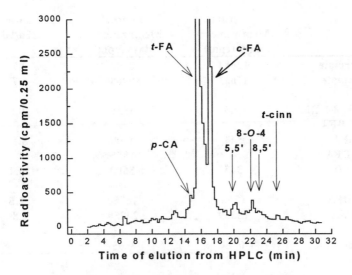

Figure 1 *Profile of radio-labelled phenolic material from sugar beet seedling fed with 14C-sucrose. AIR was prepared after two weeks, phenolic material was extracted and subjected to reverse phase HPLC. Fractions (0.25 ml) were collected and monitored for radioactivity*

Uptake and incorporation of radiolabel into the beet seedling appears to be rapid. Radioactivity was found to be uniformly distributed among the leaves and roots of the beet seedlings, with more radioactivity in the young leaves as detected with the use of an autoradiographic image analyser. The amount and distribution of radioactive label appeared to remain stable over subsequent growth of the plants for at least three months. The storage root showed an interesting pattern. Feeding young beet seedlings with 14C-cinnamic acid through the roots caused the epidermis and stele of the storage root to become labelled, with relatively little label in the cortex.

By pulse-labelling red beet suspension cultures with 14C-cinnamate, radiolabel was found to be taken up and incorporated into cells quickly. Cell walls were separated from internal membranes as mentioned above from 9 days and 22 days old suspension cultures and analysed for phenolic material. Table 1 summarizes data from internal membranes. The amount of 14C-cinnamate taken up by the cells increased almost 6 fold between 9 and 22 days and the amount of total ferulate less than 2 fold. However, the content of diferulates increased more than 12 fold, accounting for a more than 6 fold increase in the diferulate:ferulate ratio after 22 days in comparison to 9 days.

Table 1 *Suspension cultures of red beet were fed with 14C-cinnamate for 9 days, sub-cultured and incubated for further 13 days. Internal membranes were separated from cell wall fragments and phenolic material extracted with alkaline. Separation of the phenolics was performed on HPLC, fractions were collected and monitored for radioactivity*

	Internal Membranes 9 days (cpm)	Internal Membranes 22 days (cpm)	Ratio 22d/9d
Total ferulate	12084	8500	0.7
t-Cinnamate	17631	100680	5.7
8-8 aryl DFA, 8-8 DFA and 8-5 DFA	366	3835	10.5
5-5 DFA	544	4400	8.1
8-O-4 DFA	-	1855	
8-5 Benz DFA	345	5660	16.4
Total DFA	1255	15750	12.5
Total FA	13339	24250	1.8
DFAs/Total FA	0.09	0.65	

It is likely that the ferulate found in the internal membranes is mainly present attached to newly-synthesised pectins in the Golgi apparatus, prior to their export to the cell wall. The diferulates present in the internal membranes could rise by crosslinking of nascent feruloylated pectins prior to export to the wall. However, the big increase in the radioactively-labelled diferulate:ferulate ratio between 9 and 22 days suggests that dimerization is occurring much later in the metabolic sequence than the initial feruloylation. The diferulate:ferulate ratio after 22 days is similar to that found in the cell-wall itself (results not shown). Hence the diferulates in the internal membranes are more likely to be derived from pectic polysaccharides which have been deposited in the wall and then taken up into an internal membrane-bound organelle, perhaps, the vacuole, as a result.

4 CONCLUSION

Based on feeding studies of young sugarbeet seedlings with 14C-cinnamate, it appears that the distribution of radioactive label is stable over the growth period and therefore phenolic compounds as a whole are not subject to turnover. However, a more detailed analysis of suspension cultures (and also of plant material from time-course analysed for phenolics; data not shown) suggested that ferulate dehydrodimers are subject to turnover. Such turnover will have important implications for the control of texture in plant-derived foods, both physiologically and also as a potential target for genetic manipulation.

REFERENCES

1. J. P. van Buren, *J. Texture Studies*, 1979, **10**, 1.
2. G. Wende and S. C. Fry, *Phytochemistry*, 1997, **44**, 1019.
3. M.-C. Ralet, J.-F. Thibault, C. B. Faulds and G. Williamson, *Carbohydrate Research*, 1994, **263**, 227.
4. J. Ralph, S. Quideau, J. H. Grabber and R. D. Hatfield, *J. Chem. Soc., Perkin Trans. 1*, 1994, 3485.
5. M. L. Parker and K. W. Waldron, *J. Sci. Food Agri*, 1995, **68**, 337.
6. A. J. Parr, K. W. Waldron, A. Ng and M. L. Parker, *J. Sci. Food Agri*, 1996, **71**, 501.
7. Hewitt, E. 'Sand and Water Culture Methods used in the Study of Plant Nutrition', Commonwealth Agricultural Bureaux: Farnham Royal, England, 1966.
8. G. Wende, R. D.Hatfield, J. H. Grabber and J. Ralph, *Planta*, 1998, submitted.
9. K. W. Waldron, A. J. Parr, A. Ng and J. Ralph, *Phytocheml. Anal.*, 1996, **7**, 305.

WATER POTENTIAL, AN EASY TO MEASURE AND SENSITIVE INDICATOR OF MECHANICAL AND CLIMATIC STRESS DURING POSTHARVEST HANDLING OF CARROTS

W. B. Herppich, H. Mempel and M. Geyer.

Inst. f. Agrartechnik Bornim e. V.,
Abt. Technik im Gartenbau,
Max-Eyth-Allee 100,
D-14469 Potsdam,
Germany

1 INTRODUCTION

The postharvest handling chain, from mechanical harvest via washing, packing and storage to sale, exposes carrots to many mechanical and climatic stresses, leading to increased dry matter and tissue water losses. This reduces product freshness and quality, lowering the price returns. Precise knowledge of tissue water status is necessary to characterise the different impacts during handling and elucidate possible effects on the physiological activity of the product. However, biological variability of samples used, often complicates a meaningful quantification of changes in tissue water content. In this paper we will show that measuring the root water potential (Ψ) with a „Scholander" pressure bomb provides a rapid and reliable solution.

2 MATERIAL AND METHODS

2.1. Plant Material

For all experiments F1-hybrids of the cultivar 'Nanthya' were used which were grown on a farm near Potsdam (Huschke, Seeburg, Brandenburg, Germany). Fertilisation and plant protection was as typically for farming in this area. Carrots were carefully harvested by hand and leaves were cut off immediately if not mentioned otherwise. Roots were gently washed. If they were not used directly the carrots were stored in plastic bags at 16-19 °C for a short period (max. 6 h).

2.2. Methods

Tissue water potential was measured with a commercial 'Scholander'-type pressure bomb (Plant Water Status Console 3000, Soilmoisture Inc., Santa Barbara, CA, USA). After severing part of the primary root the carrots were wrapped in aluminium foil to reduce water loss during measurements. They were then enclosed in the pressure vessel of the pressure bomb. Only the end of the root protruded through the specimen holder. Then pressure was slowly increased until the xylem sap started to emerge at the exposed water conducting system. This indicated that equilibrium pressure was reached which denotes the mean bulk water potential[1].

Dry matter related water content was calculated as the difference of root fresh- and dry matter, divided by the dry matter which was obtained after over drying (48 h at 85 °C). Sample size typically was at least 10.

Respiratory CO_2 evolution was measured in an open system with an IRGA (CIRAS-1, PP Systems, Herts, UK) as described earlier[2].

3 RESULTS

Water potential measurements with a pressure bomb are sound and do not damage or stress the tissue used. After a two day storage at 18 °C and 98% rH carrots used for water potential determinations before, had obtained the same Ψ as untreated controls, only measured after the two days (Figure 1 A). Furthermore, respiration rates of these samples were not enhanced relative to that of the controls (24.0 ± 4.8 and 25.2 ± 5.8 mg_{CO_2} h^{-1} $kg_{freshmatter}^{-1}$, respectively). However, while means of water potential were always significantly lower after storage (Figure 1 A) that of dry matter related water content were not (Figure 1 B).

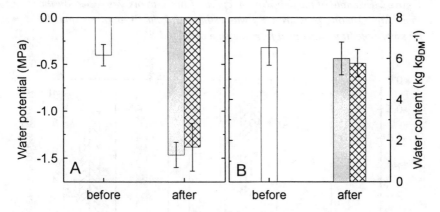

Figure 1 *Influence of storage at 18 °C and high air humidity (98% rH) on water potential (A) and water content (B) of carrots, used for water potential measurements only after storage (hatched columns), and those that had been measured before (open columns) and after storage (dotted columns)*

This discrepancy was due to a high biological variability as carrots were only roughly selected for weight and seize during harvest to mirror the situation in practice but to reduce the effects of different root weights on water loss[4]. Clearly, if the same roots were weighed before and after storage water losses of the individual carrots could be detected. This was not possible if samples were chosen randomly out of the large number of carrots typically used for the experiments.

The higher resolution obtained if water potential was used as an indicator of changes in water relations is easy to understand because Ψ decreases exponentially as tissue water deficit increased (Figure 2). This relationship can be obtained experimentally when changes in fresh matter and water potential were followed during controlled dehydration of the roots. Pressure-volume-analysis[1] also showed that bulk tissue pressure potential became zero only after a water deficit of ca 13% has been reached. This PV-analysis can also provide further valuable information about bulk root water relations[1], not dealt with here. e.

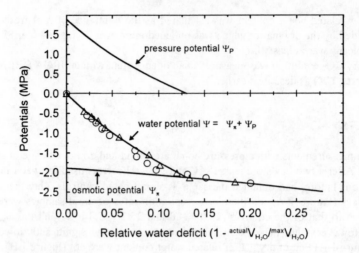

Figure 2 *Example of the dehydration curve of two carrots, harvested on the same day. Osmotic pressure (Ψ_π) was estimated from PV-analysis and the pressure potential calculated from the sum of Ψ and Ψ_π*

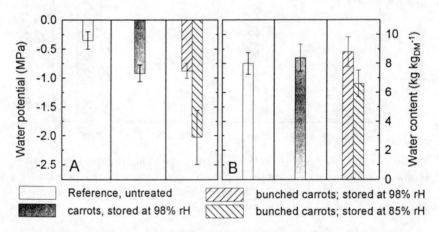

Figure 3 *Water potential (A) and water content (B) of carrots with (bunched carrots) and without leaves that had been stored for two days at 18 °C, and 98% rH or 85% rH, respectively*

Even if experiencing very high air humidity (98-99% rH) carrots suffered considerable decrease in water potential when stored at room temperature for two days (Figure 1 A, 3 A). This was virtually independent of whether the leaves remained attached to the roots (bunched carrots) or were cut off directly after harvest (Figure 3 A). Again, changes in water content could not mirror the strained tissue water status. After a storage at reduced air humidity (85% rH) changes in this parameter were significant (Figure 3 B). However, a relative air humidity of 85% may be a rare situation of high humidity during marketing.

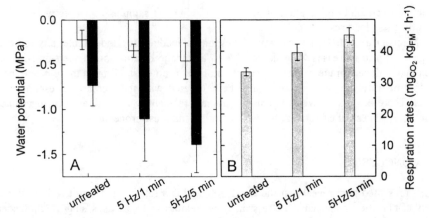

Figure 4 *Water potential (A) and respiration rates (B) of carrots that had been mechanically stressed by shaking in a stress simulator at different frequencies. Water potential was measured before (open columns) and after storage (filled columns)*

In carrots that had been mechanically stressed in a stress simulator[4] by shaking with a frequency of 5 Hz for different times (1 or 5 min), changes in water potential partially correlated with the degree of stress (Figure 4A), while differences in water content could not be resolved (data not shown). Respiration of these roots showed a very distinct response to mechanical stress (Figure 4B). Nevertheless, these results may point out that reduction of water potential can indicate and correlate with changes in physiological activity.

4 DISCUSSION

Reports on water relations in carrots are relatively rare[5,6,7] and mainly deal with the relationships between water potential and tissue toughness[6] and other physical properties[5]. This may be due to the fact that methods currently used for water potential measurements with carrots, i.e. psychrometry[5,6] or vapour equilibration[4,7], are either very time consuming[4,7] or destructive or both[4,5,6]. On the other hand, the pressure bomb provides a well accepted means to accurately and rapidly determine water potential[1]. It has even been successfully used to establish Ψ in stripes of leaf tissue[8]. Our results also prove that it is non-destructive and does not lead to 'unknown effects of repeated pressurisation'[7] if used correctly[1]. Thus it allows repeated measurements on a large number of samples within an acceptable timespan.

Water potentials obtained here with the Scholander bomb on freshly harvested carrots lay well in the range of values previously published[6] and obtained with a psychrometer. The much lower Ψ determined with the vapour equilibrium method[7] on „fresh carrots purchased from a local supermarket" may be due to the fact that these roots have been stored for some time before use. Two days storage at low room temperatures and high air humidity, which may be a realistic situation during marketing, resulted in a significant but still acceptable reduction of water potential and product freshness. However, if air humidity is moderate, e.g. 85% rH, water losses will rapidly lead to a water potential below turgor loss point. This long known fact[9] is trivial because water losses of astomatal tissues are directly related to the water vapour gradient[1]. However, the systematic quantification of the imposed stress by recording the changes in the means of the water content of a large sample size is impeded by the high variability of size and weight of the individual samples if carrots were not selected for these features.

Mechanical stress induces wound responses leading to undesirable physiological changes[10]. It also increases water losses due to a damage or abrasion of the skin surface[11] which may reduce storage capacity. Reduction of water potential is much more pronounced than that of water content and seemed to be correlated with the increase in respiration of carrots. Therefore, tissue water potential may help to identify the influence of mechanical impact on freshness of carrots during harvest and postharvest procedures.

CONCLUSION

Results presented here reveal that the water potential is a reliable and sensitive indicator of effects of climatic or mechanical stress on water relations of carrots. Using a Scholander-type pressure bomb Ψ can rapidly and easily obtained in a large number of roots during an experiment. Furthermore, this method proves to be non-destructive, thus roots can be used for further investigation after initial determination of water potential.

Acknowledgements

The authors wish to thank G. Wegner for excellent technical assistance. Part of this work was supported by a grant of the Deutsche Foschungs Gemeinschaft to M. Geyer.

References

1.   D. J. von Willert, R. Matyssek and W. B. Herppich, Experimentelle Pflanzenökologie, Grundlagen und Anwendungen, Georg Thieme Verlag Stuttgart, 1995.
2.   H. Mempel and M. Geyer. *Proceedings of the Technical Program for the fifth Symposium on Fruit, Nut and Vegetable Production Engineering, Session 2 No. 5*, Davis, CA, USA, 1997.
3.   G. Wormanns, *Bornimer Agrartechnische Berichte* 1996, **8**, 161-182.
4.   S. I. Shibairo, M. K. Upadhyaya and P. M. A. Toivonen, *Sci. Hort.*, 1997, **71**, 1-12.
5.   K. Golacki, *Zesz. Probl. Post. Nauk Rol.*, 1993, **399**, 77-81.
6.   A. McGarry, *Ann. Bot.*, 1995, **75**, 157-163.
7.   J. J. Jobling, B. D. Patterson, S. Moradi, and D. Joyce, *Postharvest Biol. Technol.*, 1997, **10**, 1-8.
8.   W. B. Herppich, B. M-T. Flach, D. J. von Willert and M. Herppich, *Flora*, 1995, **191**, 59-66.
9   J. Apeland and H. Baugerød, *Acta Hort.*, 1971, **20**, 92-96.
10.   L. R. Howard and L. E. Griffin, *J. Food Sci.*, 1993, **58**, 1065-1072.
11.   R. W. Den Outer, *Sci. Hortic.*, 1990, **41**, 201-207.

DEVELOPMENT OF MEALINESS IN APPLES UNDER STORAGE AND SHELF LIFE CONDITIONS.

V. De Smedt[1], Pilar Barreiro[2], Coral Ortiz[2], J. De Baerdemaeker[1] and B. Nicolaï[1]

1. Department of Agro-Engineering and -Economics, Katholieke Universiteit Leuven, Kardinaal Mercierlaan 92, 3001 Heverlee, Belgium
2. Lab. de Propiedades Ficicas de Productos Agricolas, Escuela T.S. de Ingenieros Agronomos, 28040 Madrid, Espana

1 INTRODUCTION

Mealiness is an important quality parameter of fruits, which is characterised by a texture deterioration of the fruits during inappropriate storage resulting in soft, dry fruit. Although the exact mechanism is yet unknown, there are indications that mealiness is related to the ratio of cell rupture to cell separation[1-3]. Earlier work in EU-FAIR-project CT95-0302 resulted in a definition for mealiness as a combination of a lack of crispiness, hardness and juiciness. These three attributes can be measured by performing a confined compression test with a texture analyser. A mealiness scale has been described in the literature combining the three measurements[4]. The objective of this contribution was to evaluate the influence of storage and shelf-life conditions on the development of mealiness.

2 MATERIALS AND METHODS

2.1 Materials

2.1.1 *Storage experiment.* In the storage experiment, the influence of size, picking data and storage time of Cox's Orange and Elstar apples was evaluated. The apples were purchased at the auction of Borgloon (Belgium). The picking date for the Cox's apples was 18/9/1997 for the early ones and 26/9/97 for the late ones. The picking date for the Elstar apples was 15/9/1997 for the early ones and 24/9/97 for the late ones. The diameter of the small and large Cox's apples was 70-80 mm and 75-85 mm, respectively; the diameter of the small and large Elstar apples was 70-75 mm and 80-85 mm, respectively. The apples were stored under ULO conditions, each variety at its optimal temperature, 3.5 for Cox's and 1°C for Elstar. The relative humidity in the storage rooms was 80% .The apples were stored for a few months. Four times during the storage period, 10 apples were removed from the storage room and several measurements were performed on the apples. The Cox's apples were measured immediately after being removed from the storage rooms, the Elstar were kept for one week at room temperature and 95 % RH before measurement. The Cox's apples were measured before storage and after 5, 10 and 15 weeks of storage. The Elstar apples were measured before storage and after 6, 16 and 24 weeks of storage.

2.1.2   *Shelf-life experiment.* In this experiment, the shelf-life of Jonagold apples was evaluated in three different shops in the city of Leuven (Belgium). The first shop was a supermarket in which the apples are stored in a cooled room in the middle of the supermarket. In the second shop, also a supermarket, the apples are not stored in a controlled room but in the same room as all the other products in the supermarket. The temperature in the shop is set 21°C in winter and is 5°C lower than the outside temperature in summer. The third shop is a specialised fruit and vegetables retail shop. The temperature of the shop is very variable as a consequence of frequent openings of the front door. In general the temperature is rather low. The fruits stayed in the shops for 18 days. Each 4 days, 20 apples were removed from the shops and several tests were performed on the apples. The temperature and the relative humidity were recorded in the three shops during the whole period with dataloggers (Escort Junior, Tech Innovatort LTD, Auckland, New Zealand).

2.2   Methods

Several mechanical tests were performed on the apples. A confined compression test was carried out with a universal test machine (UTS Testsysteme GmbH, Ulm, Germany) on cylindrical probes of 1.7 cm height and diameter. A maximum deformation of 2.5 mm was applied at a compression velocity of 20 mm/min. The deformation was immediately removed at the same velocity. The probes were confined in a disk, which had a hole with the same size as the probes. A filter paper about the size of the disk was placed beneath it in order to recover the juice extracted during the compression test. Several parameters were measured through this test including maximum force, and the deformation at maximum force. The slope of the graph (Max. force/deformation at max. force) is a measurement of the hardness of the apple. The juice content was measured by means of a filter paper (Albet n°1305 of 77.84 g/m²), which was placed under the sample during compression. The juice content was determined by measuring the area of the spot accumulated in the filter paper. The stiffness was measured by means of the acoustic impulse response technique[5]. The stiffness factor is a measure of the firmness of the fruit. A digital refractometer (PR-101, Palette Series, ATAGO CO., Ltd., Japan) was used to measure the SSC of apple juice. The accuracy of the system is +/- 0.2°Brix. For the storage experiment only, the pH of the apple juice was measured with a pH-measurement system (Orion, Ag/AgCl 91 series pH electrodes, model 91-02sc, Boston, USA). A mealiness index was calculated based on the hardness and juiciness of the apple. The index was proved to be useful to describe mealiness in Starking apples[4] but preliminary experiments indicated that it may not be applicable to Cox's and Elstar apples. As a clear relationship between mealiness and lack of juiciness and hardness also has been seen in other varieties, the latter two parameters will be used as a mealiness indicator in this contribution.

3 RESULTS AND DISCUSSION

3.1   Storage Experiment

The first measurements were performed at the beginning of the experiment and give an idea about the differences in ripeness of the apples. In general the differences between the different groups were not very high. For the Elstar apples, the late picked apples (12.6 % Brix) were significantly riper then the early picked apples (11.6 % Brix) Only for the late picked apples there was a difference in ripeness between the large and small apples (brix value: 13.3 comparing to 11.5-12 for the other groups). The early picked Cox's apples (pH 3.36) were less ripe than the late picked ones (pH 3.54). Also large apples (pH 3.5) were riper than small apples (pH 3.4). In general the Cox's apples deteriorate very fast. Especially between week 0 and week 5, all three parameters: firmness, hardness and juiciness decrease considerably. After two months of storage the large Cox's apples were of such a low quality that they were removed from storage. Firmness, hardness and juiciness were at the beginning higher for the early picked small apples than for the other apples. After 4 months of storage the differences were not significant anymore. Juiciness was very low in the large apples already at the start of the experiment indicating the low quality of the batch. Also for the small, late picked apples the juiciness was low from the start. Combining the attributes hardness and juiciness to evaluate the degree of mealiness, only the small, early picked apples were significantly less mealy than the other apples but after a few months of storage this difference was minimised. The evolution of hardness and juiciness is shown in Figure 1.

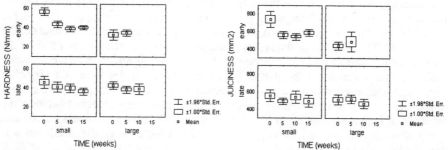

Figure 1 *Evolution of hardness and juiciness as a function of time, picking date and apple size for Cox's apples*

For the Elstar apples, the decrease of the different parameters is more gradual than in the Cox's apples, indicating the better storage potential of the apples. The firmness of the apples decreases gradually for all the groups. The variance within one group is very high, especially at the end of the storage period. The hardness of the early, small apples is higher in the beginning of the experiment but it decreases very fast in storage. The hardness of the late apples is already very low in the beginning indicating a low quality. The juice content of all the batches decreases gradually, but also here the variability within one batch is very high. Combining the factor hardness and juice content to evaluate the degree of mealiness, all batches become more mealy in storage and no significant differences were observed between the different groups. The evolution of hardness and juiciness is shown in Figure 2.

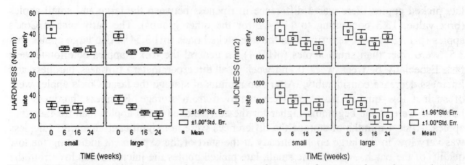

Figure 2 *Evolution of hardness and juiciness as a function of time, picking date and apple size for Elstar apples*

3.2   Shelf-life Experiment

Figure 3 gives an overview of the temperature and relative humidity conditions in the different shops. Shop 1 and 3 are rather similar, while in shop 2 the temperature is higher and the RH lower than in the other two shops. This had a clear effect on the firmness of the apples (Fig. 4). The firmness of the apples in the second shop decreased faster than in the other two shops. Tables 1 gives an overview of the effect of time and shop on the evolution of hardness and juiciness. At the start of the experiment the apples were of a very high quality (high juiciness and relatively high hardness). It can be seen that only juice content changes significantly in time and no significant differences were observed between the three shops. A high temperature and a high relative humidity induce mealiness but because the only shop (shop2) with a relatively high temperature has a lower RH, mealiness was not induced more than in the other shops.

Table 1 *Anova-table for the shelf-life experiment*

Factor	Measurement		
	Firmness	Hardness	Juiciness
Shop	15.82	not significant	not significant
Time	94.69	not significant	23.54
Shop / Time	1.91	not significant	not significant

4   CONCLUSION

The size and picking date of Elstar and Cox's Orange apples influence the development of mealiness in commercial storage. When the overall quality of the apples is low before storage, mealiness will develop very fast. The Jonagold apples were very juicy and hard before being put in shelf-life conditions. In none of the three shops apples became very mealy, and in general the apples were still good after 18 days. The apples in the second shop were less firm than in the other two shops.

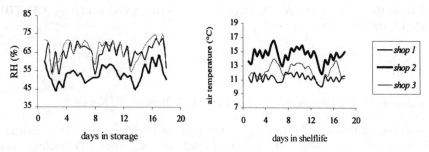

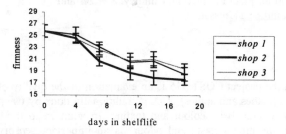

Figure 3 *RH and temperature in the three different shops*

Figure 4 *Evolution of firmness in shelf-life conditions in three different shops*

5 ACKNOWLEDGEMENTS

The financial support from the E.U. (Fair project-CT95-0302) and the Flemish Minister of Science and Technology is gratefully acknowledged. B. Nicolaï is Postdoctoral Fellow with the Flemish Fund for Scientific Research (F.W.O. Vlaanderen)

6 REFERENCES

1. R. Ben-Arie and N. Kislev, *Plant Phys*, 1979, **64**, 197.
2. V De Smedt, Submitted for publication in '*Postharvest Biol. Technol.*', 1998
3. F.R. Harker and I.C. Hallett, *Hort.Sci.*, 1992, **27**(12), 1291.
4. P. Barreiro and C. Ortiz, Second Year Report of project Fair project-CT95-0302 Mealiness in Fruits: Consumer Perception and Means for Detection, Universidad de Polytechnic de Madrid, 1997.
5. J.J. Langenakens, X. Vandewalle and J. De Baerdemaeker, *J. Agric. Eng. Res.*, 1997, **66**, 41.

APPLICATION OF A NEW CONCEPT FOR THE EVALUATION OF THE QUALITY OF FRUITS

R. Azodanlou[1,2]*, Ch. Darbellay[3], J.L. Luisier[2], J.C. Villettaz[2], R. Amadò[1]

1) Swiss Federal Institute of Technology, Institute of Food Science, Zurich, Switzerland
2) Institute of Technology of Valais, Department of Food Technology and Biotechnology, Sion, Switzerland
3) Swiss Federal Research Station for Plant Production, Conthey, Switzerland
   *To whom correspondence should be addressed

1 INTRODUCTION

The aim of our work within the project COST 915 is an evaluation of the quality of fruits (strawberries, raspberries, tomatoes and apples). This is traditionally done by using different criteria such as acidity, degree Brix, colour and texture. In addition to these physico-chemical measurements, the aroma (taste and odour) is an important sensory property of food which is perceived as a quality parameter. Aroma is caused by the interaction of sensory organs with volatiles and semivolatiles associated with the food matrix. It is a challenge for analytical chemists to study and define the complex mixtures of volatile food aroma compounds. In the course of our work, we developed a simple and suitable method for the evaluation of total aroma of foods (this part of work will be published elsewhere[1]). This method gives us the possibility to investigate the heterogeneity of fruits (tomatoes, apples and strawberries) with regard to their volatiles and to determine the ripeness of strawberries by measuring the quantity of their volatiles.

1.1 Aroma analysis

In aroma analysis, sample preparation usually involves concentrating the analytes of interest, using headspace, purge and trap, liquid-liquid extraction, solid phase extraction, or simultaneous distillation/extraction techniques. These methods have various drawbacks, including long preparation time and the use of organic solvents. The newly developed 'Solid phase microextraction', SPME, eliminates most of the drawbacks of the sample preparation. SPME is a technique of headspace sampling which concentrates analytes by adsorption onto a polymer-coated silica fibre prior to thermal desorption in the injection port of a gas chromatograph[2, 3, 4, 5].

1.2 Our concept for aroma analysis

The system we propose is a hybrid between electronic nose and SPME-GC. The idea consists of trapping the volatiles on a SPME fibre and determining the total amount of the adsorbed substances after desorption without separation in a gas chromatography detector. The feasibility of the method was proved by several applications with strawberries. For this work the optimised sampling parameters described previously by Azodanlou et al.[1] were used.

2 MATERIALS AND METHODS

2.1 Sample preparation for extraction of the volatiles

Two varieties of strawberries (*Tango* and *Evita*) were obtained from the Swiss Federal Research Station for Plant Production (Conthey, Switzerland) in autumn 1997. *Marmolada* and *Elvira*, Italian varieties, were purchased at the local market (Placette, Sion, Switzerland) in spring 1998.

The tomatoes (*Sombrero*) were also purchased from a local food store (Placette) and the apples (*Maigold*) were obtained from the Valais Cantonal Station for Fruits Production (Chateauneuf, Switzerland).

For the estimation of the heterogeneity of strawberries (*Tango* and *Elvira*), intact individual fruits of an approximate weight of 10 g were placed in a 150 ml round-necked flask with a wide opening (NS/45) at 25 °C. The sample equilibration time was 15 min and the sampling time by SPME (PDMS, 100 μm) was 5 min.

In order to investigate heterogeneity the number of fruits was increased steadily. Strawberries were placed in a tightly closed 2 l round-necked flask, apples and tomatoes were put in a 6 l flask. Sample equilibration time was 5 min. The sampling time for absorption was 5 min, each sample was analysed five times. All the fibres described below were used.

For the measurements of maturity, 200 g of strawberries (*Evita*) were placed in a tightly closed 1 l round-necked flask. The same analytical conditions as above were applied, except for sample equilibration time which was 30 min.

2.2 Analytical procedure

SPME was performed as described for gas chromatographic separations. The volatiles were collected by inserting the needle through a Teflon coated silicone septum into the headspace of the sample flask. After a defined sampling time (see above) the adsorbed substances were desorbed in the injection port of a gas chromatograph Carlo Erba HRGC 5300 (Carlo Erba S.P.A., Milano, Italy). The splitless injector port was coupled directly to the flame ionisation detector using a transfer tubing (20 cm in length, 0.1 mm i.d., N°160-2630, J&W, New Brighton, MN, USA), with a helium pressure of 150 kPa, at a flow of approximately 5 ml/min. The oven temperature was 250°C. The injection port and detector temperatures were 200 and 250°C, respectively. For quantification, the area under the peak was measured by a gas chromatographic integration program Chrom Card (Fisons Inst., Milan, Italy) in μVs.

The following SPME fibre types were used: poly(dimethylsiloxane) 100μm (PDMS) (Cat No. 5-7300-U); polyacrylate 85μm (Cat. No. 5-7304); porous fibres Carbowax/divinylbenzene 65μm (CW/DVB; Cat. No. 5-7312); bi-polar fibres: PDMS/DVB 65μm (Cat. No. 5-7310-U) and the new fibre Carboxen-PDMS 75μm (Cat. No. 5-7318), all available from Supelco Co., Bellefonte, PA, USA.

3 RESULTS AND DISCUSSION

From our preceding work (see ref 1) we knew that the measurements were reproducible. So we could investigate the differences between individual fruits. As a measure for the heterogeneity of a batch we took the %RSD of five batches.

3.1 Heterogeneity with respect to variety or the type of production

Interesting results were obtained when individual strawberries were analysed. As shown in Table 1, larger differences between individual strawberries of *Tango* variety were found than for the fruits of the *Elvira* varieties, although the sample weights and sizes were approximately equal. The quotient of signal area to weight is strongly different and no correlation between the weight of fruit and signal area could be found. These differences can be explained by the type of production and the harvest season of the strawberries. One variety of strawberries (*Tango*) were produced in a field crop in autumn, whereas the other (*Elvira*) were grown in a glass-house at the end of winter. Even if these strong differences between fruits and between varieties can be influenced from other parameters such as degree Brix or pH, the measurement of those differences will be invaluable for sensory analysis.

	Tango variety			*Elvira* variety		
	Weight (W)	Peak area (P)	P/W	Weight (W)	Peak area (P)	P/W
	10.10	70016	6932.3	11.4	16728	1467.4
	10.33	46320	4484.0	10.52	16820	1598.9
	9.13	91711	10045.0	10.98	15751	1434.5
	10.51	21152	2012.6	11.32	20276	1791.2
	10.12	49966	4937.4	10.01	15857	1584.1
	9.72	49727	5115.9	10.09	18865	1869.7
Average	9.98	54815.3	5587.9	10.72	17382.7	1624.3
SD	0.49	23862.7	2695.2	0.61	1805.6	173.8
%RSD	4.9%	43.50%	48.2%	5.6%	10.40%	10.7%

Table 1 *Aroma releases of individual strawberries*

3.2 Determining the appropriate sample size of fruits

The first results on the heterogeneity of batches leads us to investigate the appropriate sample size to characterise a batch. Enlarging sample size improves the accuracy of quantitative measurements. Using only different fruits, we made five measurements of each batch with an increasing number of fruits from one to ten. The measurements of total volatiles were performed with all available SPME fibres. As a measure of the accuracy of the measurements we used the variance or %RSD as shown for strawberries as an example in Table 2.

Number of Fruits	Weight (g)	Peak area	Specific Area	%RDS on Area	%RSD on weight
1	31.57	13486	427.2	25.6	3.3
2	59.64	20145	337.8	15.5	2.8
3	94.89	24235	255.4	9.4	1.3
4	134.92	23799	176.4	13.1	1.3
5	164.60	30168	183.3	13.3	1.1
6	182.05	28971	159.1	10.8	0.5
7	201.26	31916	158.6	10.2	1.2
8	214.86	31648	147.3	10.5	0.7
9	249.95	33513	134.1	6.6	0.5
10	284.05	38354	135.0	5.2	1.0

Table 2 *Batch size and accuracy : strawberries on CW-PDMS fibre*

With strawberries, we obtain a variance of about 20% with individual fruits per batch and 5.2% with 10 fruits per batch, respectively. Figure 1 gives graphically the development of the %RSD with increasing the number of fruits measured. As we can see, the minimum batch for a correct measurement should be about ten fruits for strawberries. With other fibre type we obtained very similar results.

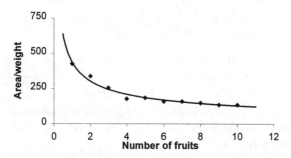

Figure 1 *%RSD as a function of batch size; strawberries; CW-PDMS fibre*

Another interesting result of these measurements is the fact that the specific area per mass is decreasing with an higher number of fruits, being almost constant at/above six fruits (Figure 2). This could arise from geometric factors either of fruits or of the measuring device.

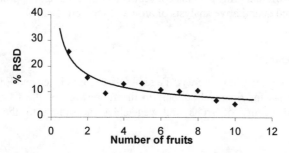

Figure 2 *Evolution of the specific area with the number of fruits*

For apples (*Maigold*) variety or tomatoes (*Sombrero*), we obtained very similar results, but with a much smaller variability. 3 apples are necessary to reach a %RSD of about 4%, whereas with one fruit it was 15%; tomatoes exhibit the smallest heterogeneity between individual fruits, giving %RSD of about 11% for one, and 3.25% for 3 fruits, respectively.

3.3 Maturity of the strawberries

With strawberries, we were able to document the differences in total volatiles between ripe and unripe fruits. We selected one batch of ripe fruits and one of unripe fruits by their colour. The total volatiles measured by peak area is notably higher when the strawberries are ripe. This technique permits to follow the ripening of strawberries (Figure 3).

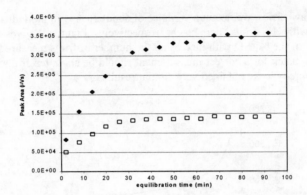

Figure 3 *Effect of maturity on the signal of total volatiles of strawberries*
The symbols denote values measured with SPME (PDMS): ◆ : *ripe fruits,*
□ : *unripe fruits.*

4 CONCLUSION

This method is convenient for measuring the intensity of aroma in fruit samples. It does not only allow to differentiate between fruits but gives also the possibilities to describe some properties such as maturity, variety, and harvest time. Heterogeneity exists between individual fruits, it requires a large sample size, which has to be determined for each fruit species, for a correct quantification of the total volatiles. The use of several SPME fibres permits to obtain different profiles. This concept is a convenient and appropriate sampling technology for rapid quantitative analysis of aroma of food products.

Acknowledgement

The authors are grateful to the Swiss Federal Office for Science and Education, and to the Canton of Valais for financial support in the framework of the COST 915 project ("Improvement of quality of fruits and vegetables, according to the needs of the consumers").

References

1. R. Azodanlou, Ch. Darbellay, J.L. Luisier, J.C. Villettaz , R. Amadò, *Z. Lebensm. Unters Forsch.* (submitted).
2. R. P. Bellardi, J. Pawliszyn, *Water Pollut. Res. J. Can.*, 1989, **24,** 179.
3. C. L. Arthur, J. Pawliszyn, *Anal. Chem.*, 1990, **62,** 2145.
4. J. Pawliszyn, *Trends Anal. Chem.,* 1995, **14,** 113.
5. B. D. Page, G. Lacroix, *J. Chromatogr.,* 1993, **648,** 190.

EVALUATION OF FLAVOUR POTENTIAL OF DIFFERENT TISSUES FROM ONION (*ALLIUM CEPA* L.)

J.R. Bacon, G.K. Moates, A. Ng, M.J.C. Rhodes, A.C. Smith and K.W. Waldron

Institute of Food Research
Norwich Research Park
Colney
Norwich NR4 7UA
United Kingdon

1 INTRODUCTION

Onion processing in the European Union produces in the region of 450,000 tonnes of onion waste per annum. The introduction of landfill taxation and environmental concerns has encouraged research into the exploitation of this waste for use as a source of food ingredients. This research focuses on the flavour potential of the different onion tissue types within this waste and the influence of bulb storage on flavour levels.

Onion flavour results after tissue disruption by the catabolism of endogenous S-alk(en)yl-L-cysteine sulphoxide flavour precursors by the enzyme, alliinase, to produce pyruvate, ammonia and a range of both volatile and non-volatile sulphur compounds giving the characteristic odour and flavour of onions. Such flavour volatiles are traditionally extracted from tissues as onion oil by steam distillation.

2 MATERIALS AND METHODS

Samples of onion bulbs of two different cultivars (Hysam and Grano de Oro) were taken immediately after curing and after six months subsequent storage at 0°C and dissected into four tissue types as in commercial peeling operations - top/bottom sections, outer leaf bases, inner leaf bases and the dry, brown skin. Each section was extracted and assayed by HPLC for flavour precursors, the S-alk(en)yl-L-cysteine sulphoxides. The amounts of pyruvate produced by total disruption of the same tissues was also determined and compared with the flavour precursor levels. The level of enzymatically produced pyruvate is generally accepted to have a close relationship with perception of pungency (1).

3 RESULTS AND DISCUSSION

3.1 Effect of Tissue Types on Flavour Precursor and Pyruvate Levels

Most of the precursors are present as *trans*-(+)-S-(1-propenyl)-L-cysteine sulphoxide (1-PECSO) with a small amount of S-methyl-L-cysteine sulphoxide (MCSO). (+)-S-propyl-L-cysteine sulphoxide was not observed in these extracts. On a fresh weight basis, both varieties exhibited the same distribution with the top and bottom sections containing the highest proportion of flavour precursors (Fig. 1). Similar trends between tissue types were observed for pyruvate levels. The top and bottom sections therefore represent a potentially good source of onion oil.

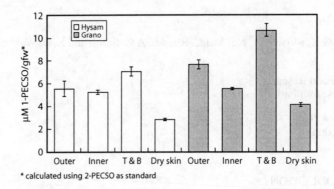

Figure 1. *Effect of variety and tissue type on flavour potential*

3.2 Effect of Variety on Flavour Precursor and Pyruvate Levels

Grano de Oro exhibited higher levels of both flavour precursors and pyruvate than Hysam. However, it should be noted that the varieties were grown at different locations with no control on the soil sulphur fertility levels and this may have a significant influence on these observed differences (2).

3.3 Effect of Storage on Flavour Precursor and Pyruvate Levels

After storage for six months at 0°C, both varieties show significant increases in levels of 1-PECSO in the inner tissues and the top/bottom sections (Fig. 2). Levels of 1-PECSO in the outer tissues and the brown skin show no significant changes. Similarly, significant increases in levels of enzymatically-produced pyruvate in the inner tissues and the top/bottom sections were observed. This would seem to suggest that, since there is no significant water loss, the onions develop a more intense flavour upon storage.

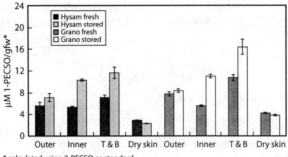

* calculated using 2-PECSO as standard

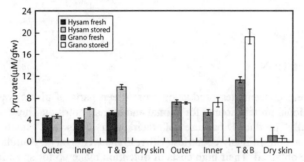

Figure 2. *Effect of storage on flavour potential*

4 CONCLUSIONS

(i)   The amount of pyruvate produced upon tissue disruption and the level of flavour precursors (S-alk(en)yl-L-cysteine sulphoxides) in the tissue types studied show similar distributions.

(ii)   The top / bottom sections of the onion contain the highest levels of flavour precursors - these sections may be a good source of onion oil.

(iii)   Levels of flavour precursors in the top / bottom sections and the inner tissues show a significant increase upon storage of onion bulbs for six months at 0°C.

(iv)   The increase in levels of the sulphur-containing flavour precursors must be as a result of a rearrangement of the total sulphur within the onion bulb to form the S-alk(en)yl-L-cysteine sulphoxides. Total sulphur is reported as unchanged on storage.

5 ACKNOWLEDGEMENT

The authors would like to thank the UK Biotechnology and Biological Sciences Research Council and the European Union (Contract No. FAIR-CT96-1184) for financial support.

References

1. S. Schwimmer and W.J. Weston, *J. Agric. Food Chem.* 1961,**9**, 301.
2. W.M. Randle, J.E. Lancaster, M.L. Shaw, K.H. Sutton, R.L. Hay, M.L. Bussard, *J. Amer. Soc. Hort. Sci.*, 1995. **120**, 6, 1075.

ANTIOXIDATIVE EFFICIENCY OF CHLOROPLAST CONCENTRATES FLOCCULATED FROM LUCERNE JUICE CHEMICALLY MODIFIED

B. Baraniak, M. Paczosa and U. Gawlik-Dziki

Department of Biochemistry and Food Chemistry
Agricultural Academy
Lublin, Poland

1 INTRODUCTION

Process of protein concentrates production from green parts of plants was elaborated to obtain the additional protein source for animal and human nutrition.These preparations are not commercially produced because of their high costs and small efficiency of cytoplasmatic fraction. In many countries (USA, France, Hungary, Australia) chloroplastic concentrates are obtained. Their high costs of obtention make up for specific physiological interactions. Researches carried out in Japan and Australia[1] proved hypocholosterolemic activity of commercial preparations Quintessence® obtained using thermal method from alfalfa juice. The contribution of concentrates in chicken's diet lowered the cholesterol content by 27.5% and triacyloglycerols by 34.3% in their plasma. Authors cited before left the traditional name for the preparations - LPC (leaf protein concentrate) - giving the new name for those obtained using the same technology - PLEX (pigmented leaf extracts). Such name has to underline the physiological meaning of compounds co-precipitated with proteins, especially the natural plant dyes. These compounds have the ability to neutralise the free radicals and thus they are of a great importance for prevention and treatment of many diseases. Their meaning in the struggle against cancer consists of preventing the premutagenic changes of DNA.[2] Hence, many authors have been focused on testing the antioxidative potential of nutrients of plant origin lately.

The purpose of the research was to study the effect of chemical modification of juice from alfalfa on antioxidative properties of chloroplastic concentrates obtained during juice fractionation using anionic flocculant Magnafloc LT-26.

2 MATERIAL AND METHODS

Green parts of alfalfa (*var.Kleszczewska*) taken at the stage before flowering were material for studies. Forage was ground in a screw press and then subjected to pressing in expeller. Fibrous material was removed from the juice using the sieve of 0.15mm = 100mesh.

The following substances were added into the juice: Ag^+, Hg^{2+} (0.5mM/dm^3), acetaldehyde (1ml/dm^3), simazine (1mg/dm^3), filtrate after thermal (55°C) proteins fractionation and using flocculant (Magnafloc LT-26) in volume ratio 1:10 as well as the juice extracted from buckwheat in volume ratio 1: 4. Chloroplastic concentrates were precipitated (at room temperature) using anionic flocculant Magnafloc L-26 in amount of $300mg/dm^3$ juice.

Precipitated chloroplastic concentrates were rotated (4000 rpm), twice washed with distilled water, dried at 50°C and after grinding stored at 6-8°C.

Substances with potential antioxidative properties were extracted using 0.4g concentrate samples and 80% acetone. Antioxidative properties using linolenic acid were estimated according to Linghert's et al. method,[3] and chlorophylls content - spectrophotometrically according to Arnon.[4]

3 RESULTS AND DISCUSSION

It is possible to prevent the leaf extracts due to modification of pressed juice against unadvantageous changes: carotenoids oxidising[5] or proteolytic enzymes activity.[6] Inactivation of proteolytic enzymes is a very important problem because peptides and amino acids liberated then do not coagulate. Protein losses due to its hydrolysis can even reach 50%.[7]

Lucerne juice modification using chosen substances caused - compared to control flocculated concentrated from the juice with no additives - both changes in obtained antioxidative potential values and changes of their dependence on the time of incubation process (table 1).

Table 1 *Antioxidative effect (0.1mg/ml linoleic acid) of chloroplastic concentrates flocculated from lucerne juice modified by chosen substances*

	Time of incubation (h)		
Substance	12	24	36
Ag^+	0.74	0.93	0.63
Hg^{++}	0.68	0.88	0.50
Acetaldehyde	0.56	0.70	0.38
Simazine	0.61	0.78	0.38
Filtrate from fractionation lucerne juice by:			
Flocculant Magnafloc LT-26	0.76	0.75	0.49
Thermal (55° C)	0.79	0.84	0.50
Buckwheat juice	0.76	0.90	0.46
Control	0.66	0.66	0.35

Chemicals introduced into the juice positively affected the change of antioxidative potential of flocculated preparations. In all preparations studied - except from that precipitated in the presence of filtrate obtained during proteins fractionation of alfalfa juice using flocculant Magnafloc LT-26 - the highest potential was achieved after 24 hours. Concentrate flocculated in the presence of silver and mercury ions showed its highest value, that with acetaldehyde addition - the lowest.

For modification of lucerne juice due to introduction of silver and mercury ions as well as buckwheat juice, potential obtained after 24-hour incubation was higher then that for concentrate coagulated in the presence of acetaldehyde.

The influence of buckwheat juice is particularly important because its addition regulates the activity of proteolytic enzymes and significantly improves the filtration properties of protein-vitamin precipitates.[6]

During extraction of compounds of antioxidative character using 80% acetone, chlorophylls were isolated from concentrates as well. Chlorophylls are magnesoporphyrins, the most common plant pigments that take part in the photosynthesis processes. In higher plants two forms of chlorophylls are present - „a" and „b" - differing with substituent at C-3 carbon. They are stable only in natural form in which they occur in chloroplasts - bonded to proteins, phospholipids or other components. The disturbance of native structures during technological processes (heating, grinding, solvents action) causes the colour changes resulting from chlorophyll transformations.[8] Chlorophylls content in protein concentrates can be modified by introducing chosen chemicals into ground material[9] or extracted juice. It also depends on the species of plant and method of concentrate coagulation.[6] Chlorophyll damage is promoted by elevated temperature, enzymes action (chlorophylases, lypooxygenases and lypases), oxygen, light and pH changes. In weak acidic environments due to exchange of magnesium into two hydrogen atoms pheophytin is formed. Both chlorophyll and pheophytin accelerate lipids oxidising on light but with no light they are inhibitors of lipid autooxidation processes.[10]

Obtained acetone extracts differed with chlorophylls contents: the greatest amount of pigments were isolated from the preparation precipitated from juice with the filtrate after thermal fractionation of lucerne proteins; the lowest - in preparation flocculated in the presence of acetaldehyde (table 2). The percentage of chlorophyll „a" in its total amount was the highest for preparation coagulated from the juice with the addition of filtrate after alfalfa proteins fractionation using flocculant Magnafloc LT-26. In all studied concentrates, the percentage of extracted chlorophyll „a" using 80% acetone was higher than in control.

Table 2 *The chlorophyll content (mg/kg d.m.) in acetone (80%) extracts of chloroplastic concentrates flocculated from lucerne juice modified by chosen substances*

Substance	Chloro- phyll „a"	Chloro- phyll „b"	Total chloro- phyll	Percent of chloro- phyll „a" in total	Amount ratio of chloro- phyll „a":"b"
Ag+	1.22	0.14	1.36	90	9:1
Hg++	0.88	0.12	1.00	88	7:1
Acetaldehyde	1.71	0.22	1.93	89	4:1
Simazine	0.87	0.11	0.98	89	8:1
Filtrate from fractionation lucerne juice by:					
Flocculant Magnafloc LT-26	1.13	0.11	1.24	91	10:1
Thermal (55° C)	2.45	1.09	3.54	69	2:1
Buckwheat juice	1.08	0.25	1.33	81	4:1
Control	0.90	0.43	1.33	68	2:1

Results obtained in the research point out that the ratio of both chlorophyll forms have the greater effect on the value of anti-oxidative potential than their content in extract. The more difference between the value and the ratio characteristic for higher plants (3:1), the greater anti-oxidative potential was shown by studied extracts (table 2). The time required for achieving the maximum anti-oxidative potential depended on the method applied for chloroplastic concentrates precipitation and on the kind of substance modifying the properties of juice extracted from alfalfa.

4 CONCLUSIONS

Concentrates flocculated from chemically modified alfalfa juice are characterised by stronger antioxidative properties than that flocculated from non-modified juice.

The value of anti-oxidative potential of protein concentrates depends to the higher extent on the relative ratio of chlorophylls „a" and „b" than on the total amount of these pigments.

References

1. H. T. Ostrowski-Meissner, Proc. World Conf. Animal Prod., pp.394, 1993.
2. P. A.Lachance, „ Food phytochemicals for cancer prevention", V.1, American Chemical Soc., Washington, D.C., 1994.
3. A. Lingnert, K. V. Vallentin, C. E. Eriksson, J. Food Proc. Preserv. 1979, 3, 87.
4. M. J. Arnon, „Chemistry and biochemistry of plant pigments", Academic Press. London, 1960.
5. D. V. Layung, M. Ohshima, H. T. Ostrowski-Meissner, H. Yokota, J. Japan. Grassl.Sci.,1995, 40, 4, 410.
6. B. Baraniak, „ Studia nad optymalizacją procesu wyodrębniania frakcji białkowych chloroplastycznych i cytoplazmatycznych z zielonych części roślin uprawnych" WAR Lublin, 1994 (in Polish).
7. J. W. Finley, C. Pallavicini, G.O. Kohler, J. Sci. Food Agric., 1980, 31, 156.
8. J. Wilska-Jeszka, Barwniki in: „Chemiczne i funkcjonalne właściwości składników żywności", WNT, Warszawa 1994 pp. 407 (in Polish).
9. B. Baraniak, M. Bubicz, M. Niezabitowska, Rocz. Nauk Rol., 1990, s.B, 106, 3/4, 31 (in Polish).
10.R. A. Larson, Phytochemistry, 1988, 27, 4, 969.

THE INFLUENCE OF TEMPERATURE ON ACTIVITY OF PROTEOLYTIC ENZYMES OF ALBUMIN FRACTION FROM GREEN PEA

B. Baraniak, K. Zienkowska and M. Karaś

Department of Biochemistry and Food Chemistry
Agricultural Academy
Lublin, Poland

1 INTRODUCTION

In several past years, special attention has been paid to the increase of a natural plant products contribution into human diet because of physiological interaction of many bioactive substances occurring in them.

Legumes are mostly annual species cultivated mainly for seeds and green forage. Seeds are directly applied in human and animal nutrition and they are used in various food-stuff and processing industry branches. Species typically edible are: bean, pea, horse bean, soya bean .

To increase the consumption of these valuable plant raw materials, a wide range of genetic, agricultural and technological studies are conducted in order to improve their nutritional values.

In Poland, pea seeds that are particularly appreciated due to high protein content are most often eaten. Proteins are the main component of protoplast and along with reservative and structural functions they play enzymatic roles that are of significant importance in seed physiological conversions. In pea seeds, similarly to other legumes, main proteins are albumins and globulins. Globulins are the major fraction and they play role as reservative proteins. Their content amounts to 66% total protein and their structure and composition is well known.[1]

The other main protein fraction is albumins that play structural and enzymatic roles in legumes. They are proteins showing high heterogenicity.[2] Percentage of albumins in pea's proteins mainly depends on its variety and cultivation conditions. All above mentioned conditions also determine the functional properties of pea's albumins.[3] This fraction properties are less known, thus the purpose of the research is to study the influence of thermal processing of green pea on activity of proteolytic enzymes connected to albumin fraction.

2 MATERIAL AND METHODS

Green pea *(var.Telefon)* recommended by seed producer as particularly valuable cultivar for processing and freezing industries was used for studies. Harvest was done in 80-85 days after sowing selecting seeds of mean size (8.5-10 mm). Selected seeds were divided into two portions: the first was subjected to the thermal processing as a fresh material; the second was frozen and subjected to the same processes after 2-week storage.

Seeds of a fresh and frozen pea were blanched at 90°C in 2, 4 or 6min. The same periods were applied to heating products in microwave cooker (Philips/Whirlpool model 606) using three power ranges: 0.5, 0.75 and 0.9kW at 2.450MHz.

Albumin fraction was extracted for half an hour and activity of proteolytic enzymes was recorded in filtrate using azoalbumin as a substrate.[4] Azoalbumin hydrolysis reaction lasted for 4 hours.

Proteolytic enzymes activity was expressed in units per 1g of sample accepting the absorbance increase by 0.1 during 4 hours as a unit.

3 RESULTS AND DISCUSSION

Pea's proteins are important components in animal and human nutrition. The biological value of seeds decreases during industrial processing[5] and the presence of anti-nutritional compounds does it as well. Among the latter there are protease inhibitors that are connected - along with enzymatic proteins - to albumin fraction. Hence, most of the studies refer to protease inhibitors of legume seeds inactivation occurring during technological processes.[6-8]

Little attention is paid to proteolytic enzymes inactivation although their action causes the modifications in protein properties and leads to active peptides formation that can affect the protein biological value.[9]

In our studies, proteolytic enzymes activity in frozen green pea was higher than in fresh pea (9.1 unit/g and 7.7 unit/g, respectively) (Table 1).

Table 1 *Influence of blanching on activity of proteolytic enzymes (unit/g fresh matter) of green pea seeds.*

		Time (minutes) of blanching		
	0	2	4	6
Fresh	7.7	6.8	5.2	3.5
Frozen	9.1	5.3	4.9	4.6

Obtained results prove high resistance of pea's proteolytic enzymes to temperature action - 6-minute blanching reduced their activity by over 50% in fresh seeds (Table 1). Analogous changes were obtained for microwave heating (0.9 kW) within 2 min or using 0.5 kW for 6 min. (Table 2).   In frozen product the inactivation was faster (Table 2) which can prove the enzymatic proteins structural changes taking place during freezing. The total protease inactivation in frozen material after 6-minute radiation of 0.9 kW can result from the damage of enzymatic molecules conformations. Non-ionizing electromagnetic radiation - microwaves - invokes in material it passes through fast dipoles re-orientation, hydrogen bonds splitting and molecular friction.[10] In our studies the changes of protease inhibitors activities were not recorded but on the basis of Yoshida's et al. research[11] their co-operation in protease inactivation can be excluded. Cited authors found that soya seeds heated in microwave cooker for 4 minutes using 0.5 kW resulted in total inactivation of presented protease inhibitors.

Table 2 *Influence of microwave treatment on activity of proteolytic enzymes (unit/g fresh matter) of green pea seeds.*

	Fresh			Frozen		
	0.5 kW	0.75 kW	0.9 kW	0.5 kW	0.75 kW	0.9 kW
2 min	6.0	4.8	3.8	7.7	5.2	0.8
4 min	5.4	3.0	2.1	7.3	3.5	0.7
6 min	3.9	2.5	2.1	2.9	1.5	0.2

4 CONCLUSIONS

1. Proteolytic enzymes connected to pea's albumins are slightly inactivated due to short-time action of higher temperatures.
2. The rate of protease activity inhibition increases along with the thermal processing time both due to blanching and in microwave cooker.
3. During pea's microwaving, the activity of proteolytic enzymes decreases along with the microwave power increase.
4. Freezing of the material causes the significant shortening of the time needed for inactivation of proteases connected to green pea's albumins.

References

1. J. Gueguen, J. Barbot , *J. Sci. Food Agric.,* 1998, **42**, 209.
2. C.Gruen, R. E. Guthrie, J. Blagrove, *J. Sci. Food Agric.,* 1987, **41**, 167.
3. B.Czukor, L. Gajzago-Schuster, Z. Cserhalami, „Agri-Food Quality", ed. G.R. Fenwick et al. The Royal Society of Chemistry, 1996, pp.131.
4. H. Thomas, *Planta,* 1978, **142**,161.
5. B. Grześkowiak, Z. Pazoła, *Rocz. AR. Pozn.,***CCXVIII**, 1990, 43. (in Polish)
6. M. Klepacka, M. Piecyk, H. Porzucek, Proc. XXVII Symposium K. T. Ch. Ż.PAN.„Postępy w technologii, przechowalnictwie i ocenie jakości żywności". Szczecin, 1996, (in Polish).
7. J. Sadowska, J. Fornal, A. Ostaszyk, *Pol. J. Food Nutr. Sci.,* 1996, **5/46**, 2, 61.

8. S. Kadm, R.R. Smithard, M. D. Eyre, D. D. Armstrong, *J. Sci. Food Agic.,* 1987, **41,** 267.
9. M. Karmać, R. Amarowicz, H. Kostyra, Proc. XXVII Symposium K. T. Ch. Ż. PAN. „Postępy w technologii, przechowalnictwie i ocenie jakości żywności''. Szczecin, 1996, (in Polish).
10. K. Surówka, *Żywność, Technologia, Jakość,* 1994, **1,** 13. (in Polish).
11. H. Yoshida, G. Kajimoto, *J. Food Sci.,* 1988, **53,** 6, 1756.

SIMPLE IN-LINE POST COLUMN OXIDATION AND DERIVATIZATION FOR THE SIMULTANEOUS ANALYSIS OF ASCORBIC AND DEHYDROASCORBIC ACIDS IN FOODS BY HPLC

Dr. A. Bognár

Federal Research Centre for Nutrition
Institute of Chemistry and Biology
Garbenstr. 13
D-70599 Stuttgart

Dr. H.G. Daood

Central Food Research Institute
Herman Ottó u. 15
H-1022 Budapest

1 INTRODUCTION

Most of the non-simultaneous HPLC procedures for the analysis of ascorbic acid derivatives are based on either the reduction of L-dehydroascorbic acid (LDHAA) to L-ascorbic acid (LAA), or oxidation of LAA and D-isoascorbic acid to LDHAA and IDHAA, and derivatization with 1,2-phenylene diamine (OPDA) to produce fluorometrically measurable 3-(1,2-dihydroxy-ethyl) furo (3,4-b) quinoxaline-1-one [DFQ] (1-4). Simultaneous HPLC determination of LAA and DHAA has been achieved by several methods based on dual wave length UV monitoring, UV and fluorescence detection or in-line post-column oxidation of LAA and IAA to LDHAA and IDHAA by $HgCl_2$ solution, followed by derivatization with OPDA to DFQ-s (5-9). However, these methods have some disadvantages regarding the pumping system and detection, uncontrolled side reaction of $HgCl_2$ and much time required for sample preparation and HPLC-separation.

On the basis of methods described in the literature a new procedure simultaneous determination of the four ascorbic acid derivatives in foods by HPLC with in-line oxidation of LAA and IAA using a short column, packed with activated charcoal has been developed. Post-column derivatization of LDHAA and IDHAA with OPDA was optimized.

2 EXPERIMENTAL

2.1 Reagents and materials. L-ascorbic acid, 1,2-phenylene diamine, ammonium dihydrogen phosphate, metaphosphoric acid, acetic acid, acetonitrile (Merck),
activated charcoal (Sigma); D-isoascorbic acid (Fluka). Foods from a local market.
2.2 Sample preparation. *Solid foods* (e.g. fruit, vegetables, sausages, liver): 5 g to 50 g of the sample were homogenised in the presence of 80 ml of 7.5 % metaphosphoric acid containing 20 % acetic acid solution for 0.5 - 1 min (3). The mixture was poured into a 200 ml graduated flask, filled up with water, mixed and filtered.
Liquid foods (e.g. fruit juices, milk): To 5 g to 50 g of the sample in a 100 ml graduated flask, 40 ml of 7.5 % metaphosphoric acid containing 20 % acetic acid solution were added, filled up with water, shaken and filtered. For HPLC analysis, the filtered sample extract was mixed with acetonitrile (1:1) and filtered by ultrafilter (0.2 μm).
2.3 Standard preparation. Stock solutions of LAA and IAA were prepared by dissolving 100 mg of each acid in 100 ml of 3 % metaphosphoric acid solution containing

3 % acetic acid. To prepare standard solutions of L-DHAA and IDHAA, 50 ml of diluted stock solution (50 μg/ml) were oxidized by shaking thoroughly with 1 g of activated charcoal for 15 sec. It was then filtered. The stock and standard solutions were diluted to different concentrations with metaphosphoric / acetic acid (3 % / 8 %) solvent.

For HPLC-analysis, standard solutions were mixed with acetonitrile (1:1).

2.4 Conditions for HPLC analysis.

Equipment (see Figure 1): 2 HPLC isocratic pumps, autoinjector, fluorimetric detector, integrator, thermostat.

Separation column: stainless steel 250 mm x 4 mm, packed with Grom SIL 120-Amino-2PA, 5 μm. *Oxidation column*: stainless steel 10 mm x 4 mm, packed with activated charcoal (Sigma) at our laboratory; it may also be obtained from Grom (Germany).

Reactor: Tefzel tubing 20 m (1.58 mm OD, 0,3 mm ID). T-connection (PEEK).

Eluent: acetonitrile + 0.05 M (NH_4) H_2PO_4 (75+25 v/v); flow rate: 0.8 ml/min.

Post-column reagent: 0.3 % 1,2-phenylene diamine in 0.15 % metaphosphoric/0.4 % acetic acid solution, buffered with 2.5 M sodium acetate to pH 5.2; flow rate: 0.3 ml/min. Reactor temperature: 70 °C.

Injection volume: 10 - 25 μl (5-200 ng of each AA and DHAA).

Detection: fluorimetric; ex.: 350 nm, em.: 430 nm.

Evaluation: external standard method.

Note: The OPDA reagent, if kept in a brown bottle, is stable for 1 day at room temperature. After completion of one analytical series it is advisable to rinse the reactor tubing and the detector with distilled water. The oxidizing column is capable of oxidizing more than 1000 samples. An index of decreased capacity in the marked peak broadening of LAA and decrease oxidation.

3 RESULTS AND DISCUSSION

The four ascorbic acid derivates were eluated within 10 min with no distinct shift in elution times after several hundred injections.
The retention times were ~5.9 min for LDHAA, ~6.4 min for IDHAA, ~8,0 min for IAA and ~9.4 min for LAA. The peak area increased linearly as a function of quantity in the range of 1 ng to 100 ng ascorbic acid derivatives per injection. The fluorescence activity of IAA and IDHAA was lower than of LAA and LDHAA. Correlation coefficients of the linear regression ranged 0.999 and 1.000, suggesting that the method is useful for sensitive and accurate analysis of vitamin C and isoascorbic acid

Figure 1: *Diagram of the HPLC system*

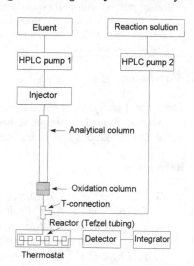

For routine analysis to determine the content of total vitamin C (LAA + LDHAA) and of total isoascorbic acid (IAA + IDHAA), the oxidizing column can be connected before the separation column. In this case, the resulting peaks are LDHAA and IDHAA.

The present method was used to analyse fruit, fruit juices, vegetables, milk, dairy products, sausage and liver (about 25 foods totally). The chromatograms of foods were free from interfering peaks. The recoveries of vitamin C (LAA and LDHAA) and of isoascorbic acid were in the range between 98 % and 104 %. Table 1 shows results of comparative studies of some selected foods. Vitamin C contents determined according to the present method are in the same order of magnitude as the results of fluorimetrical analysis (ISO) and HPLC with pre-column derivatization.

Table 1: *Vitamin C content in selected foods analysed by different analytical methods - mg/100 g food -*

Food	- present method[a) -]			HPLC[b)]	ISO[c)]
	LAA	*LDHAA*	*Vit. C*	*Vit. C*	*Vit. C*
Red pepper	212.0	3.1	215.1	215.3	215.3
Potato	32.0	1.8	33.8	33.1	33.9
Spinach	29.6	2.3	31.9	30.9	31.5
Apple	19.8	2.5	22.3	22.7	23.1
Strawberry	30.0	5.2	35.2	35.9	34.2
Orange juice	36.9	6.8	43.7	43.6	42.4
Pig liver	13.4	1.3	14.7	14.5	14.9
Sausage (Becel)	71.3	2.8	74.1	72.9	72.2
Milk (3.5 % fat)	0.5	0.4	0.9	0.8	0.7

[a)] = present method with in-line post-column oxidation and dervatization; mean value of 3-5 analysis; the variation coefficients were < ± 5%
[b)] = reversed phase HPLC, pre-column oxidation and derivatization; separation as DFQ (3); [c)] = ISO (10);

3.1 Conclusion

The present analytical method is highly specific in determining LAA, LDHAA, IAA and IDHAA in food. It is simple to handle, not susceptible to failure, perfectly suitable for serial determinations and yields reproducible results very well.

References

1.   G. M. Sapers et al., J. Chromatogr., 1990, **503**, 431-436
2.   V. Gokmen and J. Acar, Fruit Processing, 1996, **6**, 198-201
3.   A. J. Speek et al., J. Chromatogr., 1984, **305**, 53-60
4.   A. Bognár, Deutsche Lebensmittel-Rundschau, 1988, **84**, 73-76
5.   L. L. Lloyd et al., Chromatographia, 1987, **24**, 371-376
6.   R. W. Keating and P. R. Haddad, J. Chromatogr., 1982, **245**, 249-255
7.   S. Zapata and J. P. Dufour, J. Food Sci, 1992, **57**, 506-511
8.   J. T. Vanderslice and D. J. Higgs, J. Micronutr. Anal., 1988, **4**, 109-118
9.   B. Kacem et al., J. Agric. Food Chem., 1986, **34**, 271-274
10.   Reference method ISO 6557/1-1986 (E)

ORAL BREAKDOWN OF APPLES IN RELATION TO MECHANICAL PROPERTIES, CELL WALL COMPOSITION AND SENSORY PROPERTIES

W.E. Brown, D. Eves, A. Smith and K. Waldron

Institute of Food Research,
Earley Gate,
Reading, RG6 6BZ, UK.

1 INTRODUCTION

Sensory perception of food texture occurs primarily from sensations derived during oral processing of food and relates to progressive breakdown of the product during mastication. The level and type of activity of the masticatory muscles provides an indication of how the product changes in physical and mechanical properties during breakdown. This activity was measured as human subjects ate a series of apples to examine the effect of ripeness on oral processing and texture perception.

2 METHODS

Apples – Cox transferred to appropriate storage directly after harvesting.

Storage conditions	Sample 1	5 weeks controlled atmosphere
	Sample 2	5 weeks $0^{o}C$ in ice bank
	Sample 3	3 weeks $0^{o}C$, 2 weeks $8^{o}C$
	Sample 4	2 weeks $0^{o}C$, 3 weeks $8^{o}C$
	Sample 5	1 weeks $0^{o}C$, 4 weeks $8^{o}C$
	Sample 6	5 weeks $8^{o}C$

This generated a series of samples which were allowed to ripen to various extents after removal from the ice bank.

Mechanical properties - penetrometer tests were performed radially at the equator of each apple after removal of skin. A 5mm probe was used at a penetration rate of 100mm/min.

Biochemical analysis - for cell wall sugars was carried out as described by Waldron and Selvendran[1].

Sensory analysis - sensory profiling was undertaken by a trained sensory panel of 12 members.

Mastication analysis -electromyographs (EMGs) were recorded from the temporalis and masseter muscles bilaterally as described by Brown[2]. Records were obtained from 20 human volunteers each eating 2 repetitions of each apple sample. Data analysis provided, among other parameters, the number of chews (activity bursts), muscle work per chew (area under each activity burst) and peak height of each activity burst. A typical EMG record is shown in Figure 1.

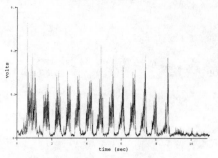

Fig. 1. Typical EMG record for mastication of apple. Each burst of activity represents a single chew, pauses between chews represent jaw opening actions.

3 RESULTS

In all analyses sample 1 (controlled atmosphere) appeared to exhibit more ripe characteristics than sample 2 (ice bank) although these differences were not significant.

Mechanical and sensory analysis indicated a progressive decline in mechanical strength and intensity of sensory texture properties with ripening of the apples (see Figure 2). Between sample differences were significant ($p < 0.05$) for all parameters except first bite juiciness.

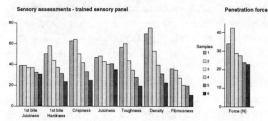

Fig. 2. Changes in intensity of sensory attributes and mechanical properties with apple ripening

Biochemical analysis of cell wall sugars indicated progressive changes in relative concentrations of galactose (decreasing), and glucose and xylose (increasing)

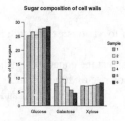

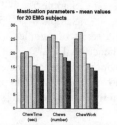

Fig. 3. Changes in relative proportions of cell wall sugars with apple ripening.

Fig. 4. Changes in mastication parameters with apple ripening.

Mastication data were averaged for the all subjects and analysis revealed that ripening resulted in progressively shorter chewing sequences which involved less total chewing work (Figure 4). These data reflect the changes seen in mechanical and sensory data. Sample differences were significant (p<0.05) for all parameters.

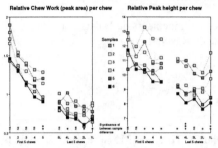

Fig. 5. graphs of chew work and peak height across chews

The first 5 chews and last five chews in each sequence were examined individually (Figure 5). Chew work declined most markedly over the first 5 chews. Rank order of samples was more or less maintained over the whole chewing sequence. Sample differences were significant only by chew 5, and at the end of the chewing sequence.

Peak height for muscle activity for each chew also declined over the chewing sequence – however sample differences were more apparent (significant) by chew 3 and at the end of the sequence.

3 CONCLUSIONS

Ripening resulted in :
- reduction in firmness in mechanical tests and in sensory assessments
- reduction in relative proportion of galactose and increase in glucose and xylose in cell walls
- reduction in the number of chews prior to swallowing, and more chewing work overall
- no significant difference in chew work in initial chews
- no significant difference in perception of juiciness in first bite
- reduction in chew work and peak height in later and final chews

Sensory and mastication differences between the apples were not all evident from the first bite. Much of the sensory information relating to perceptions was acquired subsequent to the first bite at a time when differences in the mechanical properties were evident in the chewing effort. Subjects may modify their first bite when expected to rate the samples. Although the less ripe apples required longer to chew they also required more work per chew throughout the chewing sequence, indicating that they retained their mechanical integrity even to the end of the sequence. Differences in sensory perceptions may be related to the resistance of the sample to transform to a mush at the end of the chewing sequence.

1   K.W. Waldron and R.R. Selvendran, Phytochemisty 31, 1931-1940, 1992.
2   W.E.Brown, J. Texture Studies, 25, 1-16, 1994.

ASSESSING TOMATO QUALITY

C. Darbellay[1], A. Granges[1], J.-C. Villettaz[2] and R. Azodanlou[2]

[1] Station fédérale de recherches en
production végétale de Changins
Centre d'arboriculture et d'horticulture
des Fougères, CH-1964 Conthey

[2] Ecole d'ingénieurs du Valais
Rte du Rawyl 47
CH-1950 Sion

1 INTRODUCTION

Within the frame of Swiss participation in the COST action 915, a research module is devoted to the quality of tomato fruits. This work is aimed at describing the physical and chemical components which may influence the perception of the product's quality. A special emphasis is put on gustatory value.

2 CRITERIA AND METHODS

A product's quality results from a pool of interacting factors. Breeders and traders favored so far the external features : appearance, shape, colour. There is currently a consumers' demand for better intrinsic quality, with emphasis on taste, vitamin contents, consistency etc. Methods have to be developed which allow for a more comprehensive quality assessment.

Table 1 *Method and criteria for quality assessment*

1 Appearance, external quality
Method :   visual inspection, grading
Criteria :   - shape
- colour
- size
- mean weight
2 Gustatory value, intrinsic quality
Methods :
2.1   Chemical and physical analysis
Criteria :   - Sugar contents (°Brix)
- Acidity (g citric acid)
- firmness (penetrometry)
- stringiness
- vitamin contents
2.2   Sensoric analysis (tasting)
Criteria :   - perception of sweetness
- perception of tomato smell
- perception of mealiness
- perception of firmness

The sensoric analysis was performed on one hand by a board of specialists (12-15 persons active in the tomato supply chain), and on the other hand by a « general public » panel made up of about a hundred consumers.

3 FIRST RESULTS

The first investigation campaign (1997) allowed to show several factors to be significant in the valuation of quality, notably :
- Texture
The gustatory quality of tomato is impaired by excessive flesh firmness, skin toughness and stringiness. The mealiness is particularly detrimental to gustatory value.
- Sugar contents
The consumer's appreciation is influenced favourably by higher levels. The Brix index varies from 4.8 and 5.5.
- Acidity
The relation between acid contents and quality perception by the taster is more complicated to establish. The sugar-acid ratio seems to have some significance. The acid contents varies from 4.0 to 5.5 g of citric acid / litre.
- Vitamin contents
The tomatoes are high in vitamin C (15-25 mg/100 g), lycopene (1.7-5.9 mg/100 g) and β-carotene (0.35-0.67 mg/100 g). Vitamin contents are variety-related.

There is a good correspondence between the board of experts' and the panel of consumers' valuations.

4 CONCLUSIONS

This first stage of the study allowed a first selection of criteria and the test of various approaches. A second stage (1998) will allow for a new tasting campaign aimed at setting up a quality scale for the different products.

OPERATING PARAMETERS FOR INSTRUMENTAL MEASUREMENT OF JUICINESS IN FRUITS

P. Eccher Zerbini, M. Grassi, S. Grazianetti

Istituto Sperimentale per la Valorizzazione Tecnologica dei Prodotti Agricoli (IVTPA)
via Venezian 26
I-20133 Milan, Italy

1 INTRODUCTION

Juiciness is one of the characteristics of fruit most appreciated by consumers, but as a sensory characteristic it cannot be measured instrumentally. However physical or chemical quantities correlated to sensory assessment can be measured instrumentally. Szczesniak and Ilker[1] found that sensory juiciness was more correlated to the juice expressed at the first bite of GF Texturometer, than with juice extracted in 3 or more successive bites. Assuming that juiciness is due to juice extracted with mastication, and that the extraction is not exhaustive of the juice contained in the cells, the operating parameters of compression tests were studied, which could allow extraction of amounts of juice correlated to the sensory assessment of juiciness.

2 MATERIALS AND METHODS

2.1 Fruits

Fruits of different juiciness were used: peaches cv Glohaven ripened at room temperature for 1-4 days; pears cv Comice ripened at room temperature for 1-3-5-7 days; apples cv Starking at harvest and after storage. ripened or not.

Small cylinders of fruit flesh ($\varnothing$ = 15 mm, height = 10 mm) were sampled in radial or tangential positions from the fruit.

Sensory juiciness was evaluated on slices of the same fruits by a laboratory panel. Juiciness was defined as the sensation of a progressive increase of free liquids in the oral cavity during mastication[2].

2.2 Compression and relaxation tests

Fruit flesh samples were posed with the axis vertical on filter paper and compressed between two plates with an Instron UTM at deformation rate of 50 or 100 mm/min to a compression of 50% or 80% of the original height of the cylinder; the compression remained constant for 10 seconds, allowing outflow of juice and stress relaxation of the sample. The number of replications was 10 in apples and peaches, 20 in pears, for each factorial combination of sample position in the fruit, deformation rate and % compression.

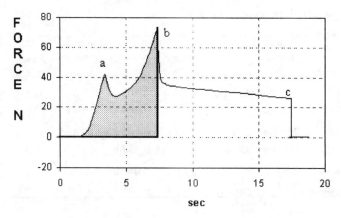

Figure1. *Force-time curve during compression and relaxation.*

Per cent juice was computed as the difference of weight of the filter paper after and before compression, referred to as the initial sample weight.

The applied force was recorded both in the compression and in the relaxation phase. From the force-deformation or force-time curve (Figure 1) the following variables were recorded: a) failure force, b) force at maximum compression, c) residual force after 10 s relaxation, deformation at failure (mm) and compression work. Failure stress, maximum stress and residual stress after relaxation were calculated by dividing the corresponding forces by the original cross-sectional area of the sample. Per cent residual stress was calculated in relation to maximum stress; it is a measure of the residual integrity of the sample after compression. Failure strain was calculated as the deformaton at failure divided by the original height of sample. The ratio force/deformation at failure was also calculated.

2.3 Statistical analysis

Compression and relaxation variables were processed by analysis of variance to determine the effect of the different operating parameters.

Correlation coefficients were calculated between per cent juice obtained by operating with different parameters, and sensory assessment of juiciness.

3 RESULTS

All the variables, except deformation at failure, were significantly affected by the orientation of the sample in the fruit in apples and pears (Table 1). With radial samples higher values were obtained for per cent juice in apples and peaches, as well as for failure strain (data not reported), while the other variables were lower than those obtained with tangential samples.

By compressing the samples to 80%, higher values were obtained for maximum and residual stress and compression work (Table 2), as expected. This level of compression was the most destructive for the samples: the per cent residual stress was lowest, while per cent juice expressed was the highest in all three fruit species. With compression of only 50% on the contrary less juice was expressed, but a better discrimination was obtained in per cent

Table1. *Effect of sample position in the fruit.*

	Position	Failure stress kPa	Max stress kPa	Residual stress kPa	Work J	Force/def at failure N/mm	% Resid. stress	Juice%
APPLES	*Radial*	321.1 a	437.2 a	175.9 a	0.305 a	29.0 a	42.6 a	33.4 a
	Tang.	382.0 b	587.3 b	244.2 b	0.364 b	34.4 b	45.0 b	29.1 b
PEARS	*Radial*	222.2 a	260.9 a	91.8 a	0.193 a	17.8 a	31.9 a	34.0 a
	Tang.	307.4 b	451.8 b	150.4 b	0.277 b	25.3 b	35.3 b	33.6 a
PEACHES	*Radial*	59.4 a	135.2 a	26.6 a	0.069 a	2.2 a	21.6 a	39.1 a
	Tang.	81.5 a	147.7 a	32.8 a	0.084 a	2.5 b	24.9 b	33.1 b

Table 2. *Effect of per cent compression.*

	Compres-sion	Maximum stress kPa	Residual stress kPa	Work J	% Residual stress	Juice %
APPLES	*50%*	287.8 a	145.7 a	0.220 a	50.0 a	20.62a
	80%	750.2 b	276.4 b	0.454 b	37.0 b	42.62b
PEARS	*50 %*	190.7 a	84.4 a	0.158 a	40.1 a	22.90a
	80 %	521.9 b	157.8 b	0.313 b	27.1 b	44.73b
PEACHES	*50%*	55.7 a	16.4 a	0.044 a	28.6 a	23.48a
	80%	226.7 b	42.9 b	0.109 b	17.9 b	48.72b

juice between fruits of different ripeness (data not shown).

Deformation rate significantly influenced some variables in apples and peaches but not one in pears (Table 3). Maximum and residual stress were lower with higher rate, indicating that the samples were more damaged at 100 mm/min; in these conditions the per cent juice expressed from apples was significantly higher than that expressed at 50 mm/min.

As regards correlation between per cent juice and sensory juiciness, in apples higher coefficients were obtained with 50 mm/min deformation rate and 50% compression. Similar results were found for pears as regards per cent compression. With peaches the only significant coefficients were those of radial samples, with both compressions at 50 mm/min rate, and with 50% compression at 100 mm/min rate.

Table 3. *Effect of deformation rate.*

Def.rate mm/min	Maximum stress kPa	Residual stress kPa	% Residual stress		Juice %
	APPLES	APPLES	APPLES	PEACHES	APPLES
50	525.4 a	225.7 a	45.8 a	25.0 a	30.08 a
100	495.8 b	193.2 b	41.9 b	21.6 b	32.38 b

Table 4. *Correlation coefficients of sensory juiciness with % juice expressed in compression tests with different operating parameters.*

Def.rate	*50 mm/min*				*100 mm/min*			
Compr.	*50%*		*80%*		*50%*		*80%*	
Position	*Rad.*	*Tang.*	*Rad.*	*Tang.*	*Rad.*	*Tang.*	*Rad.*	*Tang.*
APPLES	0.73**	0.80**	n.s.	n.s.	n.s.	0.70**	0.64**	0.66**
PEACHES	0.84**	n.s.	0.83**	n.s.	0.84**	n.s.	n.s.	n.s.
PEARS	0.65**	0.66**	0.46**	n.s.	-	-	-	-

4 CONCLUSIONS

Among those tested, the compression parameters most suitable to extract amounts of juice correlated to sensory juiciness were a compression limited to 50% and a deformation rate of 50 mm/min.

These conditions are the least destructive tested and also allow better discrimination between fruits of different ripeness.

The orientation of the sample in the fruit can affect the results: amounts of juice correlated with sensory juiciness were only obtained with radial samples in peaches, while in apples the correlation was slightly better with the tangential samples.

The fact that the best correlations with sensory juiciness were obtained with the least destructive compression conditions confirms that to perceive juiciness it is not necessary to completely destruct the structure of fruit flesh and extract the whole juice, as the perceived juiciness depends on that amount of juice which is readily expressed. One of the components of sensory juiciness in fact is the force with which the juice squirts out of the cells.

References

1. A.S. Szczesniak and R. Ilker, *Journal of Texture Studies,* 1988, **19**, 61-78.
2. P. Eccher Zerbini, A. Pianezzola and M. Grassi, *Journal of Food Quality*, 1998, in press.

GLUCOSINOLATE METABOLISM BY A HUMAN INTESTINAL BACTERIAL STRAIN OF *BACTEROIDES THETAIOTAOMICRON* IN GNOTOBIOTIC RATS

L. Elfoul [1], S. Rabot [1], A. J. Duncan [3], L. Goddyn [3], N. Khelifa [2], A. Rimbault [2]

[1] Institut National de la Recherche Agronomique, Unité d'Ecologie et de Physiologie du Système Digestif, 78352 Jouy-en-Josas, France, [2] Faculté des Sciences Pharmaceutiques et Biologiques, Laboratoire de Microbiologie, 75006 Paris, France, [3] Macaulay Land Use Research Institute, Aberdeen AB9 2QJ, UK

1 INTRODUCTION

Glucosinolates, which are thioglucosides prevalent in cruciferous vegetables, are considered to be protective against cancer. These beneficial effects are due to their breakdown products obtained after hydrolysis by plant myrosinase (thioglucoside glucohydrolase EC 3.2.3.1) [1]. Among glucosinolate derivatives, isothiocyanates are the most effective chemopreventive agents known [2]. After cooking of the vegetables, the plant myrosinase is inactivated and the intact molecules can reach the distal digestive tract where the microflora metabolise them, into still unknown compounds [3].

Our aim is to evaluate the ability of a human digestive strain of *Bacteroides thetaiotaomicron* to convert glucosinolates into isothiocyanates *in vivo* and to assess the influence of diet on this metabolism.

2 MATERIALS AND METHODS

The method involved giving gnotobiotic rats harbouring a human digestive strain of *Bacteroides thetaiotaomicron* a single oral dose of sinigrin, a glucosinolate commonly found in *Brassica*. Allyl isothiocyanate (AITC) putatively released in the digestive tract by the bacterial strain was then ascertained by its detection in the faeces and by quantifying its principal urinary metabolite, allyl mercapturic acid (N-acetyl-S-(N-allyl thiocarbamoyl)-L-cysteine).

2.1 Animals and Diets

Sixteen F344 male adult germ-free rats were inoculated with the *B. thetaiotaomicron* strain and offered either a glucosinolate-containing diet (3.9 µmol/g) followed by a glucosinolate-free diet (n=8) or *vice versa* (n=8). Dietary glucosinolates were provided by a dehulled myrosinase-free and sinigrin-free 00 rapeseed meal (Darmor *cv*).

2.2 Experimental Design

Gnotobiotic rats were dosed with 50 μmol of sinigrin by stomach tube under light ether anaesthesia. For each rat, sinigrin administration was performed once during the glucosinolate-containing diet period and once during the glucosinolate-free diet period. On separate occasions, animals were dosed with AITC directly to measure the extent to which allyl isothiocyanate is recovered as its urinary mercapturic acid derivative.

Rats were housed in individual metabolism cages to allow total collection of faeces and urine for 6 hours prior to dosing and at 6, 18, 30 and 48 hours after dosing.

The microbial metabolism of sinigrin was assessed by quantifying its faecal excretion. The release of AITC was ascertained by its detection in samples of fresh faeces collected at intervals. The degree of conversion of sinigrin into AITC was estimated by quantifying allyl mercapturic acid excreted in the urine.

2.3 Analytical Methods

Sinigrin and allyl mercapturic acid analyses were performed by HPLC, using the method of Hoelbe *et al.* [4] modified by Bjerg and Sorensen [5] and the method of Mennicke *et al.* [6] following slight modification [7], respectively. AITC was analysed by a headspace-gas chromatography method specifically developed for this experiment.

3 RESULTS

Whatever the diet, residual intact sinigrin was mostly excreted within 18 hours after dosing and the total faecal excretion was nearly 4 μmol. So, the human digestive strain of *B. thetaiotaomicron* was able to metabolise 92.6% (sem = 0.6) of the administered sinigrin, *in vivo* (Figure 1).

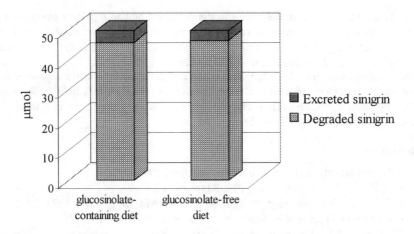

Figure 1 *Proportion of sinigrin degraded in the digestive tract of gnotobiotic rats inoculated with the B. thetaiotaomicron strain*

Trace amounts of AITC were detected in freshly collected faeces, 18 hours after sinigrin administration, whatever the diet. Peak identity was confirmed by mass spectrometry (molecular mass at m/z 99 and fragments at m/z 58 and 72).

Most allyl mercapturic acid urinary excretion occurred within 30 hours following sinigrin administration, whatever the diet. A greater proportion of sinigrin was converted into AITC during the glucosinolate-containing diet period than during the glucosinolate-free diet period (14.5% *vs.* 9.4% ; P<0.025) (Figure 2).

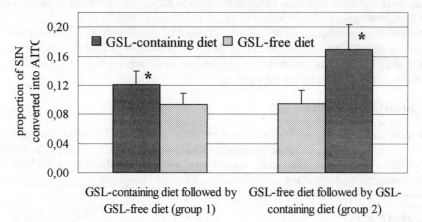

Figure 2 *Influence of diet on the proportion of sinigrin (SIN) converted into allyl isothiocyanate (AITC) in the digestive tract of gnotobiotic rats inoculated with the B. thetaiotaomicron strain (* P < 0.025)*

4 CONCLUSIONS

Our study shows that a strain belonging to the dominant human digestive microflora is able to convert, *in vivo*, a glucosinolate commonly found in *Brassica* vegetables into a specific isothiocyanate with potential health benefits (anticarcinogen). Moreover, the presence of glucosinolates in the diet enhances this metabolism.

These findings give more knowledge on the fate of glucosinolates in the digestive tract and on the bioavailability of their isothiocyanate derivatives. They may help to decide on whether potential health benefits of glucosinolates are worth being used as a target for genetic selection and engineering of *Brassica* cultivars.

References

1.   L. Nugon-Baudon and S. Rabot, *Nutr. Res. Rev.*, 1994, **7**, 205.
2.   S. S. Hecht, *J. Cell. Biochem.*, 1995, Supplement **22**, 195.
3.   S. Rabot, L. Nugon-Baudon, P. Raibaud and O. Szylit, *Br. J. Nutr.*, 1993, **70**, 323.
4.   P. Helboe, O. Olsen and H. Sorensen, *J. Chromatogr.*, 1980, **197**, 199.
5.   B. Bjerg and H. Sorensen, 'World Crops : Production, Utilization, Description', Wathelet, Dordrecht, 1987, Vol. 13, p. 125.
6.   W. H. Mennicke, T. Kral, G. Krumbiegel and N. Rittman, *J. Chromatogr.*, 1987, **414**, 19.
7.   A.J. Duncan, S. Rabot and L. Nugon-Baudon, *J. Sci. Food Agric.*, 1997, **73**, 214.

DIETARY FIBRE COMPOSITION ON DIFFERENT ONION TISSUES

R.M. Esteban, L. Jaime, A. Fernández, M.A. Martin-Cabrejas, E. Mollá, F.J. López Andréu and K. Waldron*
Dpto Química Agrícola. Facultad de Ciencias. Universidad Autónoma de Madrid.
28049 Madrid. Spain
*Institute of Food Research, Norwich Research Park, Colney Lane
Norwich NR4 7 UA. UK

1 INTRODUCTION

Onion (*Allium cepa* L.), an important bulb vegetable, has a widespread use as food, specially as a seasoning and flavouring agent in food preparations. Onions contain moderate amounts of ascorbic acid and other hydrosoluble vitamins and contribute to the intake of minerals and dietary fibre[1]. They are poor sources of calories, proteins and fats. However, they are highly valued for their flavour and also onions are said to possess beneficial effects for human health[2,3].

The varied and widespread use of onions generates substantial quantities of onion wastes derived from the industrial processing of these vegetables. Like many other food industry byproducts, onion wastes are presumably rich in dietary fibre, and they could be a potentially valuable source of fibre. In this sense, the demand for high fibre products has led to new product development with fibre from many sources. At present, high fibre diets have been recommended to improve human health, however, not only the amount of fibre but also the type of fibre[4,5] and the insoluble to soluble fibre ratio are important variables related to nutritional and sensorial quality of foods. The availability of high quality foods with high dietary fibre contents is of key importance for obtaining changes in fibre intake.

As part of a larger study investigating the possible exploitation of onion wastes, we have carried out a preliminary study on dietary fibre content of different onion tissues, which were obtained from onion bulbs as in commercial peeling operations.

2 MATERIALS AND METHODS

Onions (*Allium cepa* L. cv. Grano de Oro) were grown in Spain and supplied by the British Onion Producers Association, UK. The weight of the onions varied from 194 to 256 g and the size ranged from 73 mm to 84 mm diameter. The onions were cut into different tissue types: (1) top and bottom, (2) brown dry outer skin, (3) outer two fleshy layers, and (4) the remaining inner fleshy leaf bases. Top and bottom tissues were obtained slicing off 5-10 mm of the top and bottom ends of the onions, and the rest of tissue samples were obtained by handpeeling. In addition, samples of the whole onion, the tops and the bottoms were also individually analysed. The separated tissues were frozen in liquid nitrogen after cutting and stored at -30°C. Subsequently, they were freeze-dried, milled to pass through 0.5 mm mesh, and stored until analysed.

2.1 Dietary Fibre Determination

Total, insoluble and soluble dietary fibre were determined using the enzymatic-gravimetric method of Lee *et al.*[6]. The principles of the method are the same as those for the AOAC dietary fibre methods 985.29 and 991.42, including the use of the same three enzymes (heat-stable α-amylase, protease and amyloglucosidase) and similar enzyme incubation conditions. In the modification, minor changes have been made to reduce analysis time and to improve assay precision: (1) MES-TRIS buffer replaces phosphate buffer; (2) one pH adjustment step is eliminated; and (3) total volumes of reaction mixture and filtration are reduced.

3 RESULTS AND DISCUSSION

The results are shown in Figure 1. Regarding the insoluble dietary fibre (IDF), brown skin is the tissue which exhibited the highest content, suggesting that this material is an important source of that fibre fraction. The IDF content in mixed top-bottom tissue was also high (41.4%, on a dry weight basis). Nevertheless, the analysis of the separated tissues showed that top ends of onions had higher content of IDF than bottom ends.

Inner layers showed the lowest content of IDF. Although inner layers and outer two fleshy layers apparently showed the same physical characteristics, they had different content of IDF. The results obtained in this study indicated an increasing trend in IDF levels from the inner to the outer parts of the onion.

The IDF content was obtained as the gravimetric residue remained after the enzymatic treatment and substration of ash and protein contents. The contribution of these components to the gravimetric residue will be influenced by physiological characteristics of each onion tissue. Thus, bottom was the tissue which exhibited the major contribution of ash (10% referred to gravimetric residue weight). It might be due to the nutrient uptake process in this tissue.

The soluble dietary fibre (SDF) contents were significantly lower than IDF. Top-bottom tissue showed the highest content of this fibre fraction, due to the high level of SDF in the bottom ends of onion (10.2% on a dry matter basis). In contrast, brown skin exhibited the lowest content of SDF.

The differences in total dietary fibre (TDF) between tissue types were mainly due to differences in IDF content, since the amounts of SDF were generally closer. IDF:SDF ratio decreased from 39.2 in brown skin to 1.4 in inner layers.

The study of tissue yields showed that brown skin contributed a minor average weight of onions, whereas inner tissues contributed the major weight. In contrast, brown skin contributed most to the fibre yield of whole onion, due to its high TDF content (on a dry weight basis).

From the above results, it is concluded that onion and specially certain tissues such as brown skin and top-bottom show very high contents of dietary fibre, being very interesting for their potential use as fibre source.

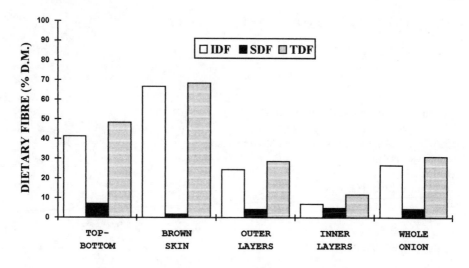

Figure 1. *Dietary fibre content in onion tissues*

References

1. G.R.Fenwick, "The Genus Allium", Part 2, *CRC Critical Reviews in Food Sciences and Nutrition*, 1985 **22**, 273.
2. N.F. Lewis, B. Y. K. Rao, A.R. Shah, G.M. Tewari and C. Bandopadhyay, *J. Food Sci. Technol.*, 1977, **14**, 35.
3. K. Sambaiah and K.Srinivasanr, *J. Food Sci. Technol.*, 1991, **28**, 35.
4. R.R. Selvendran and J.A. Robertson, "Physico-chemical Properties of Foods and Effect of Processing on Micronutrient Availability", EEC Brussels, 1994, p. 11.
5. R.J. Redgwell and R.R. Selvendran, *Carbohydr. Res.*, 1986, **157**, 183.
6. S.C. Lee, L. Prosky and J.W. DeVries, *J. AOAC Int.* 1992, **75(3)**, 395.

Acknowledgements
This work was funded by the European Communities FAIR4 Programme, Contract N° FAIR-CT96-1184.

VEGETABLES AS SOURCES OF VITAMIN K IN FINLAND

T. Koivu, V. Piironen and P. Mattila

Department of Applied Chemistry and Microbiology
University of Helsinki
P.O. Box 27
00014 University of Helsinki
Finland

1 INTRODUCTION

Vitamin K has been historically identified for its role in blood coagulation, in which it functions as an essential cofactor in the posttranslational synthesis of prothrombin and other clotting proteins. It is currently known that vitamin K dependent proteins also include proteins affecting bone metabolism (1). On the other hand, dietary vitamin K has an effect on the efficiency of anticoagulant drugs.

Green vegetables are generally regarded as good dietary sources of vitamin K; especially the dark-green vegetables (spinach, lettuce and broccoli) are excellent contributors of dietary phylloquinone. In the present study we estimated the average daily intake of vitamin K from vegetables, fruits and berries in Finland.

2 METHODS

The vitamin K contents of vegetables, fruits and berries used for estimation of dietary vitamin K intake were obtained from our study done during the years 1996-1997 (2). The sampling covered 38 different items available in Finland and also variation in the phylloquinone contents was investigated. Phylloquinone was quantified by the reverse-phase HPLC with dual-electrode electrochemical detection.

Data on the average consumption of vegetables of the Finnish population was derived from the Finnish Horticultural Products (1993).

3 RESULTS

The phylloquinone contents of vegetables, fruits and berries analysed are summarized in Table 1. The best sources of phylloquinone were dark-green vegetables: parsley, dill, spinach and Brussels sprouts (mean content > 200 µg/100 g). The amounts of phylloquinone were also moderately high in other green vegetables (> 100 µg/100 g). The phylloquinone levels were considerably lower in fruits, berries, and red and yellow vegetables (generally < 20 µg/100 g).

Table 1 *Phylloquinone in vegetables, fruits, and berries*

Phylloquinone (μg/100 g)

0 - 10	onion, peppers, potato, redbeets, swede, tomato, apple, plum, lingonberry, strawberry
10 - 20	blueberry, raspberry, red currant, grapes, carrot, cucumber
20 - 30	cauliflower, avocado, peas
30 - 50	black currant, kiwifruit, iceberg lettuce
50 - 100	leek, white cabbage, Chinese cabbage
100 - 200	broccoli, lettuce in pot, leaf lettuce
>200	Brussels sprouts, dill, spinach, parsley

The use of vegetables, fruits and berries has been fairly low in Finland; the average consumption figures were in 1993 169, 119 and 29 g/day, respectively. The various cabbages, cucumber, tomato, and carrot were the most commonly used vegetables.

On the basis of the above consumption figures, the average daily intake of vitamin K from these foods was estimated to be 37 μg, when the proportion of vegetables accounts for about 90% (Figure 1). Among the vegetables the most significant sources are various cabbages (34%), lettuces (15%) and root crops (15%), especially carrot. The average intake estimated here contributes approximately 50-60% of the current RDA (3). However, a fairly high variation in the consumption of vegetables among individuals is to be expected. In consequence, the daily dietary intake of phylloquinone may vary considerably.

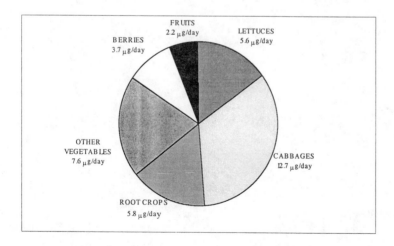

Figure 1. *Vegetables, fruits, and berries as sources of phylloquinone*

In conclusion, green vegetables are excellent sources of phylloquinone; for example, eating 50 g of pot-grown lettuce per day will fulfil over 50% of the daily RDA. The end part of the RDA is easily satisfied by eating also one carrot (70 g) and 50 g of black currant.

References

1. M. J. Shearer, *Lancet,* 1995, **345,** 229.
2. T. J. Koivu, V. I. Piironen, S. K. Henttonen and P. H. Mattila, *J. Agric. Food Chem.,* 1997, **45**, 4644.
3. NRC, 'Recommended Dietary Allowances', 10th ed., National Academy Press, Washington, 1989.

IMPORTANT AROMA COMPOUNDS IN TOMATO DETERMINED BY INSTRUMENTAL AND SENSORY ANALYSIS

A. Krumbein and H. Auerswald

Institute for Vegetable and Ornamental Crops
Grossbeeren/Erfurt e. V.
Theodor-Echtermeyer-Weg 1
14979 Grossbeeren
Germany

1 INTRODUCTION

Flavour is an important quality criteria of vegetables. The flavour of tomato is mainly attributed to its aroma volatiles, sugar and acid content[1]. The following questions are not adequately clarified regarding the aroma volatiles.
- Which aroma volatiles are essential to tomato flavour?
- Which aroma volatiles and descriptive taste attributes are responsible for consumer preference (taste) of tomato fruits?

2 MATERIAL AND METHODS

2.1 Materials

The tomato cultivars 'Supersweet 100', 'Pronto', DRW3126F1 and LYC 1045/90 were grown in soilless culture in a greenhouse. Fruits were harvested at the same stage of ripeness, marked by the same colour.

2.2 Sample Preparation Techniques and GC Analysis

2.2.1 Dynamic Headspace Method on Tenax TA. A modified dynamic headspace method of Buttery et al[2] with small Tenax trap was used. After blending 500 g of tomatoes for 30 s and holding the mixture for 180 s, 500 ml saturated calciumchloride solution were added and the mixture blended for 10 s. An internal standard (2-octanone) was then added and the mixture blended for another 10 s. The mixture was placed in a 3 l flask containing a magnetic stirrer, and purified air (150 ml min^{-1}) passed through the mixture and out of the flask through a Tenax trap (200 mg). The isolation was carried out for 150 min, then the trap was removed and volatiles extracted with 3 ml acetone. Samples were concentrated with nitrogen flux to a volume of 50 µl.

2.2.2 GC-MS. Instruments: HP5890 Serie II plus MSD 5972A; splitles, injector temperature 250 °C; Supelcowax 10 column 30 m x 0.25 mm i. d./0.25 µm; 1 ml helium min^{-1}; temperature programme: 3 min at 40 °C; from 40 °C at 1 °C min^{-1} to 60 °C, 2 min at

60 °C; from 60 °C at 5 °C min^{-1} to 180 °C, 10 min at 180 °C. The relative peaks were normalized with the peak area of the internal standard.

2.2.3 GC-olfactometry. Instruments: HP6890 with FID and a sniffing port; 2 ml hydrogen min^{-1}; GC parameters: the same as GC-MS. Sniffing was performed by five assessors. Using aroma extraction dilution analysis (AEDA) for determination of flavour dilution factors (FD factor) the extracts were diluted in ascending order with solvent in a geometrical line[3].

2.3 Sensory Analysis

Hedonic preference measurements of taste were carried out by 100 housewives using unstructured scale with the anchor points 0 (unpleasant) and 100 (pleasant). In parallel, the tomato fruits were evaluated by a trained descriptive panel (9 panelists), using unstructured scale too, anchor points 0 (not) and 100 (very strong).

2.4 Statistics

For significance investigations Mann-Whitney U test was used at the 5 % level. Correlations were determined by Spearman`s correlation coefficient R.

3 RESULTS AND DISCUSSION

54 volatiles were identified by GC-MS in the dynamic headspace extract of fresh tomatoes. Only a limited number, however, are essential to tomato flavour. The GC-olfactometry yielded 50 odour active compounds in the dynamic headspace extract. According to aroma extract dilution analysis (AEDA) the most important aroma volatiles with FD factors greater than 1000 were (Z)-3-hexenal, hexanal and 1-octen-3-one with fresh green, grassy and mushroom-like notes, respectively. Odour-active compounds with FD factors greater than 100 could be focused on green notes (1-penten-3-on, 2-methyl-4-pentenal), potato notes (methional, unknown compound) and an unpleasant note (3-methyl butanal). Further important compounds with FD factors greater than 10 could be identified as linalool (flowery), phenylacetaldehyde (honey), benzaldehyde (almondy), methylsalicylate (oily), (E,E)-2,4-decadienal (sharp), (E)-2-hexenal (green) and (E,Z)-2,6-nonadienal (cucumbery).

The consumers preferred the cultivars 'Pronto' and 'Supersweet' in taste with acceptance values of 73.7 and 72.0, respectively. LYC 1045/90 was rejected by the consumers (acceptance value of 37.6). DRW 3126 was evaluated more positively than LYC 1045/90 (acceptance value of 54.2). The preference of the consumers could not be explained by acid and sugar contents of the fruits only[4]. The aroma volatiles have to be included.

Relations between consumer preference in taste, descriptive taste attributes and aroma volatiles were found (table 1). There were positive significant correlations between the consumer preference in taste and the results of the descriptive panel in the taste attributes "sweet" and "fruity". The sweet taste attributes were associated with the most potent odour active compound (Z)-3-hexenal. The cultivars 'Supersweet' and 'Pronto', which were preferred by the consumers, had the highest content of this volatile. Furthermore, the attributes "sweet" and "fruity" correlated positively to 1-penten-3-one, 1-octen-3-one and (E,E)-2,4 hexadienal and negatively to hexanal and 2-isobutylthiazol. The bitter taste was

Table 1   *Relations between consumer preference, descriptive taste attributes and aroma volatiles*

	descriptive taste attributes			
	sweet*	sweet**	fruity	bitter
	R	R	R	R
consumer preference	0.76	0.75	0.81	-0.82
aroma volatiles				
(Z)-3-hexenal	0.70	0.80		-0.74
1-penten-3-one	0.85	0.83	0.70	
1-octen-3-one		0.71	0.73	
(E,E)-2,4-hexadienal	0.82	0.77	0.79	
hexanal	-0.77	-0.87	-0.78	
2-isobutylthiazol			-0.89	
ß-phellandrene		-0.81		0.80

* first impression before chewing, **   second impression during chewing

rejected by the consumers. ß-phellandrene, which had the highest content in the cultivar LYC 1045/90, followed by cultivar DRW 3126, correlated to the bitter taste.

References

1.   Petro-Turza, *Food Rev. Int.*, 1986 - 1987, **2,** 309
2.   R. G. Buttery, R Teranishi and L. C. Ling, *J. Agric Food Chem.*, 1987, **35**, 540
3.   F. Ullrich and W. Grosch, *Z. Lebensm. Unters. Forschung*, 1987, **184**, 277
4.   H. Auerswald, C. Kornelson and A. Krumbein, 'Flavour perception-aroma evalution', Proceedings of the 5 th Warthburg Aroma Symposium, Potsdam 1997, 469

ENDOGENOUS MARKERS FOR ORGANIC VERSUS CONVENTIONAL PLANT PRODUCTS

M. Lucarini, M. Carbonaro, S. Nicoli, A. Aguzzi, M. Cappelloni, S. Ruggeri, G. Di Lullo, L. Gambelli, and E. Carnovale.
Istituto Nazionale della Nutrizione
Via Ardeatina 546,Roma-Italy.

1 INTRODUCTION

In Italy organic production is regulated by the EC Legislation (n° 2092/91 and updatings) and by relative Italian Laws which strictly define all the parameters regarding soil, fertilisation, product treatments, essentially excluding the aid of chemical-synthetic pesticides and largely without the use of readily soluble mineral fertilisers. The control is assigned to official organisms selected by the Ministry of Agriculture (MIPA). The control also involves Regions and Consumer Associations.

There is increasing interest in Italy, like in other European countries, in organic products, mainly from the consumers that consider these products as more safe and of higher nutritional quality than conventional products. However many aspects in this field need to be clarified (yield, biological contamination, safety, product shelf-life)[1,2].

A project, funded by MIPA, started in 1996 focused on comparison between organic and conventional production, including chemical, microbiological, toxicological, rheological and sensorial evaluation. In this preliminary study apple and tomato were chosen among plant products, considering their quantitative role in the Italian organic production. The aim of our work was to evidence possible differences between the two systems of production, conventional and organic, and to try to identify endogenous markers developed by plant tissue as defence mechanism against external attacks, like changes in fibre composition and changes in their antioxidative system.

2 MATERIALS AND METHODS

Apple and tomato from the same cultivar, production area, harvest time, differing only for the system of production, conventional or organic, were examined. Apples (Golden cultivar) and tomatoes (S. Marzano-like cultivars) were produced in organic or conventional farms in the Toscana region, middle of Italy, and were harvested respectively at the beginning of October and at the end of August. Apples were stored at 4°C; before analysis unpeeled samples were frozen at -30 °C and then homogenised using a waring blender.

Tomatoes were stored at -30°C before sample preparation. Unpeeled freeze samples were homogenised using a waring blender and then freeze-dried at -20°C before analysis.

Samples were analysed according to the following analytical plan: moisture, protein, ash, (according to AOAC methods 1990[3]), total, soluble and insoluble fibre[4], mineral and trace elements (by atomic absorption on a Varian SpectrAA 400, after microwave mineralization), total polyphenols[5], malondialdehyde[6] (MDA). Free sugars, extracted in water for 30 min by sonication, were filtered and injected in a Dionex Biol LC instrument; the column was CarboPac PA1 (0,4x25 cm), the eluent was NaOH 160 mM, the flow rate 1,0 ml/min and detector was PAD (pulsed amperometric detector).

Ascorbic acid was determined by HPLC after extraction in 5% m-phosphoric acid for 15 minutes under stirring, filtered and then injected in a Water chromatogram system; the column was Hypersil ODS C18 (0,46x25 cm Sigma Aldrich), the mobile phase was

0,015M sodium phosphate at pH 2,8 at flow rate 0,8 ml/min. The UV detector was at 254 nm.

Phenolic acids were extracted by 96% ethanol. After centrifugation the surnatant was evaporated under vacuum in a rotavapor. The ethanol residue was hydrolysed with NaOH 1N for 30 minutes at 100°C; the solution was acidified and phenolic acids were extracted by diethyl ether. The organic layer was evaporated under N_2 and the mobile phase was added to the residue, filtered and then injected in a Water chromatogram system, equipped with a photodiode array detector. The column was Alltima C18 (0,46x25 cm Alltech Associates), the mobile phase was 0,01M sodium citrate buffer at pH 5,4 at flow rate 1 ml/min. Phenolic acids were identified by both their retention time and UV spectra in comparison to external standards.

In tomatoes carotenoid extraction was performed according to a published procedure[7]. Conditions for HPLC separations were: C18 column Inertsil (0,46x25 cm G.L. Sciences Inc), mobile phase was $CH_3CN/CH_2Cl_2/CH_3OH$ (70/18/10 v/v/v) at flow rate 1ml/min and a Waters 996 photodiode array detector was used. Data were examined by Student's *t* test analysis.

3 RESULTS AND DISCUSSION

3.1 Apples

Results relative to both apple samples are shown in **table 1**. Total dietary fibre was higher, by 17%, in organic than in conventional apples with a significantly higher insoluble fraction.

Fructose was the main sugar averaging about 50% of total sugars both in conventional and organic products. Organic apples showed a lower content in soluble sugars (both total and single constituents) and a higher content in malic acid than conventional ones. Because sugars and malic acid content are strictly related to the stage of ripening[8,9], the differences observed seem to evidence a delay in ripening of organic apples.

Considering compounds with antioxidant activities ascorbic acid was lower in organic samples than in conventional, while an opposite behaviour was shown by total polyphenol compounds; in particular caffeic acid and p-coumaric acid were almost doubled in organic samples.

The caffeic acid content could be correlated with chlorogenic acid content. In fact, for every milligram of chlorogenic acid hydrolysed approximately 0,5 mg of caffeic acid should be generated[10]. Chlorogenic acid and phenolic compounds are an important part of the general plant defence mechanism against phytopathogens. Studies by Lattanzio et al.[11] show that chlorogenic acid was generally higher in infected apples than in the uninfected ones.

It may be interesting to point out that organic apples showed a higher contamination by fungi.

As concerns minerals higher value for copper in conventional apples may derive by the high number of treatments with copper against fungi applied to these products.

Table 1 *Nutrient composition of apple (% fresh weight)*

	Conventional Apple	Organic Apple	LS
Moisture, g	84.0 ±0.14	82.9±0.04	***
Protein, g	0.27 ±0.01	0.22 ±0.01	**
Ash, g	0.29 ±0.04	0.32 ±0.04	
Soluble Fibre, g	0.73 ±0.09	0.83 ±0.06	
Insoluble fibre, g	1.84 ±0.11	2.17 ±0.11	*
Total dietary Fibre, g	2.57 ±0.20	3.00 ±0.17	*
Glucose, g	1.33 ±0.08	0.97 ±0.04	**
Fructose, g	5.60 ±0.36	5.19 ±0.08	
Sucrose, g	4.43 ±0.10	4.14 ±0.08	*
Malic Acid, mg	534 ±39.1	708 ±91.0	*
Citric Acid, mg	13 ±4.4	11 ±0.6	
Tartaric Acid, mg	117 ±12.1	114 ±12.1	
Ascorbic Acid, mg	8 ±1.2	6 ±0.6	*
Total Polyphenols, mg	35.6 ±3.94	56.3 ±6.99	**
Caffeic Acid, mg	15.82 ±3.50	37.51 ±3.89	**
p-Coumaric Acid, mg	2.54 ±1.10	6.04 ±1.64	**
Ferulic Acid, mg	0.80 ±0.69	0.36 ±0.03	
Sodium, mg	0.8±0.01	3.5 ±0.01	***
Potassium, mg	143.5 ±0.49	158.9 ±0.39	***
Calcium, mg	4.82 ±0.51	4.52 ±0.01	
Magnesium, mg	8.10 ±0.17	7.51 ±0.26	*
Zinc, mg	0.13 ±0.08	0.11 ±0.04	
Iron, mg	0.14 ±0.02	0.13 ±0.01	
Copper, mg	0.08 ±0.00	0.04 ±0.00	***

Note. The tabulated values represent the Mean ± Standard Deviation of a minimum of three determinations. LS, level of significance: ***$P<0.001$; **$P<0.01$; *$P<0.05$.

3.2 Tomatoes

The results for tomatoes are reported in **table 2**.

Organic tomatoes showed a lower dietary fibre content with a reduction in the soluble/insoluble ratio. They were also characterized by both a higher sugar content and a higher malic acid content.

The behaviour of antioxidant compounds in tomatoes was different in comparison to apple; antioxidant activity differed significantly between compounds. Polyphenol compounds, which are not the major responsible, did not show important differences among the two kind of samples: total content did not differ, while single phenolic acids were lower in organic tomatoes.

Table 2 *Nutrient composition of tomato (% fresh weight)*

	Conventional Tomato	Organic Tomato	LS
Moisture, g	94.4±0.19	93.9±0.30	
Protein, g	1.09 ±0.01	1.13 ±0.04	
Ash, g	0.60 ±0.01	0.64 ±0.04	
Soluble Fibre, g	0.37 ±0.02	0.34 ±0.01	
Insoluble fibre, g	1.67 ±0.05	1.44 ±0.12	*
Total dietary Fibre, g	2.04 ±0.07	1.78 ±0.13	*
Glucose, g	1.13 ±0.04	1.32 ±0.03	**
Fructose, g	1.54 ±0.07	1.80 ±0.03	**
Total Sugars, g	2.66 ±0.11	3.13 ±0.05	**
Malic Acid, mg	83 ±10.8	92 ±2.5	
Citric Acid, mg	318 ±17.9	356 ±22.1	
Total Polyphenols, mg	21.8 ±1.31	22.3 ±0.37	
Caffeic Acid, mg	2.16 ±0.79	1.93 ±0.71	
p-Coumaric Acid, mg	0.71 ±0.16	0.38 ±0.13	*
Ferulic Acid, mg	0.37 ±0.06	0.30 ±0.14	
Lutein, mg	0.15 ± 0.01	0.19 ±0.01	**
Lycopene, mg	7.94 ±0.29	8.06 ±0.71	
β-carotene, mg	0.47 ±0.03	0.34 ±0.02	**
Ascorbic Acid, mg	18 ±1.0	22 ±0.6	**
Sodium, mg	5.5 ±0.22	3.4 ±0.04	***
Potassium, mg	176.20±3.75	363.9 ±3.61	***
Calcium, mg	10.65 ±0.25	9.47 ±0.91	
Magnesium, mg	16.20 ±0.00	16.82 ±0.97	
Zinc, mg	0.14 ±0.00	0.20 ±0.10	
Iron, mg	0.57 ±0.03	0.39 ± 0.02	
Copper, mg	0.13 ±0.01	0.11 ±0.00	*

Note. The tabulated values represent the Mean ± Standard Deviation of a minimum of three determinations. LS, level of significance: ***$P<0.001$; **$P<0.01$; *$P<0.05$.

Lutein, β-carotene and lycopene are the main components of the antioxidant defense of tomatoes and their content is strictly dependent on many factors (fertilisation, ripening process, extent of light exposure). Lycopene showed the highest variability (partially due to analytical variability). Also differences in ascorbic acid (that are in the opposite direction than β-carotene and phenolic acid) are difficult to interpret, even at low level. Considering mineral composition differences for potassium were found and they are related to fertilisation treatments of conventional products. Polyphenols and phenolic acid levels in tomatoes did not appear to be greatly affected by their different growing conditions.

4 CONCLUSIONS

In this preliminary study we found significant differences between organic and conventional products as far as apples are concerned. The significantly higher level in phenolic compounds found in organic apples are in agreement with physical and microbiological evaluation of apple (data not shown here) that reported in organic apples a higher level of fungi infestation, a higher presence of "bitter point".

Although the concentration of phenolics obtained by the Folin-Ciocalteu method should be interpreted with caution because nonphenolic material can interfere with the assay, the levels of phenolic acids may provide an endogenous marker for organic versus conventional plant products. In fact the relative content of phenolic acids could be relevant because it is well-known that they were more effective against phytopatogens when used as a mixture then when used individually[12].

As far as tomatoes are concerned it is important to underline that both organic and conventional products showed great variability of physical defects and contamination by fungi; for this study we choose for both kind of products samples of the first class, that means without any defects. This may be the reasons for the flattened responses in tomatoes and in opposite trend with respect to apples. Although apple and tomato behaviour was different, considering their different level of biological contamination (as resulted from collateral studies) we suggest that antioxidant levels may be an interesting marker of stress-induced changes in plant tissue.

Further studies (already started) need to confirm these preliminary results.

References

1. Woese K., Lange D., Boess C., Bogl K.W., *J. Sci Food Agric.*, 1997, **74**, 281-293.
2. Pither R and Hall M.N., Technical Memorandum No 597, MAFF Project No 4350, June 1990.
3. AOAC *"Official Method of Analysis", 15th ed.*; Association of Official Analytical Chemists: Arlington, VA.,1990.
4. Prosky L., Asp N.-G., Schweizer T.F., De Vries J.W., Furda I., *J. Ass. Off. Anal. Chem*, 1988, **71**, 1017-1023.
5. Singleton V.L., Rossi J.A. jr.,Am. J. Enol. Vit, 1969, **167**, 144-148.
6. Tarladgis G. B., Watts M.B., Younathan T.M., *J. Ass. Off. Anal. Chem.*, 1960, **37**, 44-48.
7. Tonucci L.H., Holden J.M., Beecher G.R., Khachik F., Davis CS., Mulokozi G. *J. Agr. Food Chem.*, 1995, **43**, 579-586.
8. Lintas C., Paoletti F., Cappelloni M., Gambelli L., Monastra F., Ponziani G., *Adv. Hort. Sci.*, 1993, **7**, 165-168.
9. Blanco D, Moran M.J., Gutierrez M.D., Moreno J., Dapena E., Mangas J., *Zeitschrift Lebensmittel Untersuchung Forschung*, 1992, **194** (1), 33-37.
10. Spanos G.A., Wrolstad R.E., *J. Agr. Food Chem.*, 1992, **40**, 1478-1487.
11. Lattanzio V., De Cicco V., Di Venere D., Lima G., Salerno M., *Ital. J. of Food Sci.*, 1994, **1**, 23-30.
12. Ghanekar A.S., Padwal-Desai S.F., Nadkarni G.B, *Potato Res.*, 1984, **27**, 189-199.

NEW SOURCES OF ANTHOCYANINS

M. Máriássyová, S. Šilhár and M. Kovác

Food Reseach Institute Bratislava,
Priemyselná 4, SK-820 06 Bratislava

1 INTRODUCTION

Interest in natural colouring agents from various fruit and vegetable materials has increased in recent past, a question regarding the safety of artifical colours has been raise. Included among the natural colouring materials which have been investigated are anthocyanins from many fruits. Especially the anthocyanins processed from waste from manufacture of grape is applied as a colourant, but recently experiments to evaluate the possibility for application of other fruits such elderberry[1-4] and black chokeberry[5-7] has been carried out.

The aim of this paper is to determine the quantity and quality of pigments and level of fruit yield by extraction from different cultivars of elderberry (Sambucus nigra L.), black chokeberry (Aronia melanocarpa L.) and edible honeysuckle (Lonicera edulis). The anthocyanins content in fruits is influenced by several factors, such as variety, weight, time of harvesting and the weather during vegetation[6].

2 EXPERIMENTAL PROCEDURES

The fruits were obtained from Pomology Research Institute at Bojnice. Harvesting was generally conducted in June to October. In the initial phase of this study, an ethanol-HCl extraction was used to evaluate anthocyanin concentration in berries, saft and pomace. Following the investigation of extraction solvent effectiveness, a standard reaction procedure was developed for elderberry, honeysuckle and chokeberry.

Qualitative and quantitative anthocyanins composition in black chokeberry, elderberry and honeysuckle were determined by HPLC and UV/VIS-methods[8].

The pigment concentrates were made in plant pilot scale. Fresh berries were mashed in a mill, pressed to separate juice and pomace extracted in a discontinual extractor. Next step is sedimentation and filtration. The sediments were collected, EtOH was removed. Obtained concentrate was utilised in colouring jams and similar food products. After sedimentation and filtration the ethanol was removed from extracts in vacuum evaporator at a temperature below 40 °C.

3 RESULTS AND DISCUSSION

3.1 *Quantitative analysis.*The content of anthocyanins was influenced by the cultivars (Table 1). The pigment content is much higher from experimental cultivation than from normal horticultural production.

Table 1 Content of anthocyanin in saft and press cake of different fruits.

cultivar	content of anthocyanins		
	berries (g/kg)	saft (g/l)	pomace (g/kg)
honeysuckle (Lonicera edulis)			
2-303-82/15	6,1 - 6,3	5,0 - 5,2	7,9 - 8,1
Gerda	4,0 - 4,2	2,1 - 2,3	5,1 - 5,4
Lazurnaja	3,6 - 3,9	1,7 - 1,8	4,7 - 4,9
elderberry (Sambucus nigra)			
wild	6,1 - 7,0	1,3 - 1,5	7,2 - 8,3
Haschberg	12,7 - 14,5	2,3 - 2,6	15,1 - 17,2
Sambo	4,1 - 4,7	0,7 - 0,9	4,9 - 5,4
Novošľachtenec 7	7,7 - 9,1	1,5 - 1,7	9,2 - 10,9
chokeberry (Aronia melanocarpa)			
Nero	5,3 - 7,6	1,5 - 2,3	7,5 -10,5
Granatinaja	1,2 - 1,4	0,5 - 0,7	1,8 - 2,1
Burka	2,8 - 3,5	0,6 - 0,8	4,2 - 5,2

3.2 *Qualitative analysis.* In chokeberry (Aronia melanocarpa) the four main pigments are cyanidin-3-galactoside (64 %), cyanidin-3-glucoside (4,5 %), cyanidin-3-arabinoside (29 %) and cyanidin-3-xyloside (3,5 %). As about three quarters of the anthocyanins in chokeberry are retained in the fruit skins. In honeysuckle (Lonicera edulis) cyanidin-3-glucoside, cyanidin-3,5-diglucoside, but also peonidin-3,5-diglucoside are present. We found the ratio of these three anthocyanins are 89:4:7. The main two anthocyanins in elderberry (Sambucus nigra) are cyanidin-3-sambubioside and cyanidin-3-glucoside, in small amounts are cyanidin-3-sambubioside-5-glucoside and cyanidin-3-glucoside-5-glucoside. Here we found a ratio of 44:47:8:1

3.3 *Colour concentrate production.* Several studies on extraction of anthocyanins have been conducted. For the processing conditions used, there were about equal amounts of juice (elderberry 55-60%, chokeberry 45-50%, honeysuckle 65-70%) and pomace. About 80% of the anthocyanins are extracted from pomace during first extraction step, reaching the maximum of colour units after at least 40 to 60 minutes. To obtain most of anthocyanins, ratio of pomace to ethanol 1:1, temperature about 25 °C, time of first extraction 3 - 4 hours, second extraction 6 - 8 hours is adequate.

When elderberry, chokeberry and honeysuckle concentrates are applied as colourant for soft drinks, jams, syrups or confectionery a satisfactory colour stability is obtained.

4 CONCLUSION

From examined cultivars for the pigment concentrates production, the most suitable are:
elderberry - cultivar Haschberg,
chokeberry - cultivar Nero,
honeysuckle - cultivar 2-303-82/15.

References

1. K. Kaack, *Tidsskr. Plantea.*, 1990, **94**, 423.
2. M. Drdák and P. Daucik, *Acta Alimantaria*, 1990, **19**, 3.
3. K. Bronnum-Hansen, F. Jacobsen and J. M. Flink, *J. Food Technol.*, 1985, **20**, 703.
4. A. Porpáczy and M. László, *Acta Alimentaria*, 1984, **13**, 109.
5. W. Plocharski and J. Zbroszcyk, *Fruit Sci. Reports*, 1989, **26**, 38.
6. K. Kaack and F. Kühn, *Tidsskr.Plantea.*, 1992, **96**, 183.
7. H. Lehmann, *Flüss. Obst*, 1990, **57**, 746.
8. A. W. Strigl, E. Leitner and W. Pfannhauser, *Z. Lebensm. Unters. Forsch.*, 1995, **201**, 266

RED BEET AS SOURCE OF PIGMENTS

M. Máriássyová, S. Šilhár, S. Baxa and M. Kovác

Food Reseach Institute Bratislava,
Priemyselná 4, SK-820 06 Bratislava

Betanin pigments extracted from red beet (Beta vulgaris) roots provide a natural alternative to synthetic red dyes. Betalains have been successfully used in commercial food colouring operationsfor a number of years. Betalains are derivatives of betalamic acid and can be classified into two groups: the red-violet betacyanines and yellow betaxanthins[1].

This paper is concerned with chemical analysis of different cultivars of red beet aimed at the suitable kind for the production natural pigment concentrates with high content of the pigments.

The pigment concentrates we produced by ethanol (40 - 80 %) extraction of pigments from whole cut beet or from pomace.

The content of betacyanines, sugars, acids and dry matter in cultivars Rubin, Rote Rüben, Betina, Lauka, Libero and Red Round, in saft, in pomace and in extracts were determined (Table 1). For determination of pigment, a spectrophotometric method was used which does not require separation of individual pigments and enables betacyanines and betaxanthines to be determined[2]. Determined sugar concentrations are 150-250 times greater than pigment concentrationss in the beet root. The commercial product would be significantly enhanced by lowering sugar levels in the beet root through selection of cultivars[3] and different methods of extraction of pigments[4,5]. The differencies in content of sugars in examined cultivars are small. The pigment content in root is from 426 to 691 mg/kg of fresh weight. From this point of view the cultivars Rubin, Lauka and Libero are more suitable for production of betacyanine concentrates from whole roots. When the pigment concentrates are produced from pomace, only the cultivars Rubin and Libero are suitable from an economical point of view.

Table 1 Compositon of different cultivars of red beets.

	cultivar					
	Rote Rüben	Red Round	Betina	Lauka	Rubin	Libero
Whole red beet						
betanin (mg/kg)	530	426	534	651	634	691
sugar (g/100g)	11,1	12,3	12,2	13,2	12,7	10,9
dry mater (g/100g)	17,6	13,4	16,6	15,4	14,2	12,7
Saft						
betanin (mg/kg)	645	518	680	700	729	934
sugar (g/100g)	7,2	7,2	7,2	7,3	7,3	7,2
acid (as citric g/l)	1,9	1,5	1,8	2,1	1,9	2,3
Pomace						
betanin (mg/kg)	416	268	374	368	489	571
sugar (g/100g)	5,7	5,3	5,8	6,1	5,8	6,1
dry mater (g/100g)	21,6	17,5	19,8	18,9	18,4	17,5
Extract						
betanin (mg/kg)	169	125	159	160	254	269

References

1.   T. J. Mabry and A. S. Dreiding, Recent Advances in Phytochemistry. New York Academic Press, 1968.
2.   T. Nilson, Lantbrukshogskolans Ann., 1976, **36**, 179.
3.   I. L. Goldman, K. A. Eagen, D. N. Breitbach and W. H. Gabelman, J. Am. Soc. Hort. Sci., 1996, **121**, 23
4.   M. Abeysekere, S. R. Sampathu and M. L. Shankanarayana, J. Food Sci. Technol., 1990, **27**, 336.
5.   E. Sobkowska, J. Czapski and R. Kaczmarek, Int. Food Ingredients, 1991, 3, 24

PHENOLIC COMPOUNDS IN "GAZPACHO" AS FUNCTION OF THEIR INGREDIENTS

Martinez-Valverde, I.[a]; Periago, M.J.[a]; Provan, G.[b]; Chesson, A.[b] and Ros, G.[a]

[a]Department of Food Science and Nutrition. "Bromatología". Faculty of Veterinary, Murcia University. Espinardo, Murcia 30071, SPAIN.
[b]Rowett Research Institute. Greenburg Road, Bucksburn. Aberdeen AB21 9SB, Scotland, UK.

1 INTRODUCTION

Phenolic compounds are secondary plant metabolites that have important roles in providing flavor and color characteristics in fresh and processed plant foods[1]. Moreover, phenolic compounds possess outstanding antioxidant and free radical scavenging properties, suggesting a protective role for human health. The *in vitro* antioxidant activity of polyphenols has been amply researched, and there is vast literature documenting their ability to act as primary as well as secondary antioxidants through sequestration of metal ions and by scavenging reactive oxygen species[2]. It is well known that phenolic compounds are present in all vegetables and its content depends on many factors as variety, maturity degree, season, etc. So, in processed vegetable foods it is very interesting to select those raw material with a high content in phenolic compounds, to increase the total phenolic content in the final product. Gazpacho is a cold soup consumed in Spain, prepared with different vegetables such as tomato, onion, garlic, red pepper and cucumber. The aim of the present study is to know the content of soluble and insoluble phenolic compounds in gazpacho and their ingredients.

2 MATERIAL AND METHODS

2.1 Material

Gazpacho as well as their vegetable ingredients (tomato, onion, red pepper, garlic and cucumber) were analysed in this study. All the samples were freeze-dried after their harvesting and manufacture, and then ground to a fine flour, which was stored with desiccant until analysis. All samples were analysed in triplicate for each paramenter studied.

2.2 Methods

The concentration of total soluble phenolic compounds (SPC) present in dry ground samples was determined spectrophotometrically using Folin-Denis method[3]. The content of insoluble phenolic compounds (IPC) was evaluated in 10 mg of the residue obtained after the methanolic extraction of the SPC with the Folin-Denis method. The residues were dried in an oven at 40°C overnight before weighing the samples, following the modified method described by Iiyama and Wallis[4]. Total phenolic compound (TPC) were calculated as the sum of SPC and IPC. All values were expressed as g/Kg of dry matter.

3 RESULTS AND DISCUSSION

Tables 1 and 2 show the results obtained in this study. Table 1 shows the moisture value for gazpacho and their ingredients, whereas Tables 2 shows the phenolic compounds distribution. SPC in gazpacho was 4.25 g/Kg of dry matter. Related to their ingredients the highest content of SPC were showed in red pepper (4.79 g/Kg), followed by tomato (4.59 g/Kg) and onion (3.71 g/Kg), whereas the lowest content was determined in garlic (2.25 g/kg). The IPC were higher in all samples analysed compared with the content of SPC. Gazpacho showed the highest content (76.89 g/Kg) which was provided by different ingredients. About all of them garlic showed a content of 58.85 g/Kg, whereas tomato showed only 21.12 g/Kg. The next column on the Table 2 gives the TPC. As was expected gazpacho showed the highest content in TPC (81.14 g/Kg) followed by garlic (61.10 g/Kg), whereas red pepper, onion and cucumber showed a similar value around 40 g/Kg. The result obtained showed that IPC are main compounds to determine the TPC composition in vegetable foods.

4 CONCLUSION

As conclusion of this previous study we can summarize that, if we evaluate the content of phenolic compounds in the vegetables that compose Gazpacho, we could chose the raw material to select those with the highest content increasing the content in the final product to obtain a food with those beneficial physiological properties related to phenolic composition. However, more studies should be developed to understand the effect of agricultural conditions and to extend the research to acknowledgement of each type of phenolic compounds.

References

1. G.A. Spannos, and R. E. Wrolstad, *J. Agric. Food Chem.* 1992, **40**, 1478.

2. M. Serafini, A.Giselli and A. Fero-Luzzi, *Europ. J. Clin. Nutr.* 1996, **50**, 28.

3. K. Iiyama and A.F.A. Wallis, *J. Sci. Food Agric.* 1990, **51**, 145.

4. T. Swain and W.E. Hillis, *J. Sci. Food Agric.* 1959, **10**, 63.

Table 1 *Moisture observed in Gazpacho and in their ingredients expressed as percentage*

Sample	Moisture (%)
Gazpacho	90.35
Red pepper	93.21
Tomato	94.73
Onion	94.23
Garlic	69.25
Cucumber	95.54

Table 2 *Content of total phenolics compounds in raw material and final product*

Sample	SPC (g/kg)	IPC (g/kg)	TPC (g/kg)
Gazpacho	4.25±0.02	76.89±6.64	81.14
Red pepper	4.79±0.03	37.86±2.91	42.35
Tomato	4.59±0.02	21.12±0.109	25.71
Onion	3.76±0.05	40.62±5.85	44.38
Garlic	2.25±0.01	50.85±9.94	61.1
Cucumber	3.06±0.02	38.17±1.77	41.23

UTILISATION OF EXTRUDED HARD-TO-COOK BEANS (Phaseolus vulgaris) IN THE PREPARATION OF MOIN-MOIN

C. M-F. Mbofung[†], A.C. Smith and K.W. Waldron

Institute of Food Research,
Norwich Research Park,
Colney,
Norwich NR4 7UA, UK
[†] University of Ngaundere, Cameroon

1 INTRODUCTION

The hard-to-cook (HTC) defect is a problem which has prevented effective utilisation of stored *Phaseolus* beans in tropical countries. In addition to increased cooking times, the beans also exhibit higher levels of toxic antinutrients including lectins[1]. The present study was undertaken to investigate the potential for incorporating extruded HTC-red kidney bean flour into cowpea flour for the preparation of moin-moin, a pudding made from cowpea paste. A previous report has considered the extrusion cooking of HTC kidney bean flour.[2]

2 MATERIALS AND METHODS

2.1 Materials

Hard-to-cook red kidney beans (*Phaseolus vulgaris*) were bought from a local market in Ngaoundere, Cameroon. White Cowpea or black-eye beans (*Vigna unguiculata*) were bought from A. Miah and Co., Norwich, U.K. Beans were decorticated by soaking in water, drying in a fluid-bed dryer, and impact milling gently. Seed coats were removed by aspiration. Samples were milled using a Condux V2/S Disk mill and an Alpine 160Z pin mill.

2.2 Extrusion

A laboratory single screw extruder was employed (Brabender). Decorticated red kidney bean flour was extruded at 180°C at a moisture content of 30% on a dry solids basis (dsb) at a feed rate of 70g/min. Extrudates were dried at room temperature for 5 days and then milled as described above.

2.3 Production of Composite Flours

The composite flours were obtained by thoroughly mixing (w/w) the cowpea and extruded red kidney bean flour in different proportions: 100:0, 90:10, 80:20, 70:30, 60:40, 50:50, 0:100. Cowpea flour (100:0) was used as the control.

2.4 Preparation of Moin-moin

20 g flour was mixed with 60 ml tap water. The mixture was whipped for 0, 1 or 5 min and poured into metal tins (diameter 5.6 cm and height 4.3 cm). The tins were placed on wire gauge in boiling water. Duplicate samples were heated for 60 min and then cooled immediately by placing in cold water.

2.5 Determination of solubility, water absorption, moisture and protein

The moisture content of the grains and flours were determined using a hot oven box at 105°C for 5 hr and the infra-red moisture meter (Mettler LP16) at 130°C for 1 hr respectively. All determinations were performed in duplicate. Crude protein contents of the flour and moin-moin samples were determined by the Keljdahl method. Total nitrogen was multiplied by 6.25. The water-solubility index (WSI) and water absorption index (WAI) were measured in duplicate by the method of Anderson *et al.*[3]

2.6 Bulk Density of flour and Moin-moin samples (g/ml)

Bulk densities of flour and moin-moin were calculated as weight of sample per unit volume of sample.

2.7 Textural Measurements

The maximum force required to cut through moin-moin was measured by a cutting test for individual samples, using a universal testing instrument (model 1122 Instron Inc., Canton, MA, 02021) that was fitted with a 50 ± 0.05 Kg load cell and a stainless steel flat blade (width 18.1 mm, height 18.95 mm, thickness 0.67 mm and length 59.4 mm). The full scale was 5 N and the cross-head and chart speeds were both set at 10 mm/min. The test was done on single moin-moin blocks of thickness 18.95mm, placed on a flat metal plate and the blade descended to within 1 mm of the metal plate. The average force was reported in N. The compression force was measured with the same instrument: single discs of moin-moin (thickness 15 mm and diameter 19.9 mm) were placed on a flat metal and compressed using a metal press (thickness 1.73 mm and diameter 19.9 mm), which descended to 1 mm of the metal plate. The cross-head and chart speeds were set at 10 mm/min, while the full scale load was 50 N. All measurements were performed in duplicate.

3 RESULTS AND CONCLUSIONS

The results in Table 1 show that cowpea flour had a greater solubility in water than the extruded bean flour. Hence, the WSI decreased with increasing amounts of extrudate in the mixtures. In contrast, WAI showed the opposite trend.

Sample	MC	BD	WAI	WSI
Cowpea flour (c)	5.3	0.95	2.9	35.5
Kidney bean flour (kb)	4.9	0.97	4.2	20.0
Extruded bean flour (eb)	10.8	1.20	5.3	18.4
Cowpea:extruded bean				
90/10	5.7	1.01	3.5	31.6
80/20	5.9	1.05	3.6	28.9
70/30	5.9	1.08	3.7	27.8
60/40	6.0	1.14	3.9	24.0
50/50	6.2	1.17	3.9	21.0

Results are means of duplicate values

Table 1. *Moisture content (% w.w.b.), bulk density (g/ml), WAI and WSI (%) of extruded kidney bean and cowpea flour mixtures*

The bulk density of moin-moin was not affected significantly by the incorporation of extruded bean flour (Fig. 1). Furthermore, the maximum cutting force was not altered significantly up to 20% incorporation (Fig. 2). The maximum compression force was reduced markedly from over 24% to approximately 10% after incorporation of extruded bean flour at a level of only 10%. Informal sensory analysis indicated that the supplemented Moin-moin was acceptable at levels of up to 20%, both in terms of taste and textural quality. However, above this level, the extruded bean flour imparted a slight burnt taste resulting from the extrusion process.

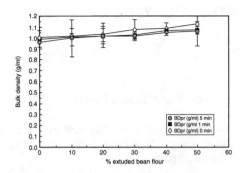

Figure 1. *Bulk density (g/ml) of moin-moin from extruded kidney bean/cowpea flour blends whipped at different times*

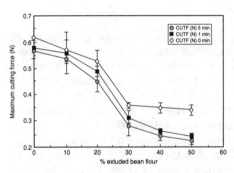

Figure 2. *Maximum cutting force of moin-moin from extruded kidney bean/cowpea flour blends whipped at different times*

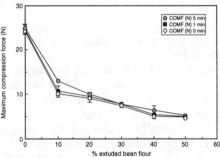

Figure 3. *Maximum compression force of moin-moin from extruded kidney bean/cowpea flour blends whipped at different times*

4 ACKNOWLEDGEMENTS

The authors wish to thank Mr. T. Tunkap for assistance with experimental work and to acknowledge the financial support of the UK Biotechnology and Biological Science Research Council, and the European Commission (STD-3 project No. TS3*-CT92-0085).

5 REFERENCES

1. M. A.Martin-Cabrejas, R. M. Esteban, K. W. Waldron, G. Maina, G. Grant, S. Bardocz and A. Pusztai. *J. Sci. Food Agric.*, 1995, **69**, 429.

2. C. Karanja, G. Njeri-Maina, M. Martin-Cabrejas, R.M. Estenam, G. Grant, A. Puszti, D. M. R. Georget, M. L. Parker, A. C. Smith and K. W. Waldron. Proceedings "Agri-Food Quality", eds. G. R. Fenwick, C. Hedley, R. L. Richards and S. Khokhar., Royal Society of Chemistry, 1996, 279.

3. R. A. Anderson, J. F. Conway, V. F. Pfeifer and E. L. Griffin. *Cereal Sci. Today*, 1969, **14**, 372.

D. Pérez-Conesa, M.J. Periago and G. Ros

Department of Food Science and Nutrition, "Bromatología e Inspección de Alimentos". Veterinary Faculty, Murcia University. Espinardo 300071, Murcia, Spain.

1 INTRODUCTION

Weaning food is usually started, with homogenized foods, granulated in small particles that don't need to be chewed, or with a thicker texture (bigger particles) to stimulate infant chewing before swallowing[1]. Cereals, incorporated as flours, are the common food recommended by paediatricians, to start the weaning food at the age of 4-6 months[2]. They have a high energetic content per volume unit (80 kcal/100 g), and are mixed with human milk or formula to equilibrate the infant diet[3]. Cereals are technologically processed by roasting and using enzymatic treatment for better digestibility and texturisation of the products, with a final heating to inactivate the enzyme. Processing will produce some changes in the chemical components that will determine the nutritional value. The aim of the study was to evaluate the effect of the technological treatment on the nutritional protein value of "multicereal" infant cereal.

2 MATERIAL AND METHODS

2.1 Material

Infant cereals of the so-called "multicereal" (a mix of 8 cereals: wheat, millet, barley, rye, sorghum, rice, oat and maize), were sampled and classified according to their technological stage of processing: mix of raw flours (A), mix of roasted flours (B), mix of drum dried flours (C) and commercial infant cereals (D). These four samples were processed and provided by local infant foods producer (Hero España S.A). Afterwards, they were stored at room temperature in plastic containers, ready to be analysed.

2.2 Methods

The protein nutritional value was determined according to the techniques recommended by the AOAC (1990)[4]. Crude protein (CP) was estimated using Kjeldahl method (N x 5'7), and true protein (TP) and free amino acids (FAa), using the trichloroacetic precipitation and ninhydrin reagent methods respectively[5]. *In vitro* protein digestibility (IVPD) was carried out using a multienzymatic technique according to the method of Satterlee *et al.* (1982)[6]. Amino acids profiles were determined following the method of Moore and Stein (1963)[7] after hydrolysis with 6N HCl at 110 °C for 24h in vacuum sealed tubes. The data were statistically evaluated by ANOVA and Tukey's test with a significant level of P < 0'05.

3 RESULTS AND DISCUSSION

Table 1 presents changes in protein nutrition quality parameters and analysis of variance affected by processing. The analysis of variance showed that for all the parameters considered, statistical significantly differences (P < 0'05) were observed.

Crude protein was always higher than true protein (2'28-6'74 times higher) with the consequent overestimation of CP for infant cereals in all processing steps. In addition, at the time that CP increases (from A to C), TP decreases as processing forwarded (A to D). This fact is a consequence of the heat treatment by protein denaturalization, decreasing the true protein content in the flour. The trend described by TP was related with FA which increased significantly (P < 0'05) as a function of processing (C and D). The protein hydrolysis due to the heat treatment and consequently the amino acids liberation from the peptidic chain, increasing quickly the amount of free amino acids which are digested easier by the infant gut[8]. IVDP increased significatly (P < 0'05) from A (76'03%) to C (88'69%), and slightly at the commercial product probably due to the mixture with other ingredients with lower digestibility (sugar, minerals and vitamins). In general heat treatment makes better protein digestibility due to the protein denaturalization, exposing new places to the enzymatic attack. Heat also helps protein digestibility by inactivation of enzymatic inhibitors or antienzymatic factors,

Table 1 *Changes in protein nutritional value parameters as affected by processing*

Protein nutritional parameters	Processing stage				Significance[2]
	A[1]	B	C	D	Processing
Crude Protein (%)	8'37[b3]	8'77[b]	10'16[a]	9'10[a]	*
True Protein (%)	3'66[a]	1'98[a]	1'66[b]	1'35[b]	*
Free Aa. (mg/100g)	95'73[c]	84'78[c]	168'89[b]	213'93[a]	***
IVDP (%)	76'03[b]	78'84[b]	88'69[a]	86'55[a]	***

[1] Four stages of processing: A= mix of raw flours; B= mix of roasted flours; C= mix of drum dried flours; D= commercial infant cereals.
[2] Statistical differences by analysis of variance: NS=P > 0'05; *, P < 0'05 ; **P < 0'01; ***, P < 0'001
[3a-c] Different letters in the same row show statistical significantly differences (P < 0'05).

decrease the fat-protein and starch-protein complex[8-9]. The effect on specific amino acids is not displayed since most of them did not show significantly changes (P < 0'05) what present from nutritional lost. As a general statement, commercial cereals can be considered a good source (a serving contains 10-19 % of the daily value for the nutrient)[10] of lysine, and they've got high content (a serving contains 20 % or more of the daily value for a nutrient)[10] of the following essential amino acids: phenyl-alanine, histidine, isoleucine, leucine, methionine, threonine, tryptophan, valine.

References

1. FAO/OMS, Comisión del Codex Alimentarius, 1982, Roma, Italia.
2. J.A. Molina and J. Maldonado, 'Tratado de Pediatría', Expaxs, Barcelona, España, 7th ed, 1993, Chapter 56, 655.
3. Comitte on International Nutrition Programs, 'Food and Nutrition Board', National Academy Press, Washington, USA, 1983.
4. AOAC, 'Official Methods of Analysis', Washington DC, USA, 15th ed, 1990.
5. E.O. Awolumate *J. Sci. Food and Agric.* , 1983, **34** , 1351.
6. L.D. Satterlee, J.C. Kendrick, H.F. Marshall, D.K. Jewell, R.D. Ali, M.H. Heckman, F. Stein, P. Larson, R.D. Phillips, G. Saward and P. Slump, *JAOAC*, 1982, **65**, 789.
7. S. Moore and W.H. Stein, *Meth. of Enzym..*, 1963, **6**, 819.
8. K. Dahlin and K. Lorenz, *Food Chem.*, 1993, **48**, 13.
9. J.A. Maga, K. Lorenz and O. Onayemi, *J. Food Sci.*, 1973, **38**, 173.
10. N.H. Mermelstein. *Food Tech.*, 1993, **47**, 81.

EVOLUTION OF DEXTRIN CONTENT IN INFANT CEREALS DURING PROCESSING

M. J. Periago, M.J. Bernal and G. Ros

Department of Food Science and Nutrition "Bromatología e Inspección de Alimentos".Veterinary Faculty, Murcia University, Espinardo, 30071-Murcia, Spain.

1 INTRODUCTION

Infant cereals are the first food included in the weaning diet different from human milk or formula[1]. The most abundant nutrients of the cereal grains are the soluble carbohydrates, mainly starch in a proportion of 70-80%, which provide the total energy to the weaning diet and developed an important role in the food industry, due to the physico-chemical and functional properties[2]. The digestibility of starch have been associated with the proportion of amylose/amylopectin, which is in cereals more or less constant with a value of 25/75. For this reason starch cereals are considered highly digested starch[3], but the efficient digestion has been shown to be critical for babies since they do not have the mature amylolytic activity of pancreas up to 18 months[1]. The starch of dry infant cereals is a pregelatinized starch, which allows an instant solubilization of the ready to eat product[4]. So, the processing of cereals included three main steps: 1) roasting, 2) enzymatic treatment, and 3) drum drying, with the objective of increasing the digestibility of starch and of obtaining a product with better texture, and more acceptable for the consumers.

The enzymatic treatment included the addition of α-amylase, which hydrolysed the amylose and amylopectin chains (unions 1-4 glucosidic) giving dextrin of low molecular weight[2]. After this treatment and as a result of the higher content of dextrin, the infant cereals carbohydrates are more digestible.The aim of the present study has been to understand the effect of technological treatment on the dextrin content of four commercial infant cereals (multicereals, wheat, mix of cereals without gluten and rice with carrots).

2 MATERIALS AND METHODS

2.1.Samples

The samples used in this study were four commercial infant cereals manufactured by Hero Spain, S.A. (Alcantarilla, Murcia). The samples selected showed different cereal composition, and were grouped in infant cereals with gluten and without gluten. The infant cereals with gluten were commercially called "multicereals" and "wheat", whereas the infant cereals without gluten were called "mix of cereals without gluten" and "rice with carrots". The four samples were classified according to the technological stage in: raw flour(A), roasted flour(B), drum dried flour(C) and commercial product(D).

2.2.Determination of dextrin

To determine dextrin content a method was developed based on the solubilization with water and subsequently colorimetric determination. One gram of sample was mixed with 30 ml of distilled water and placed in a water bath at 37ºC for 1 hour, shaking the samples continuously. After that, the samples were centrifuged   at 2900 g during 10 min. The supernatant was obtained and 5 ml were made up to 50 ml with distilled water. The colorimetric reactions were developed with 5 ml of diluted sample and 100 ml of iodo reagent, reading the absorbance at 450 nm in a spectrophotometer of molecular absorption Hitachi U-2000. Dextrin were quantified according to the standard curve prepared with different concentrations of dextrin.

3 RESULTS AND DISCUSSION

Table 1 shows the evolution of dextrin content in the four commercial infant cereals according to the manufactured process. In a general view, we can observe that in all samples of infant cereals studied the content of dextrin increased significantly during industrial processing, as a result of the heat treatment and the effect of α-amylase on the amylopectin and amylose chains. So, the dextrin content increases slowly after the roasted process of the flour, since the heating could lead to a partial hydrolysis of starch. This effect can be clearly observed in the samples "wheat" and "mix of cereals without gluten", whereas it is insignificant in samples "multicereals" and "rice with carrots". The enzymatic treatment following the drum dried led to obtaining samples with high content of dextrin with a mean values of 25.61, 41.01, 28.22 and 21.89 g/100 g. However in the commercial product the dextrin content decreased per 100 g of sample, as a result of the addition of other constituents to the ready to eat instant cereals, such as vitamins, minerals and flavours. Among all samples studied, the wheat infant cereals gave the highest content of dextrin with a mean value of 35.5 g/100 g, which could be due to the effect of the natural amylases present in the grain of wheat[5]. The other samples showed values around 20 g/100 g, giving the "mix cereals without gluten" a high value as result of the content of corn in their formulation, since a great proportion of soluble carbohydrates in this cereal grain are constitued by dextrin[2]. We may conclude that "wheat" infant cereal could be better digestible for weaning babies, providing more energy and nutrients, but it will be interesting to carry out studies *in vitro* or *in vivo* to determine the starch digestibility.

Table 1 *Changes in dextrin content in infant cereals*

Infant cereals	Processing Stage				Significance
	A	**B**	**C**	**D**	
Multicereals	0.0 ± 0^a	1.2 ± 0.6^a	25.6 ± 0.3^b	21.5 ± 0.8^b	***
Wheat	0.0 ± 0^a	12.2 ± 0.1^b	41.0 ± 0.14^c	35.6 ± 0.1^c	***
Mix of cereals without gluten	0.6 ± 0^a	7.1 ± 0.3^b	28.2 ± 1.6^d	22.2 ± 1.4^c	***
Rice and carrots	4.1 ± 0^a	4.1 ± 0^a	21.4 ± 1.5^c	17.0 ± 0.5^b	***

Stages of processing:
  A= raw flour
  B= roasted flour
  C= drum dried flour
  D= commercial product

*** $P < 0.001$
$^{a-d}$Letters not sharing the same superscript within the row are significantly different for $P < 0.05$.

References

1. Department of Health, Weaning and the weaning diet. 1995 Report on Healt and Social Subjects Nº 45. London. HMSO.
2. Kent, N.L. 1987. Tecnología de los cereales. Introducción para estudiantes de Ciencia de los Alimentos y Agricultura. Ed. Acribia S.A., Zaragoza, España.
3. Dreher, M.L.; Dreher, C.J. and Berry, J.W. 1984. Starch digestibility of foods: a nutritional perspective. CRC Critical Reviews in Food Sciences and Nutrition, 20 (1), 47-71.
4. Cheftel, J.C. and Cheftel, H. Introducción a la Bioquímica de los Alimentos. Ed. Acribia S.A., Zaragoza, España.
5. Belitz, H.D. and Grosch, W. 1988. Quimica de los Alimentos. Ed. Acribia S.A., Zaragoza, España.

INFLUENCE OF FERMENTATION ON COLOUR QUALITY OF RED BEET JUICES

Peter Šimko [a] – Milan Drdák [b] – Milan Kováč [a] - Jolana Karovičová [c]

[a] Food Research Institute, Bratislava, Slovak republic
[b] Technical University, Brno, Czech republic
[c] Slovak Technical University, Bratislava, Slovak republic

1 INTRODUCTION

Production of red beet concentrate and powder are studied very extensively. Fermentation of fresh red beet juice is a suitable process for reducing the amount of saccharides to yield high pigment content and pure concentrate. The yeasts *Candida utilis*[1] and *Saccharomyces cerevisiae*[2] were used in fermentation process in past. During the fermentation with *Aspergilus niger* and *Saccharomyces oviformis*[3], saccharides were utilised effectively, even the content of vulgaxantin was lowered. The aim of this work was to follow the influence of fermentation with different strains of *Saccharomyces cerevisiae* on red beet pigment composition.

1.1 Red beet juice

Red beet was crushed, pressed, and centrifuged. The pasteurised juice was inoculated with $1 . 10^7$ vital cells/ml after having adjusted the pH to 5.8 with 5 % HCl solution. The fermentation was maintained at 26 °C in closed dark vessels and monitored after 40, 60, and 180 hrs for contents of individual pigment compounds as well as saccharides – sucrose, glucose and fructose.

1.2 Strains

In the fermentation, a strongly fermentative strains of mesophilic yeast *Saccharomyces cerevisiae* (Oenoferm, Erbslöch Geisenheim), five new isolated strains (marked for working purpose as 6C, 11C, IO90, 3A, 74 F) characterised as ethanol resistant, suitable for secondary wine fermentation, and *Saccharomyces oviformis* (Tokay 76 D) were used. *Saccharomyces cerevisiae* Oenoferm was revitalised by the instructions of producer and other were grown on malt extract broth.

The total colour of fermented juices was determined spectrophotometrically, individual colour compounds and saccharides by HPLC.

2 RESULTS & DISCUSSION

Successful production of red beet (*Beta vulgaris* var. *rubra*) pigment concentrate using fermentation requires rapid initiation and progress of the fermentation. Selection of a suitable yeast is based on this basic requirement. The suitable yeast for red beet juice fermentation, which has a high content of sucrose, has to start rapidly sacharides degradation and effect to the pigment decomposition by enzymes should be lowered to a minimum.

The samples were monitored for the compound contents after 40, 60 and 180 hr. The last reading was made only for determination of equilibrium state among the forms of pigments. The portions of individual pigments, mainly betanine, isobetanine, betanidine and isobetanidine (Table 1) were expressed as relative percentage on the basis of the total area of peaks in the chromatograms. Degradation products were shown as individual peaks in the chromatograms.

In general, different percentages of individual compounds were detected after fermentation. Also different portions of degradation products and different levels of cleavage of glucoside from betanine formed during fermentation. From the data, we could not determine if it was due to enzyme formed during fermentation or it was a result of chemical hydrolysis after fermentation. The pH values after 180 hrs were as follows: Oenoferm – 4.6, 11C – 4.3, IO90 – 4.7, 3A – 4.2, 6C – 4.3, Tokay 76D – 4.2, 74 F – 4.2. It is interesting to compare the decrease of red beet pigments determined spectrophotometrically and expressed as betanine concentrations. In table 2, the measured values and retention R [%] are shown. With regard to requirement to be finished within 40 – 48 hrs, the strains 11C, 6 C and Oenoferm or IO90 and 3A were comparable. The activity of the strains Oenoferm, 11C, 6C, 76 D and 3A could also be compared on the basis of pigment retention and average percentage of initial pigment content.

The percentage of pigment in the total solids of pigment concentrate or powder was greatly effected by the resultant saccharide concentration (shown in Table 1). The average values determined refer to the different rates of sucrose inversion caused β-D-fructofuranosidase and to the different rates of utilisation of glucose and fructose.

On the basis of results we recommend as the most suitable strains for fermentation of red beet juice Oènoferm, IO90 and 74 F. This recommendation is a compromise between the relation of pigment retention and saccharide utilisation.

References
1. J.P. Adams, J.H. von Elbe, and C.H. Amundson, *J. Food Sci.*, 1976, **41**, 78.
2. Z.K.Chkeidze, USSR patent No. 449 922
3. H.Pourrat, B.Lejeune, F.Regerat, and A.Pourat, *Biotech. Lett.*, 1983, **5**, 381.

Table 1 Percentage of individual red beet pigments and concentrations of saccharides during fermentation

Strain of Saccharomyces cerevisiae	Time of fermentation [hrs]	Percentage of individual red beet pigment					Concentration [g . l⁻¹]		
		Betanine	Isobetanine	Betanidine	Isobeta-nidine	Degradation products	Fructose	Glucose	Sucrose
Oenoferm	40	36.32	6.69	45.69	8.03	3.25	9.8	8.8	+
	60	11.59	4.29	68.33	14.23	3.54	6.0	5.1	-
	180	2.83	3.92	72.48	14.77	4.37	5.1	4.7	-
11 C	40	41.46	8.78	38.13	8.36	3.25	23.1	10.4	5.0
	60	8.51	2.85	69.64	14.07	4.93	11.6	5.6	1.6
	180	7.84	2.09	70.60	14.37	5.05	7.6	3.9	-
IO90	40	21.58	4.84	59.94	10.65	3.01	9.6	5.4	3.4
	60	10.63	3.75	70.37	11.74	3.52	8.5	6.0	0.3
	180	3.84	1.09	75.82	14.29	5.31	6.6	2.5	-
6C	40	31.21	6.00	53.15	7.09	2.25	18.0	11.1	5.0
	60	18.80	3.86	63.64	10.41	3.27	9.1	4.9	0.2
	180	10.49	2.80	65.70	14.70	6.26	8.3	4.8	-
3A	40	27.24	5.26	55.52	9.25	2.70	18.1	10.3	7.5
	60	14.73	4.06	67.96	10.42	2.82	11.9	11.7	1.3
	180	11.20	2.68	68.00	10.77	7.37	5.4	10.2	-
Tokay 76 D	40	33.24	6.79	48.87	8.12	2.96	10.2	9.1	1.4
	60	18.64	4.10	60.11	13.27	3.85	7.6	7.6	+
	180	25.68	6.76	47.67	8.26	8.63	6.3	5.1	-
74 F	40	21.41	4.55	60.32	9.92	3.71	17.3	10.0	6.5
	60	13.65	3.69	66.04	12.88	3.72	8.2	4.8	1.0
	180	18.70	4.87	54.01	10.30	12.20	7.1	1.7	-
Pasteurized juice		81.18	17.12	-	-	1.69	4.7	5.0	54.1

+ Traces
- Not detected

Table 2 Changes in total colourness during fermentation
expressed as betanine concentration and its percentage retention

Strain of Saccharomyces cerevisiae	Time of fermentation [hrs]	Concen-tration of red beet pigment C[mg . l $^{-1}$ betanine]	Retention [%]
Oenoferm	40	374.5	62
	60	343.2	55.4
	180	227.8	36.8
11 C	40	374.5	62
	60	349.0	56.3
	180	224.9	36.3
IO90	40	357.8	57.7
	60	314.8	50.8
	180	199.5	32.2
6C	40	371.5	59.9
	60	310.9	50.2
	180	177.9	28.7
3A	40	351.0	56.6
	60	305.0	49.2
	180	180.9	29.2
Tokay 76 D	40	346.1	55.8
	60	205.1	49.2
	180	104.6	16.9
74 F	40	339.2	54.7
	60	300.2	48.4
	180	84.1	13.6
Pasteurized juice		619.8	100

INVESTIGATION OF *ASPARAGUS OFFICINALIS* L. FLAVOUR BY SENSORY METHODS AND GAS CHROMATOGRAPHY

D. Ulrich, E. Hoberg and D. Standhardt

Federal Centre for Breeding Research on Cultivated Plants, Institute for Quality Analysis,
Neuer Weg 22/23,
D-06484 Quedlinburg,
Germany

1.INTRODUCTION

Asparagus belongs to the favourite vegetables in Europe. In Germany the yearly consumption is about 1 kg per capita. In addition to its high nutritional value it is considered a delicacy. It can be assumed that quality, especially sensory attributes, will become the most important competition factor for breeders and producers in the case of saturated markets.

Considerable studies on the flavour of cooked asparagus have been carried out in the past (1,2,3). Nevertheless only little information exists concerning the quantitative contribution of the flavour compounds to the typical cooking flavour and their dependence on variety and climate. A comprehensive system of sensory and instrumental methods was developed to obtain basic information on the components determining asparagus flavour.

2. MATERIALS AND METHODS

Eight asparagus cultivars (CALET, GIJNLIM, HORLIM, HUCHELS ALPHA, HUCHELS AUSLESE, RECORD, THIELIM, and VULKAN) were harvested in 1996 and 1997 in the experimental station of the German Federal Office of Plant Varieties (BSA) in Neuhof (North Germany) and of the Research Station Geisenheim in Ingelheim (South Germany). Every year the sampling was repeated after 14 days, and in 1997 there were two repetitions for every sampling. The asparagus spears were cut randomly, not in a defined quality. The fresh asparagus was shipped cool, peeled and analysed one day after sampling. The same sample was used both for instrumental and sensory analyses.

2.1 Sensory Evaluation

After collecting a choice of altogether 18 sensory features their quantification was trained. In dependence of the concrete parameter the characters were defined both verbally and with reference substances.

The sensory evaluation was carried out by a trained panel consisting of 15 members with a maximum of five samples per session, two sessions a day. The asparagus spears were washed and divided into 2-3 cm long pieces. Sample was boiled only with water in a pressure cooker for 7 minutes. The panelists quantified the features on a non-graduated 10 cm long linear scale. The acceptability was evaluated with 'very high' [5], 'high' [4], 'middle' [3], 'low' [2] and 'very low' [1]. Every variety was tested eight times in an experimental period.

2.2 Flavour Analysis

The asparagus samples were prepared for further analyses using the simultaneous steam distillation-extraction (SDE). The apparatus originally described by Likens and Nickerson was modified for using solvents with a specific weight higher than water (e.g. 1,1,2-trichloro trifluoro ethane). The asparagus samples were cut into 1 cm long pieces and cooked directly in the SDE-apparatus. The Likens-Nickerson extracts were concentrated from 30 ml to 100 µl in a modified Kuderna-Danish concentrator fitted with a three-ball Snyder column.

Gas chromatography-olfactometry: Hewlett-Packard 5890 A, injector 180 °C; split 10 ml; column: HP-INNO wax 30 m x 0.32 mm ID / 0.5 µm; 2 ml hydrogen / min; temperature programme: 50 °C - 3 grd/min - 180 °C; FID   250 °C; split of carrier gas: 50 % to FID and 50 % to sniffing port. For the aroma extract dilution analysis the extracts were stepwise diluted with solvent in a geometrical line. The sniffing panelists were five specially trained members of the sensory panel. Gas chromatography-mass spectrometry: Hewlett-Packard MSD 5972 with HP 5890 Series 2 plus, data base NBS75K and Wiley138, column: INNO-wax 60 m x 0.25 mm ID / 0.5 µm; 1 ml helium / min; temperature programme: 50 °C - 2 grd/min - 180 °C.

3. RESULTS

3.1 Sensory Evaluation

The typical asparagus flavour is mainly determined by the key compound 'dimethylsulfide'. It gives the dominant sensory sensation (4). The essential sensory parameters are: smell - typical, sweet-like, pungent, musty, sulfurous, earthy, untypical; taste - sweet, bitter, tasteless; mouth sensation - typical, astringent, metallic, untypical; aftertaste - bitter; texture - crisp, tough, stringy. The evaluation of the eight varieties resulted in significant sources of variability (at the error level of 5 %) for 14 sensory characters (5). The results also offered, that some odour parameters significantly differ between the varieties. Consequently the sensory profiles of the varieties are not only formed by the three odour parameters sweet-like, pungent, and untypical, but also by the taste parameters sweet and bitter, as well as by the mouth sensation astringent, crisp, and stringy.

3.2 Flavour analysis

The results of the aroma extract dilution analysis (AEDA) of the variety 'VULKAN' are shown in figure 1. The FD-chromatograph is compared with the total ion chromatograph obtained by GC/MS.

4. CONCLUSION

The presented sensory method is available for discrimination of asparagus varieties. This reflects the genetic background. But the environment influences the synthesis of flavour determining compounds. Therefore the measurement by instrumental methods is required for objective assessment of sensory quality.

A special method for sample preparation is used to extract the volatile compounds. GC instruments equipped with MS, FID, NPD and sniffing port are applied to identify the volatiles. The six most intensive sniffing sensations with flavour dilution factors (FD-factors) between 128 and 4096 correspond to the following chemical substances: 2,3-butanedione, 3-methyl thiopropanal, and four different substituted alkyl pyrazines. For higher conformity of sensory and instrumental results the substance identification will have to be carried out especially for nitrogen- and sulfur-compounds.

The coordinated application of the above mentioned methods and the integration of all results will lead to new conclusions for further research and selection on asparagus quality.

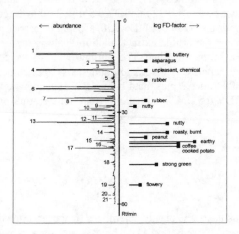

Figure 1 *Total ion chromatograph (TIC) and the result from aroma extract dilution analysis (AEDA - FD chromatograph) of an asparagus extract, variety 'VULKAN' 1997. Chemical substances (TIC): (1) 2,3-butanedione; (2) S-compound; (3) 2,3-pentanedione; (4) hexanal; (5) 3-penten-2-one; (6) pyridine; (7) 2-pentyl furan; (8) 1-pentanol; (9) 2-methyl pyrazine; (10) S-compound; (11) (E)-2-heptenal; (12) 2,6-dimethyl pyrazine; (13) 1-hexanol; (14) N-compound; (15) (E)-2-octenal; (16) 3-methyl thiopropanal; (17) furancarboxaldehyde; (18) benzaldehyde; (19) benzeneacetaldehyde; (20) nonadienal; (21) decadienal. The key compound dimethylsulfide is not detectable with this method.*

5. REFERENCES

1. R.Tressl, M. Holzer and M. Apetz, J. Agric. Food Chem. 1977, **25**, 455-459
2. R.Tressl, M. Holzer and M. Apetz, J. Agric. Food Chem. 1977, **25**, 459-463
3. R. Fellous and A. Puill, E.P.P.O.S. 1980, **63**, 83-87
4. E. Hoberg, D. Ulrich and D. Standhardt, XXXIII. Vortragstagung der Deutschen Gesellschaft für Qualitätsforschung (Pflanzliche Nahrungsmittel) e.V., 'Krankheitsresistenz und Pflanzenschutz', 23. - 24. 3. 1998, Dresden
5. E. Hoberg, D. Ulrich and D. Standhard, IX. International Asparagus Symposium, Washington State University Tri-Cities, USA, 15. - 17. July 1997

Session VII Quality Improvements and Functional Foods

FUNCTIONAL FOODS: CHALLENGES FOR PLANT FOOD DEVELOPMENT

K. Poutanen, R. Puupponen-Pimiä and K.-M. Oksman-Caldentey

VTT Biotechnology and Food Research,
POB 1500, FIN-02044 VTT,
Finland

1 INTRODUCTION

Development of health-promoting or functional foods is one of the important megatrends in food product development, and the markets for products with health benefits are estimated to expand. Modern nutrition science increases the knowledge of diet-health interactions, and biomarkers are being searched for studying the efficacy of different fractions in plant raw materials[1]. Developing novel food products which also meet the consumers´ demand for convenience and sensory appeal is a great challenge and opportunity for the modern food industry.

Functional food products should contain in adequate concentrations one or several compounds which have positive cellular or physiological effects in the body, and thus justify functional or health claims. The most common ingredients suggested and/or used in functional foods are listed in Table 1. In EU health claims are not allowed, and the use of composition claims or functional claims is regulated nationally. In Japan a category of FOSHU (Foods For Specific Health Use) has existed since 1991, and is regulated by Ministry of Health and Welfare[2]. In USA several types of health claims are allowed under NLEA (Nutrition Labeling and Education Act 1990) many of which refer to a diet rich in fruits and vegetables (Table 2).

Table 1 *Typical ingredients in functional foods*

• Dietary fibre	• Plant phenols
• Prebiotics	• Bioactive peptides
• Probiotics	• Minerals (calcium)
• ω-3-fatty acids	• Vitamins and carotenoids
• Phytosterols	(*e.g.* vitamins C, E and β-carotene)

Table 2 *Allowable health claims in the USA*

Recommended diet	Protective effect against
High in calcium	Osteoporosis
High in fiber-containing grain products, fruits and vegetables	Cancer, cardiovascular disease
High in fruits or vegetables (high in dietary fiber or vitamins A or C)	Cancer
Low in fat	Cancer
Low in saturated fat and cholesterol	Cardiovascular disease
Low in sodium	High blood pressure

2 PLANT FOODS: A VERSATILE SOURCE OF FUNCTIONAL COMPONENTS

Unrefined foods of plant origin have long been associated with a healthy diet. Progress in research into the different plant components has revealed the presence of compounds with either established or potential health benefits (Table 3). Dietary fibre, vitamins acting as antioxidants, minerals, ω-3-fatty acids, and several sulphur compounds present in garlic and related *Allium* plants already are traditional ingredients with positive nutritional properties, although more is being learnt about their role in reducing risk of various diseases[3].

Prebiotic oligosaccharides are a relatively new concept for modulating colon microflora and in that way causing several positive physiological functions. Plants also synthesize numerous primary and secondary metabolites which have not hitherto been considered nutritionally important, and have only recently been shown to have potential effects in protecting against various diseases. Bioactive peptides and proteins present in various plants *e.g.* protease inhibitors in soya, form an interesting group of health promoting primary metabolites. Ubiquituous plant sterols and phenols, such as flavonoids and phytoestrogens, as well as glucosinolates and their breakdown products specific for cruciferous vegetables, are examples of plant secondary metabolites. All these various components of the plant raw material are at present a target of intensive research internationally and offer a good base for development of functional foods and ingredients.

2.1 Prebiotics

Non-digestible oligosaccharides are widely present in various vegetables, and their positive effects on beneficial lactic acid bacteria in the intestinal tract are currently the object of intensive investigation. The concept of prebiotics was introduced only a few years ago (Table 4), but especially fructo-oligosaccharides are already in expanding commercial use. Fructo-oligosaccharides of variable chain length, generally DP<10, are being produced from inulin, a β-1,2-fructan with DP up to 60. Depending on the oligo-saccharide structure the products also have technologically beneficial functional properties. Plants with a high inulin content include chicory roots and Jerusalem artichoke (15-20%), parsley and leek (10-15%), garlic (15-25%), and onion (2-6%)[4]. The most common functional claim connected to fructo-oligo-saccharides and other oligo-saccharides, such as soya oligosaccharides, galacto- and xylo-oligosaccharides, is their bifidogenic character *i.e.* the ability to increase the amount of *Bifidobacteria* in the human gut. The other physiological functions of oligosaccharides, such as their role in cell proliferation and cancer, as well as in lipid metabolism, are currently being widely studied.

Table 3 *Health-promoting compounds in vegetables*

Well-known and beneficial health effects	*New-comers, health effects under study*
Vitamins and carotenoids • B complex, C-, E-, A and K vitamins • β-carotene	Prebiotic oligosaccharides • Fructo-oligosaccharides • Galacto-oligosaccharides • Xylo-oligosaccharides
Minerals • Calcium • Magnesium • Iron • Selenium	Bioactive peptides and proteins Phytochemicals • Flavonoids and phenolic acids • Phytoestrogens: lignans and
ω-3-Fatty Acids	isoflavonoids • Phytosterols
Dietary Fiber • Water soluble fibers • Water insoluble fibers	• Glucosinolates and their breakdown products
Sulphur compounds in *Allium* plants	

Table 4 *The concept of probiotics, prebiotics and synbiotics according to Gibson and Roberfroid* [5]

A **probiotic** is a live microbial feed supplement which beneficially affects the host animal by improving its intestinal microbial balance.
A **prebiotic** is a non digestible food ingredient that beneficially affects the host by selectively stimulating the growth and/or activity of one or a limited number of bacteria in the colon, and thus improves host health.
A **synbiotic** is a mixture of pro- and prebiotics

2.2 Phytoestrogens

Many plants contain precursors of biologically active hormone-like compounds. Two groups of diphenolic phytoestrogens, lignans and isoflavonoids, are the subject of intensive research, and their role in sex hormone performance, protein synthesis, growth factor action, malignant cell proliferation, cell differentiation and cell adhesion and angiogenesis is being studied. Plant phytoestrogens are glycosylated diphenolic compounds which are deglycosylated and converted to the biologically active hormone-like compounds by intestinal bacteria in the human colon. Isoflavonoids, such as genistein and daidzein, are found especially in soybean products, mung beans and peanuts, whereas lignans, such as secoisolariciresinol and matairesinol, are particularly abundant in flaxseeds, and also present *e.g.* in whole grain cereals, beans, some vegetables and berries, and in tea[6].

The potential of phytoestrogens as cancer-preventive compounds, especially for breast, prostate and colorectal cancer, is an interesting hypothesis, and some epidemiological evidence, as well as *in vitro* and animal experiments point in this direction. However, no *in vivo* data with pure compounds are as yet available. The role of phytoestrogens in prevention of cardiovascular disease and osteoporosis is another interesting research topic, and they also seem to have a role in releaving menopausal symptoms in women[6].

2.3 Bioflavonoids

Flavonoids form a large group of plant phenolic secondary metabolites that occur naturally in vegetables and fruits, and in beverages such as tea and wine. More than 4000 individual compounds have been identified. The most important groups of flavonoids are anthocyanins, flavonols, flavones, cathechins and flavanols. Quercetin, which is a flavonol, is a common bioflavonoid in various vegetables. The flavonoids are potent antioxidants, free radical scavengers and metal chelators; they inhibit lipid peroxidation and exhibit various physiological activities (including anti-inflammatory, anti-allergic, anti-carcinogenic, antihypertensive and anti-arthritic activities). Vegetables are good sources of flavonoids. Onions, kale, broccoli and endive contain flavonoids over 50 mg/kg, and lettuce and tomato between 10-50 mg/kg[7]. Bioflavonoids are currently the object of intensive investigation in Finland and elsewhere. Many epidemiological studies have indicated that high flavonoid consumption is associated with reduced risk of chronic diseases like cardiovascular disease[7]. However, data on health effects of flavonoids are still limited. More research is needed on flavonoid absorption, metabolism and biochemical action.

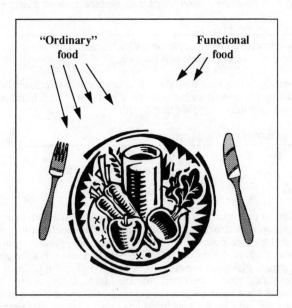

Figure 1 *Functional foods offer the consumer new alternatives to build up health-promoting diet*

3 THE FUTURE OF FUNCTIONAL FOODS

The basis for development of functional foods lies in research on nutrition, public health and medical sciences. For the creation of successful products, however, an even more multidisciplinary approach is required, and food technologists, chemists, physicians, plant physiologists, molecular biologists, consumer scientists, informatians and many others have to join the team. The first challenge is to create a safe, palatable and stable food with desired physiological functions. Another important issue is the achievement of adequate and reliable documentation of the systemic effects, and formulation and information dissemination of the claims. In this context it is essential that rules will be set for claims and labelling. The establishment of a regulatory framework will assist food manufacturers in marketing their products, and make it more clear to consumers what to expect of foods.

The relationship between diet and health will remain an intensive research even beyond the year 2000. The scientific developments will have a major impact on the food industry. It is evident that more biomarkers for diseases and disorders are needed to link the findings of epidemiologists, nutrition physiologists and food technologists to a useful chain of information. Different food categories will be created, such as enhanced variants of traditional foods, disease-specific foods, risk group specific foods, foods to reduce the effects of aging, and performance foods[8]. Consumers will diversify not only in their sensory preferences but also in their nutritional requirements.

Being rich in known and potential health-positive compounds, plants are ideal raw materials for functional foods. As one of the important functions of food for the consumer is to be the source of hedonism, the winning foods will be both delicious and healthy. The target should be a nutritious, balanced and palatable diet (Fig.1), and food manufacturers will in the future multiply the consumer´s possibilies to assemble this.

References

1. J.D. Potter and K. Steinmetz, Your mother was right - eat your vegetables: The role of plant foods in cancer. In: "Dietary Fibre in Health and Disease", D. Kritchevsky and C. Bonfield, eds, Eagan Press, St. Paul, Minnesota, USA, 1995, p. 191.

2. T. Ichikawa, Functional foods in Japan. In: "Functional foods: Designer foods, pharmafoods, nutraceuticals", I. Goldberg, ed., Chapman & Hall, Inc. New York, 1994, p. 453.

3. K. Griffiths, H. Adlercreutz, P. Boyle, L. Denis, R.I. Nicholson and M.S. Morton, "Nutrition and cancer", Isis Medical Media, Oxford, UK, 1996, Chapter 4, p. 25.

4. C.E. Westerdijk, *Agro Food Ind. Hi-Tech.*, 1997, **Jan/Feb**, 5.

5. G.R. Gibson and M.B. Roberfroid, *J. Nutr.*, 1995, **125**,1401.

6. H. Adlercreutz and W. Mazur, *Ann. Med.*, 1997, **29**, 95.

7. M.G.L. Hertog, D. Kromhout, C. Aravanis, H. Blackburn, R. Buzina, F. Fidanza, S. Giampaoli, A. Jansen, A. Menotti, S. Nedeljkovic, M. Pekkarinen, B.S. Simic, H. Toshima, E.J.M. Fesfens, P.C.H. Hollman and M.B. Katan, *Arch. Intern. Med.*, 1995, **155**, 381.

8. J.T. Winkler, The future of functional foods. In "Functional foods: The consumer, the products and the evidence", M.J. Sadler and M. Saltmarsh, eds, The Royal Society of Chemistry, Cambridge, 1998, p. 184.

UNCONVENTIONAL VEGETABLE CLUSTERBEAN FOR THE SEMI-ARID TROPICS

M.L. Saini

Department of Plant Breeding,
CCS Haryana Agricultural University,
Hisar - 125 004, India

1. INTRODUCTION

Clusterbean (*Cyamopsis tetragonoloba* (L.) Taub.) is commercially grown in India, Pakistan, United States and to a limited extent in Australia, Brazil and South Africa. Being a drought tolerant legume, it is grown in India since ancient times in arid and semi-arid conditions as vegetable, seed, fodder and green manure crop. Here, green and tender edible pods are used as a favourite vegetable in rural areas and also in many parts of the country, especially in south. Immature pods are dried and fried like potato chips. These can be cooked like Frenchbean. Green pods are dried, salted and preserved for future use. Both the pubescent and glabrous varietal types are used for vegetable purpose. Ainslie reported use of clusterbean pods as vegetable in 1813. However, in 1809 Maria Graham in a book, about life in Bombay had cited the use of clusterbean as vegetable. She had described them as some plants producing long pods which when cut into small pieces are so exactly like Frenchbeans, that it is difficult to differentiate the two. About their availability she had observed that they are plentiful throughout the year.

2. PHOTO-INSENSITIVE TYPE FOR VEGETABLE

Under north Indian conditions, clusterbean is generally photo-sensitive i.e. it comes into flowering and fruiting when sown in *kharif* season only. When grown in spring (summer), there is only vegetative growth with little or no flowering. Thus, photo-sensitivity imposes a limitation on clusterbean cultivation in the northern belt of the country. A number of genotypes have been reported as photo-insensitive by various workers in vegetable types like, Pusa Sadabahar[3], Pusa Naubahar[3,4] and Sel 28-1-1[4]. Use of these photo-insensitive genotypes would therefore be highly desirable. With such varieties in hand, it will be possible to augment clusterbean seed and vegetable production by use of surplus water after *rabi* harvest.

3. EFFICIENT PLANT TYPE FOR VEGETABLE

Pods per cluster and pod length are postively correlated. Heritability and genetic advance estimates are highest for yield[5]. Moderate to high heritability values for pod length and seed weight have been reported[6] which indicates that improvement in either would be effective. Among components of green pod yield, number of pods per plant, clusters per plant, pod length and 100-pod weight are important. Green pods per plant and 100-pod weight were reported to be the major components of green pod yield[7-9].

4. VEGETABLE CLUSTERBEAN VARIETIES

Pardeshi is a vegetable type clusterbean grown for years in Gujarat. This variety is branched, produces long pubescent pods about 12-15 cm long, and grows to a height of about 1.8 m[10-12].

Sofia has been grown for years in Gujarat as a multipurpose crop. The pods are used as vegetable and also sown as a shade plant for ginger. When the pods are plucked, the side leaves are stripped off the plant and left on the ground as a green manure for ginger. Sofia is sparsely branched and grows to a height of 1.8-3.00 m [11,12]. Anand Agricultural Institute, Gujarat during 1956-57 released four vegetable varieties, namely, G.3B-2796, G. 4-29B-17, G. 4-298-25 and G.4-300-26 which yielded 3476, 4495, 4516 and 2953 lbs/acre green pods, respectively[13-14].

Pusa Mausmi, a vegetable clusterbean variety developed at the Indian Agricultural Research Institute (IARI), New Delhi was selected from the progeny of local variety. Pusa Mausmi is sparingly branched and produces glabrous fleshy and shining pods about 10 cm long. The variety is recommended for rainy season[15-17].

Pusa Sadabahar, a vegetable type clusterbean variety developed at IARI, New Delhi, was selected from the progeny of local variety Jaipuri. Pusa Sadabahar is a non-branching variety that produces pubescent pods about 12 cm long. This variety is photo-insensitive and can be grown either in summer or in rainy season[15,16].

Pusa Naubahar, a vegetable type clusterbean developed at IARI, New Delhi, was selected from a cross between Pusa Sadabahar and Pusa Mausmi. The cross could effectively combine the photoinsensitivity of Pusa Sadabahar and smooth shining character of Pusa Mausmi. Pusa Naubahar is unbranched and produces glabrous pods about 15 cm long, which can be picked 40 days after sowing and yield 6,000 kg ha[-1] of green pods[17-18].

Punjab 4 is a vegetable variety developed at Fodder Research Station, Sirsa, by selection from a local variety. Punjab 4 is a tall, non-branching variety with pubescent pods that flowers early and continues to form pods for a long period during the growing season[19].

Durgabahar is a vegetable variety developed at the Agriculture Research Station, Durgapura, Jaipur. It is a cross between RGC401 a white flowered pubescent variety and Pusa Naubahar. Durgabahar is high yielding, unbranched, white flowered, glabrous and photo-insenstive variety released in 1984. It bears long (13-14 cm), smooth and fleshy pods at every node from the base of the plant and is moderately resistant to bacterial blight and powdery mildew[20].

5. VEGETABLE PRODUCTION TECHNOLOGY

5.1 Soil

In India, Pakistan and United States, clusterbean is confined mainly to arid and semi-arid regions[21]. Although clusterbean grows well in a variety of soils, it thrives best in sandy, sandy loam[22-25] and coarse-textured alluvial soil[26] with well drained subsoils. Clusterbean does not grow well on heavy black soils[27]. Soils with pH of 7.0 are better suited for clusterbean production than are acidic soils[28]. Clusterbean plants cannot withstand excess moisture or water logging during the growing season[29]. Singh and Sikka[30] recommended 100 kg superphosphate per ha. The fertilizer should be drilled about 15-23 cm deep when the land is being prepared for sowing. 10 kg N ha[-1] and 20 kg P_2O_5 ha[-1] applied to clusterbean significantly increased green pod yield by 38 per cent[31].

Quality of seed of Pusa Naubahar was not affected by fertilizer rate but was higher for seeds harvested after removal of two fresh pod harvest[32]. The seed yield of Pusa Sadabahar was highest at low spacing 10 cm (2.0 t ha[-1]) and treatment with cycocel 2000 ppm (2.1 t ha[-1])[33].

5.2 Date of planting

Vegetable varieties of clusterbean are sown at the beginning of the monsoon in late June or early July. In Gujarat and northern India two crops of clusterbean are grown : one at the onset of the monsoon as *kharif* crop and other under irrigated condition during hot summer premonsoon months of February to early June[30, 34-36].

In southern India, it is grown under irrigation throughout the year, being cultivated largely in kitchen garden[35-37]. As a vegetable crop clusterbean is irrigated in Gujarat and northern and southern India[27, 36, 37]. In addition clusterbean is grown in many countries as a kitchen garden vegetable by inhabitants of Indian and Pakistani origin[38].

5.3 Row spacing and seed rate

Before sowing, the land is ploughed once or twice and harrowed. The seed is either drilled 5-7.5 cm deep in rows 45-60 cm apart or dribbled at a distance of 60 x 60 or 60 x 30 cm in rows. The seed rate varies 3-12 kg ha[-1]. After germination, the crop is thinned and hoed once or twice.

5.4 Irrigation

In some areas of Gujarat as a pre-monsoon summer crop, clusterbean seeds are sown in rows 45-60 cm apart in mixture with okra and various gourds. The first irrigation is given within 15-20 days after sowing with subsequent irrigation at an interval of 7-10 days.

5.5 Harvesting

Harvesting occurs 50-80 days after planting with the tender green pods being picked at an interval of two or three days for several weeks. A yield of 2,000 kg of green pods per ha can be obtained from the pre monsoon crop, whereas it is 2,500-3,000 kg green pods per ha if crop is grown as vegetable during *kharif* season.

6. QUALITY OF CLUSTERBEAN PODS

The pods are quite nutritious vegetables which contain moisture 82.5 per cent, protein 3.7, fat 0.2, fibre 2-3, carbohydrate 9.9, mineral matter 1.4, calcium 0.13, phosphorus 0.25 per cent, iron 5.8 mg/100g and vitamin 49 mg/100g[39]. The chemical composition of Pusa Naubahar clusterbean pods wall and seed at different stages of growth is given in Table 1. Data in the table reveal that clusterbean pods are quite nutritious as the

Table 1   *Chemical composition of Pusa Naubahar, clusterbean pod wall and seed at different stages of growth*

	Days after anthesis							
	Podwall				Seed			
	15	30	45	75	15	30	45	75
Crude protein (%)	15.94	17.69	17.94	10.25	25.56	28.25	31.38	32.63
Gum (%)	-	-	-	-	6.42	14.25	23.15	23.80
Total soluble sugar (%)	-	-	-	-	13.56	11.21	9.03	7.37

protein content in pod wall at different stages of growth varies from 10.75 to 17.94 per cent, whereas in seed it ranges from 25.56 to 32.63 per cent. Gum content in seed varies from 6.42 per cent after 15 days of anthesis to 23.80 per cent after 75 days of anthesis whereas total soluble sugar ranges from 7.37 to 13.56 per cent. During the entire period of pod development, sucrose constitutes the major portion of the free sugars in the seed (both endosperm and cotyledons) and in the pod wall[40]. From above results, it can be concluded that pods harvested after 30 days of anthesis have high protein and total soluble sugar content but low gum content and this is right stage of pods harvesting.

Clusterbean has medicinal value for various diseases. Leaves of clusterbean are eaten to cure night blindness. Ashed plant is mixed with oil and is used as a poultice on cattle boils. Seeds are use as chemo-therapeutic agent against smallpox and as laxative also. Boiled clusterbean seeds are used as poultices for plague, enlarged liver, head swellings and on swelling due to broken bones. Clusterbean has also been found to lower blood cholesterol and non-insulin requiring diabetics have shown that clusterbean reduces that rise in blood glucose and serum insulin after the meal. These properties can be further improved through breeding.

Acknowledgement

The author is thankful to Dr. U.N. Joshi for helping in biochemical analysis.

References

1. W. Ainslie, Materia Medica of Hindoostan and Actisans and Agriculturalists Nomenclature, Madras, Government Press, 1813.
2. M. Graham, *Journal of a Residence in India* Edinburgh, Archibald Censtable, 1812.
3. H.B. Singh and S. Sikka, *Indian Fmg.*, 1955, **12**(9), 23.
4. S.P. Mittal, B.S. Dabas, T.A. Thomas and D.P. Chopra, *Proc. First ICAR Guar Research Workshop*, Jan. 11-12. *Central Arid Zone and Research Institute Jodhpur*, 1977, 58.
5. R.N. Vashistha, M.L. Pandita and A.S. Sidhu, *Haryana J. Hort. Sci.*, 1981, **10**, 131.
6. R.E. Stafford and G.L. Barker, *Pl. Breed.*, 1989, **103**, 47.
7. U. Menon, M.M. Dubey and R.P. Chandola, *Rajasthan J. Agric. Sci.*, 1973, **4**: 67.
8. U. Menon, M.M. Dubey, K.C. Kala and R.P. Chandola, *Indian J. Hered.*, 1973, **5**, 7.
9. U.Menon, M.M. Dubey and K.C. Kala, *Indian J. Hered.*, 1977, **9**, 21.
10. G.R. Ambekar, 'Bombay Department of Agriculture Bulletin', 1927, **146**: 1.
11. T.F. Main, Bombay Department of Agriculture , 1909-10 *Annual Report*, 1910.
12. P.L. Patel, G.C. Patel and R.M. Patel, *Bansilal Amritlal Agric. Coll. Mag.*, 1951, **4**, 3.
13. U. Menon, *Monograph Secies-2. Depart. of Agric. Rajasthan*, 1973, 51.

14.   R.S. Paroda, and M.L. Saini, *Forage Res.*, 1978, **4A**: 9.
15.   H.B. Singh and S.M. Sikka, *Indian Fmg.*, 1955, **5**: 8.
16.   B.P. Pal, S.M. Sikka and H.B. Singh, *Indian J. Hort.*, 1956, **13**: 64.
17.   S. Premsaker and K. Venkataraman, *Phytomorphology*, 1969, **19**, 303.
18.   S.P. Mital, T.A. Thomas, and B. Srivastava, *Indian Fmg.*, 1963, **13**, 13.
19.   H.C. Malek and P.C. Batra, *Indian Fmg.*, 1956, **5**, 28.
20.   U. Menon, Rekha Mathur and M.L. Mathur, *Guar Newsletter*, 1982, **3**, 30.
21.   D.K. Misra, S. Bhan and R. Prasad, *Allahabad Farmer*, 1968, **42**, 239.
22.   G.R. Ambekar, 'Bombay Department of Agriculture Bulletin', 1927, **146**, 1.
23.   G.R. Ambekar, 'Bombay Department of Agriculture Bulletin', 1928, **150**, 1.
24.   S.P. Raychandhuri, *Bull. Nat. Inst. Sci.* India, 1952, **1**, 266.
25.   W.S. Read, 'Agriculture and Livestock in India', 1936, **6**: 11.
26.   H.H. Mann, *Bombay Department of Agriculture*, 1927, **145**: 1.
27.   B.M. Pugh, 'Guara', Vanguard Press, Allahabad, 1958.
28.   S. Bhan and P. Prasad, *Indian Fmg.*, 1967, **17**, 17.
29.   A.S. Faroda, *Gosamvardhana*, 1968, **16**, 17.
30.   H.B. Singh and S.M. Sikha, *Indian Fmg.*, 1955, **5**, 8.
31.   G.S. Reddy and S. Venkateshwalu, *J. Oilseed Res.*, 1989, **6**, 300.
32.   M.B. Madalageri and M.M. Rao, *Seeds and Farms*, 1989, **15**, 7.
33.   S.V.S. Rathore, K. Puneet and D.R. Singh, *Haryana J. Hort. Sci.*, 1990, **19**, 219.
34.   P.L. Patel, G.C. Patel and R.M. Patel, *Bansilal Amritlal Agric Coll. Mag.*, 1951, **4**: 3.
35.   S.S. Purewal, *Indian Council of Agricultural Research Farmers Bulletin*, 1957, **36**: 1.
36.   A. Subramaniam and S. Premsekar, *Madras Agric. J.*, 1960, **47**: 134.
37.   J. Mollison, 'A textbook of Indian Agriculture', *Advocate of India Press, Bombay*, 1901.
38.   I.H. Burkill, 'A Dictionary of the Economic Products of the Malay Pennisula', University Press, Oxford, 1935.
39.   Wealth of India, CSIR, Govt. Press, Delhi, 1950.
40.   R.Singh, A. Aggarwal, S.S. Bhullar and J. Goyal. *J. Experimental Bot.*, 1990, **41**, 101.

BERRY SEED OILS: NEW POTENTIAL SOURCES OF FOOD SUPPLEMENTS AND HEALTH PRODUCTS

A. Johansson, T. Laine and H. Kallio

Department of Biochemistry and Food Chemistry
University of Turku
FIN-20014 Turku, Finland

1 INTRODUCTION

The oil content and chemical composition of seeds of most plants is affected by geographical location and climatic conditions of the growth site[1,2]. Temperature, in particular, has been shown to affect the fatty acid composition of plants; the degree of unsaturation and, along with that, the fluidity of lipids usually increases with decreasing growth temperatures. It is supposed that this mechanism protects the plant against cold injury and promotes the optimal function of membrane-bound enzymes. In most higher plants, triacylglycerols formed during lipid biosynthesis are accumulated in embryo or endosperm tissue of the developing seed. Palmitic (16:0), stearic (18:0), oleic [18:1(n-9)], linoleic [18:2(n-6)] and α-linolenic [18:3(n-3)] acids predominate the triacylglycerols, however, some oil seeds synthesize unusual, often species- or genus-specific, fatty acids which occur almost exclusively in the tricylglycerol fraction[3]. Climatic temperature has been observed to affect mainly oleic, linoleic and, if present, linolenic acids[2]. Only minor effects have been detected in the proportions of saturated fatty acids.

Seed oil composition correlates with its nutritive value. Seed oils of wild berries contain appreciable amounts of nutritionally important fatty acids. For example, γ-linolenic [18:3(n-6)] and stearidonic acids [18:4(n-3)] which are significant components of the seed oils of all currant species, have important roles as precursors of other long-chain polyunsaturated fatty acids and local hormones in humans[4]. Seed oils of berries have the potential of being utilized as food supplements or health products. Raw material suitable for seed oil production is produced, and at present wasted, in large quantities as the press residue from, for example, the manufacture of marmalades, juices and liqueurs.

Compositional screening study of the seed oils of four berry species was carried out. The aim was to determine whether there is variation according to the growth latitude in the seed oils of crowberry and cloudberry. Another aim was to examine the phenotype/genotype based variation in wild currant species. To achieve this a detailed study on the natural variation of the seed oil fatty acids of alpine currant and northern redcurrant was performed.

2   MATERIALS AND METHODS

2.1 Berry samples

Berry samples of cloudberry (*Rubus chamaemorus*) and crowberry (*Empetrum nigrum*) were collected in Finland in areas from 60.5 to 69.5° N Lat. and 22 to 29° E long. The fourteen samples of each species represented three collecting areas; southern Finland, the Province of Oulu and the Province of Lapland.

Berry samples of alpine currant (*Ribes alpinum*, 55 samples) and northern red currant (*Ribes spicatum*, 51 samples) were collected in the South-West Finland and in the northern Lapland, respectively. The collecting sites of both berries were divided into nine geographically distinct growth areas.

2.2 Methods

Seeds were isolated from berries and dried to a moisture content of 5-6 %. Seed oil of cloudberry and crowberry was extracted by $CHCl_3$–MeOH $(2:1)^{5,6}$; yield was determined gravimetrically. Fatty acids of triacylglycerols were analysed as methyl esters by gas chromatography after NaOMe-catalysed transesterification[7].

A rapid one-step extraction/methylation method (H_2SO_4–MeOH, $1:99)^8$ was applied to the analysis of seed oil content and composition of alpine and redcurrant berries. Triheptadecanoylglycerol was used as an internal standard. The fatty acid methyl esters formed were analysed by gas chromatography. The oil content of the seeds was calculated from the yield of fatty acid methyl esters in relation to the internal standard.

Comparisons between the different geographical groups were carried out using a one-way analysis of variance. Specific differences were identified using the Newman-Keuls test.

3   RESULTS AND DISCUSSION

3.1 One-step extraction/methylation method

Berry seeds of a cultivated redcurrant variety were used as reference material when the oil yields for the $CHCl_3$–MeOH and rapid one-step methods were measured (Table 1). The extractions were done in triplicate by both methods. Reproducibility of the two methods was equal. When compared with the yield of $CHCl_3$–MeOH method, the yield of one-step method was about 80 %. However, when the fatty acid compositions of oils produced by the two different methods were compared, highly similar fatty acid compositions could be seen (Table 2).

3.2 Cloudberry

The oil content of cloudberry seeds varied from 9.1 to 12.4 % of the dry weight. The southern cloudberry seeds had lower ($p<0.05$) oil contents compared with the more northern counterparts. Cloudberry seeds collected in the southern Finland were significantly heavier ($p<0.05$) than those collected in the Provinces of Oulu and Lapland. In addition, the oil content correlated negatively with the seed weight. Studies have shown

Table 1 *The extraction yields for the CHCl₃–MeOH and the one-step H₂SO₄–MeOH methods. Results are weight-% (wt%) ± standard deviation (SD).*

Extraction method	Yield / wt% ± SD
CHCl₃–MeOH	12.5 ± 0.9
H₂SO₄–MeOH	9.9 ± 0.9

Table 2 *Fatty acid composition of redcurrant seed oil produced by CHCl₃–MeOH extraction or by one-step H₂SO₄–MeOH method. The results are weight-% ± standard deviation.*

Fatty acid	CHCl₃–MeOH	H₂SO₄–MeOH
16:0	4.6 ± 0.1	4.8 ± 0.0
18:0	2.0 ± 0.0	3.0 ± 0.2
18:1(n-9)	17.1 ± 0.2	17.8 ± 1.0
18:1(n-7)	0.5 ± 0.1	0.6 ± 0.0
18:2(n-6)	41.3 ± 0.1	40.7 ± 0.4
18:3(n-6)	5.9 ± 0.1	5.6 ± 0.4
18:3(n-3)	25.1 ± 0.3	24.5 ± 0.4
18:4(n-3)	3.2 ± 0.0	3.0 ± 0.0
Others	0.3	0.0

that the brightness of the growth site affects the size of cloudberry seeds and berries; berries grown in shaded habitats are bigger and have heavier seeds[9,10]. In the present study, the light conditions were not taken into account when collecting the berries. However, the berries collected in the northern Finland have grown in long-day conditions which may have affected the average seed sizes.

The most abundant fatty acids in cloudberry seed oils were oleic, linoleic and α-linolenic acids, accounting for 92-93 % of the total acids (Figure 1). Cloudberries from Lapland contained the highest proportion of linoleic and the lowest proportion of α-linolenic acid, whereas oleic acid was not affected by the growth site of the berry.

3.3 Crowberry

The weight of crowberry seeds from three collecting areas was equal. In contrast to cloudberry, the oil content of the most southern berries was the highest. There was no relationship between the seed weight and oil content in crowberry. As in cloudberry, the most abundant fatty acids in the seed oil of cloudberry were oleic, linoleic and α-linolenic acids, accounting for 96 % of the total acids (Figure 1). Crowberries from the southern Finland contained more α-linolenic and less linoleic acid than berries from other collecting areas. The proportion of oleic acid was highest in crowberries from the middle Finland, the Province of Oulu.

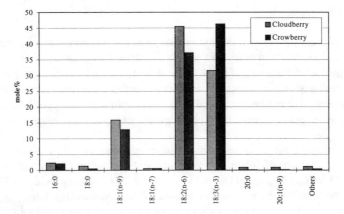

Figure 1 *Mean fatty acid compositions of the seed oils of cloudberry and crowberry.*

3.4 Alpine currant

The oil content (i.e. yield of the one-step extraction/methylation method) of alpine currant seeds varied between 3.9 and 7.7 weight-%. There was no correlation between the growth area and oil content.

Seeds of wild alpine currants contained on the average 8.6 % γ-linolenic acid (range 6.9-12.8%) and 3.0 % stearidonic acid (range 2.1-4.0 %) (Figure 2). Geographical variations were small in alpine currant. Some random differences were seen in the proportions of palmitic, oleic and linoleic acids, whereas the proportions of α-linolenic, γ-linolenic and stearidonic acids remained constant throughout the South-West Finland and the Archipelago of Turku. There was a positive correlation between the proportions of γ-linolenic and stearidonic acids in alpine currant ($p < 0.05$).

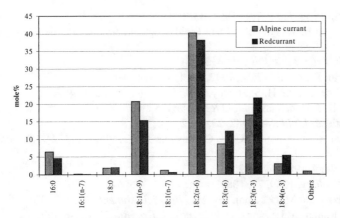

Figure 2 *Mean fatty acid compositions of alpine currant and northern redcurrant.*

3.5 Northern redcurrant

As in alpine currant, the seed oil content in wild northern redcurrant was unaffected by the collecting location, the mean being 15.9 weight-% (range 10.1-30.4 %).

The proportions of γ-linolenic (12.3 %) and stearidonic (5.4 %) acids were clearly higher in redcurrant growing in the north than in more southern alpine currant. The relative amount of γ-linolenic acid has been reported to increase from south to north[11]. In addition, a lower γ-linolenic acid content is characteristic of seeds developed at higher temperatures[12].

The growth areas of northern redcurrant can be divided into two distinct groups according to the proportions of polyunsaturated fatty acids. In the more northern group, the content of γ-linolenic acid was high (13.0 %) whereas that of α-linolenic acid low (20.9 %) compared with the more southern group (11.5 and 22.8 %, respectively). Interestingly, the proportion of stearidonic acid parallels closely the proportion of α-linolenic acid. Thus, in contrast to alpine currant, there is tendency of a negative relationship between the contents of γ-linolenic and stearidonic acids in the seed oil of northern redcurrant.

4 CONCLUSIONS

Slight latitudinal differences existed in the seed oil content and fatty acid composition of crowberry and cloudberry; for example, the proportion of linoleic acid increased and that of α-linolenic decreased towards the northern collecting locations. Rapid, one-step extraction/methylation method can be reliably applied to the analysis of natural variation of seed oil content and fatty acid composition. Northern redcurrant showed a great deal of variation in the proportions of polyunsaturated fatty acids. Whether the variance is due to differences in genotype or to the different growth conditions, should be tested by cultivating the same varieties in more controlled conditions.

References

1. J. L. Harwood, *Prog. Lipid Res.*, 1994, **33**, 193.
2. C. Hitchcock and B. W. Nichols, 'Plant Lipid Biochemistry', Academic Press, London, 1971, p. 152.
3. D. J. Murphy, *Prog. Lipid Res.*, 1994, **33**, 71.
4. D. F. Horrobin, *Prog. Lipid Res.*, 1992, **31**, 163.
5. J. Folch, M. Lees and G. H. S. Stanley, *J. Biol. Chem.*,1957, **226**, 497.
6. P. Ways and D. J. Hanahan, *J. Lipid Res.*, 1964, **5**, 318.
7. W. W. Christie, *J. Lipid Res.*,1982, **23**, 1072.
8. M. L. Dahmer, P. D. Fleming, G. B. Collins and D. F. Hildebrand, *JAOCS*, 1989, **66**, 543.
9. J. Ågren, *Ecology*, 1988, **69**, 962.
10. J. Ågren, *J. Ecology*, 1989, **77**, 1080.
11. G. Lotti, R. Izzo and M. G. Landi, *Riv. Soc. Ital. Sci. Aliment.*, 1978, **7**, 361.
12. Z. Yaniv, C. Ranen, A. Levy and D. Palevitch, *J. Exp. Bot.*, 1989, **40**, 609.

THE IMPACT OF BREEDING AND GENETICS ON BERRY FRUIT QUALITY

Rex M. Brennan, Sandra Gordon and Brian Williamson

Scottish Crop Research Institute
Invergowrie
Dundee DD2 5DA, Scotland

1 INTRODUCTION

The improvement of fruit quality is a major objective in most fruit breeding programmes, to gain advances from wild types that formed the first cultivars. The introgression of diverse germplasm in breeding programmes has offered increases in fruit quality, both physical and chemical aspects, and the opportunity to provide specific improvements for most markets. However, the multiple requirements of any breeding programme result in a balance between fruit quality and agronomic traits such as pest resistance and abiotic stress tolerance. Increasingly, the use of biotechnological methods represents an invaluable alternative means of fruit quality manipulation without alteration of the agronomic background.

2 FRUIT BREEDING

Most of the soft fruit crop genera were domesticated comparatively recently,[1,2] and in many cases, not too far removed from the original wild type – indeed, in the case of blackcurrant, an accession from the Swedish wild, Öjebyn, is still widely grown in Scandinavia and Central Europe.[3] Initial breeding efforts concentrated on the development of consistency of cropping and larger berry size, and it is only recently that breeding for improved quality has been expanded, so that in most European countries, newer, high quality cultivars such as the blackcurrant 'Ben Alder' are taking over the acreage. In the immediate past, the development of privately funded germplasm improvement programmes has enabled grower co-operatives and processors to tailor the breeding of new high quality genotypes to their own specific requirements.

Breeding for fruit quality has different emphases depending on the eventual market for the fruit – fresh and processing requirements are often quite different. From this, the improvements can be divided into the following categories:

i.   Physical characters – fruit size, evenness of ripening, skin strength, suitability for mechanical harvesting, juice yield and shelf-life.

ii.   Chemical characters – colour, vitamin content, sugar levels, nutritional components.

The major pigments found in soft fruits are anthocyanins, mainly delphinidins and cyanidins.[4] The breeding effort, certainly in *Ribes*, has been towards an increase in the proportion of delphinidins, which gives the expressed juice a more stable colouration.

Ascorbic acid levels are a major breeding objective in *Ribes* breeding programmes,[5] and there are large genotypic differences in content, ranging from *ca.* 50 mg/100 g in Scandinavian cultivars such as 'Hedda' to over 300 mg/100 g in some of the breeding lines in the SCRI programme (Brennan, unpublished). Other soft fruits have considerably lower levels of ascorbate, and screening for this character is consequently of greatly reduced importance. However, recent developments in the field of functional foods and nutraceuticals has made the whole area of anti-oxidative metabolites one that may require breeding attention in the future. While ascorbic acid and alpha-tocopherol are the best known antioxidants, many phenolic compounds, including anthocyanins, can function in a similar way. Ongoing studies at SCRI to examine total anti-oxidative capacity of juice from a range of exotic *Rubus* species sourced from China via the National Clonal Germplasm Repository, Oregon, USA, have found species with many times the total antioxidant activity of most commonly grown fruits (Deighton *et al.*, unpublished). There could, therefore, be considerable potential for bringing these species into breeding programmes in the future.

iii.   Sensory characters – flavour, aroma, aftertaste, mouthfeel, etc.

Research at SCRI and the Hannah Research Institute, Ayr, Scotland, using a trained sensory panel have shown that considerable genotypic variation exists in these characters, both in *Ribes*, and to a lesser extent in *Rubus*,[5,6] and these differences remain clear through processing into cordials and after freezing.[7] By relating these differences back to a standard product, the most appropriate genotypes for processing or for parental types can be selected.

The genetic control and inheritance of many of these characters are unclear at present, although most are clearly quantitative traits. The heritability of quality traits in fruit represents one of the main challenges to fruit breeders, who, in a long-term strategy, must focus on the genotypes and parents most likely to produce the desired results. Considerable diversity exists within the genetic resources available to the plant breeders, and more emphasis than ever is placed on the development of cultivars likely to be successful in increasingly competitive markets. However, the use of different species in *Rubus* breeding, for example,[8] in order to gain resistance to pest and diseases, can affect quality dramatically.

3 BIOTECHNOLOGICAL METHODS FOR IMPROVING FRUIT QUALITY

In addition to conventional breeding processes, many fruit crops can now be genetically modified using transgenic protocols. Further improvements in the future through the introgression of genes for upregulation of biosynthetic pathways or for reduction of fungal infections are therefore likely.

i. Ripening-related genes

By differential screening, cDNA clones of five genes from *Ribes nigrum* have been isolated which show greatly enhanced transcript levels in fully ripe fruit compared with green fruit.[9] This implies that they may have important roles in fruit ripening, and manipulation of the expression levels of these genes in transgenic plants may help to elucidate the molecular mechanisms of some of the ripening processes. Two of these clones (pRIB1 and pRIB7) were found to be fruit-specific.

ii. PGIP genes for *Botrytis* control

Ripe fruit of raspberry and strawberry are highly susceptible to spoilage due to *Botrytis cinerea*. The fungus infects the flowers as soon as they open and establishes a quiescent infection that does not cause grey mould rot until the fruits ripen. *In vitro* studies have shown that *B. cinerea* produces endo-polygalacturonases (PGs) which degrade polygalacturonic acid, an integral part of the pectin molecule in the primary walls and intercellular matrix of the fruit tissues. A polygalacturonase-inhibiting protein (PGIP) purified from immature green raspberry fruits inhibits *Botrytis* endo-PGs, an activity which declines sharply with the onset of ripening at a stage corresponding to an increase in susceptibility to infection.[10] A PGIP cDNA from *Rubus* is being used for transformation of fruit species under the control of a constitutive promoter with the objective of enhancing resistance to one of the most serious fungal threats to fresh fruit quality.[11]

iii. Other genes

Work to incorporate invertase genes into strawberry is already in progress;[12] the aim is to use these genes, encoding cell wall and vacuolar forms from potato, to modify the sugar balance in terms of the hydrolysis of sucrose to hexoses.

Transgenic plants containing the anthocyanin pathway-specific transcriptional activators R and C1 from maize are currently undergoing analysis for improved or augmented pigment production.[13]

4 GENOTYPE X ENVIRONMENT FLUCTUATIONS IN FRUIT QUALITY

Many quality-related components show considerable variation around the optimum harvest date and in different environments. The breeding of improved quality cultivars has led to a need for better characterisation of these fluctuations so that the breeding advances in quality can be maximised to the processor and consumer. Although much research effort is currently going into the development of higher quality genotypes through classical and biotechnological means, the environmental factors that may reduce quality in these genotypes remain poorly understood.

Acknowledgements

The support of the Scottish Office Agriculture, Environment and Fisheries Department, SmithKline Beecham Research and Development Fund and Scottish Soft Fruit Growers Ltd is gratefully acknowledged.

References

1.   D.L. Jennings, 'Raspberries and blackberries: their breeding, diseases and growth', Academic Press, London, 1988.
2.   J. Smartt and N.W. Simmonds, 'Evolution of Crop Plants', Longman Scientific, Harlow, Essex, 1995.
3.   R.M. Brennan, 'Currants and gooseberries'. In: 'Fruit Breeding Vol. II: Small Fruits and Vine Crops', Wiley and Sons, New York, 1996.
4.   J.B. Harborne and E. Hall, *Phytochem.,* 1964, **3**, 453.
5.   R.M. Brennan, E.A. Hunter and D.D. Muir, *Food Res. International*, 1997, **30**, 381.
6.   R.E. Harrison, D.D. Muir and E.A. Hunter, *Food Res. International*, 1998a, (in press).
7.   R.E. Harrison, R.M. Brennan, E.A. Hunter, S. Morel, and D.D. Muir, *Acta Hort.*, 1998b, (in press).
8.   V.H. Knight, *Acta Hort.*, 1993, **352**, 363.
9.   M. Woodhead, M.A. Taylor, R. Brennan, R.J. McNicol and H.V. Davies, *J. Plant Physiol.*, 1998, (in press).
10.   D.J. Johnston, V. Ramanathan and B. Williamson, *J. Exp. Bot.*, 1993, **44**, 971.
11.   V. Ramanathan, C.G. Simpson, G. Thow, P.P.M. Iannetta, R.J. McNicol and B. Williamson, *J. Exp. Bot.*, 1997, **48**, 1185.
12.   C. Bachelier, J. Graham, G. Machray, J. Du Manoir, J.-F. Roucou, R.J. McNicol and H. Davies, *Acta Hort.*, 1997, **439**, 161.
13.   A.M. Lloyd, V. Walbot and R.W. Davis, *Science*, 1992, **258**, 1773.

FOOD CHEMICAL PROPERTIES OF A NEW POTATO WITH ORANGE FLESH

G.Ishii [1], M.Mori [2], A.Ohara [2] and Y.Umemura [2]

[1] Crop Quality Physiology Laboratory
[2] Potato Breeding Laboratory
 Hokkaido National Agricultural Experiment Station
[1] Hitsujigaoka,062-8555 Sapporo, Japan
[2] Shinsei,082-0071 Memuro,Japan

1 INTRODUCTION

We have targeted to raise carotenoid and anthocyanin contents or to improve soluble protein in potato breeding. A new potato with high carotenoid was obtained by crossing some varieties(Figure.1). This diploid potato "SH-575" with orange flesh has rich carotenoid and soluble protein content. The purpose of this study was to compare the food chemical properties of a new variety "SH-575" with those of "Irish cobbler", which is a leading variety in Japan.

2 MATERIALS AND METHODS

2.1 Plant material and enzyme assay

Seed tubers of "Irish cobbler" and "SH-575" were planted in open fields in early May in 1997. Potatoes were harvested in late Aug.or early Sept., and stored in a dark room until use. Ten or five tubers were subjected to measurement. Extraction of protein and determination of polyphenol oxidase(PPO) activities were conducted as described by S.K.Sim *et al* [1]. After homogenization and centrifugation, the browning degree of samples was expressed as absorbance at 600 nm using the supernatant thus obtained under 0 °C for 5 days.

2.2 Chemical and physical analysis

Sliced surfaces of potato tubers were measured by a color analyser(Minolta,CR-300). Sugar, organic acid,ascorbate,carotenoid and free amino acids were determined by a HPLC(Shimadzu LC-10A) according to the standard procedure for extraction and preparation of components. The chemiluminescence(CL) of active oxygen species from *tert*-butyl hydroperoxide(*t*-BuOOH) was measured with or without carotenoid by a filter-equipped photon counting-type spectrometer(CLD-110I,Tohoku Electronic Ind.) in the presence of hemin and luminol. The photons counted at the respective wavelengths from 350 to 600 nm were computed totally as spectral intensities(counts/sec). The CL of water soluble proteins were also measured as above spectrometer connected to the HPLC and injector. A solution of 50% MeOH diluted with 50mM phosphate buffer containing 1.5% hydroperoxide and 5% MeCHO (pH7.0)was used as a mobile phase; the flow rate of 1 mL/min and temp. of

Table 1 *Comparison of "SH-575" and "Irish cobbler"*

		SH-575	Irish cobbler
Yield(kg/10a)		3,027	4,873
Starch(%)		16.6	13.8
Tuber weight(g)		58	84
Flesh color (L*,a*,b*)		71.9,-1.1,54.9	78.0,-3.1,15.9
Discoloration 3hrs after slicing (ΔE)		2.60	8.24
Carotenoid as zeaxanthin (μg/100gf.w.)		530-741	not detected

```
┌USW-1(diploid,Katahdin)
┌W822229-5
│   └S.phureja(114)
"SH-575"
│   ┌SD219-5(diploid,S.andigena)
└P10173-5
    └SD375-5(diploid,S.andigena)
```

Figure 1 *Genetic background*

Table 2 *Comparison of quality component of "SH-575" and "Irish cobbler"*

	Tyrosine	Chlorogenic acid	Citrate	Malate	Ascorbate	Suc.	Glc.	Fru.
			(mg/100gf.w.)					
SH-575	1.74	49.3	249	15	15	189	36	29
Irish cobbler	3.71	39.4	203	29	16	231	236	68

Table 3 *Enzymatic browning and PPO activity in the central pith zone of tubers*

	Degree of browning (Abs600nm)	Polyphenol oxidase (U/mg protein)	
		(tyrosine)	(catechol)
SH-575	0.650	0.187	47
Irish cobbler	1.170	0.396	217

1U=0.01 ΔA475nm(tyrosine),410nm(catechol)/min at 25 °C

Table 4 *Chemiluminescence(CL) intensity of soluble protein of potato tubers in the presence of acetaldehyde and hydroxyl radical*

CL intensity of soluble protein(relative photon counts/sec)

SH-575	(C)	214	Irish cobbler	(C)	192
SH-575	(B)	124	Irish cobbler	(B)	108
SH-575	(M)	123	Irish cobbler	(M)	97
SH-575	(P)	108	Irish cobbler	(P)	121
Bovine serum albumin(10μ g,10μL)					41
Gallic acid (5mM,10 μL)as a standard					1000

Each CL shows the relative value to the CL of 10 μL of 5 mM gallic acid(1000).
C:cortex zone, B:vascular bundle ring zone,M:middle zone between B and P,P:central pith zone.

23°C were according to the method of Yoshiki *et al.*[2] Hydroxyl radical(*OH) was generated through the Fenton reaction by adding 25mM FeCl$_2$ to the sample injector. Ten μL of sample was also directly injected into the sample injector with FeCl$_2$. The scavenging activities of protein solution were measured by comparing with the CL intensity of gallic acid or bovine serum albumin(BSA,standard protein,Bio-Rad), respectively.

3 RESULTS AND DISCUSSION

This orange potato has some advantageous traits in both tuber quality and active oxygen scavenging activity. The main carotenoid of "SH-575" tuber was identified as zeaxanthin occupying around 40% of total HPLC peak area(Table 1). The HPLC profile showed a similar pattern as reported by Brown *et al.*[3] "SH-575"potatoes contain more soluble protein(data not shown) and citrate content, but less glucose content than those of "Irish cobbler" ."SH-575" has also a less browning flesh between vascular bundle ring and central pith zone than that of "Irish cobbler" after cutting, probably due to lower level of tyrosine and PPO activity(Table 2&3). The pigment extract exhibited a power of scavenging active oxygen species from *t*-BuOOH(data not shown). A radical scavenger was shown to exhibit a very weak light emission(CL) in the presence of acetaldehyde and active oxygen.[2] We, therefore, examined the scavenger activity of potato tuber using CL assay. The water soluble protein showed a stronger activity of scavenging *OH than that of BSA on a basis of equal amount of protein(Table 4). The antioxidant activity was suggested to be mainly derived from patatin protein with high lysine content[4,5],although BSA was shown to be degraded with *OH.[6]

This work was supported by a budget from the Bio Renaissance Program(BRP 98-IV-A-1) of the Ministry of Agriculture, Forestry, and Fisheries(MAFF) in Japan.

References

1. S.K.Sim,S.M.Ohmann and C.B.S.Tong,*American Potato J.*, 1997,**74**,1.
2. Y.Yoshiki,K.Okubo,M.Onuma and K.Igarashi,*Phytochemistry*,1995,**39**,1411.
3. C.R.Brown,C.G.Edwards,C.-P.Yang and B.B.Dean,*J.Amer.Soc.Hort.Sci.*,1993,**118**,145.
4. J.R.Bohac,*J.Agric.Food Chem.*,1991,**39**,1411.
5. M.S.Al-saikhan,L.R.Howard and J.C.Miller,Jr,*J.Food Sci.*,1995,**60**,341.
6. K.J.A.Davies,*J.Biol.Chem.*,1987,**262**,9895.

VITAMIN C: A VARIABLE QUALITY FACTOR IN SEA BUCKTHORN BREEDING

S. T. Karhu and S. K. Ulvinen[1]

Agricultural Research Centre of Finland,
Horticulture, Toivonlinnantie 518,
FIN-21500 Piikkiö, Finland

[1]*present address:*
Boreal Plant Breeding
Myllytie 8,
FIN-31600 Jokioinen, Finland

1 INTRODUCTION

Sea buckthorn *(Hippophaë rhamnoides)* is a deciduous shrub with small, edible berries. Because of the medicinal and nutritional values reviewed recently[1] its fruits are considered most valuable, and its cultivation is expanding. The best known trait of sea buckthorn is probably the high vitamin C (ascorbic acid) content of the berries which varies, however, greatly between populations from different areas and between different subspecies[2].

Sea buckthorn is a dioecious species. Therefore, in the breeding work we need models to predict the vitamin C content of progenies of a given male progenitor when the vitamin C level of the female progenitor is known. We also need to know the variation in vitamin C content of berries within full-sib progenies. In this case study, we tested whether it is possible to estimate the effect of a male progenitor on the vitamin C content of seedling descendants in a way that could be used in the breeding work.

2 DETERMINATION OF VITAMIN C CONTENT

The vitamin C content of berries was determined in plants within a breeding programme, including Russian cultivars as well as wild strains and their hybrids as progenitors. In order to test whether it is possible to design models to predict the vitamin C content of progenies, ten hybrid seedlings within four progenies of different female progenitors but the same male progenitor were included in the analysis. The female progenitors were *cvs.* Maslichnaya, Dar Katuni, Vitaminnaya and a hybrid selection (German wild strain × Finnish strain) of *H. rhamnoides* ssp. *rhamnoides*. The male progenitor was a hybrid of a wild strain of *H. rhamnoides* ssp. *caucasica* from Elburz Mountains (Iran) and a Finnish wild strain.

Newly-ripened berries were collected in 1992 from hybrid plants growing on a field in South-Western Finland. Fruit maturity was observed on the basis of changes in colour, the collection time varying from August 24 to September 25. Three samples of 50 berries were collected from each individual and kept at -40°C for about 2 months before analysis.

The berry samples were thawed and weighed, and the seeds removed and weighed. The total vitamin C content of berry flesh was determined by high-performance liquid chromatography (HPLC), using a method essentially same as described elsewhere[3] but modified by using a 5μm × 125 mm Spherisorb column (Hewlett Packard) with 0.08 M KH$_2$PO$_4$, pH 7.8, and methanol (20:80 v/v) as the mobile phase.

3 MODEL FOR PREDICTING VITAMIN C CONTENT

Within the full-sib seedling progenies there were differences in vitamin C content greater than 100 mg/100 g of berries. On the other hand, the vitamin C level of the siblings correlated with that of their female progenitors (Figure 1). In the given case, this relationship was curvilinear, and the regression equation $\log Y = 1.502 + 0.003X$, where Y and X stand for vitamin C content of the sibling and that of the female progenitor, respectively, was significant at $P \leq 0.0001$ ($R^2 = 0.53$).

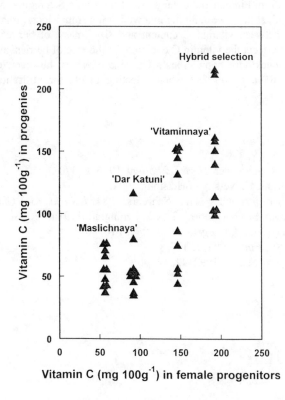

Figure 1 *Relationship between total vitamin C content (mg/100 g of berries) of four female progenitors and that of their seedling descendants. In each of the progenies, produced by crossing the female cultivars with a hybrid of wild strains from Iran and Finland, ten siblings were included in the analysis.*

Thus it was possible to design a model to predict the inheritance of vitamin C content of berries in hybrid progenies. The rather low coefficient of determination (R^2) was due to the notable variation in the vitamin C content within progenies, partly due to the small sample size and low number of regressors, too. The measured contents of Russian varieties were highly comparable with those reported in the descriptions of the varieties[4]. This showed that the ripening time was correctly estimated and the variation of the vitamin C content was very unlikely to be related to the physiological maturity of berries[5,6].

The only full-sib female plant of the male progenitor used in the test crossings had a vitamin C content of 93 mg/100 g of berries. Testing of the equation with some descendants of another male progenitor originating from the Alpine area, whose female full-sib plant had a much higher vitamin C content, 600 mg/100 g of berries, led to underestimated values of vitamin C content of the descendants. Thus, the prediction models should be related to the level of vitamin C content of the particular population of the male progenitor. The vitamin C content per 100 g of berries was about 200 mg in the natural populations from Finland and other areas of the Baltic Sea region, 52 to 147 mg in Russian cultivars and, at the lowest, 34 mg in a wild strain from Elburz, Iran.

The correlation between vitamin C content and size (mass) of berries was low. The genetic background affects the vitamin C content, not the size of berries, as also indicated in the Finnish sea buckthorn populations[7]. The size of seeds, however, correlated with berry size (r = 0.66), a useful fact when selecting plants for utilisation of chemical components of the seeds.

References

1. S. C. Li and W. R. Schroeder, *HortTechnology,* 1996, **6,** 370.
2. Y. Yao, PhD Thesis, University of Helsinki, 1994.
3. A. J. Speek, J. Schrijver and W. H. P. Schreurs, 1984, *J. Agric. Food Chem.* **32,** 352.
4. M. N. Plekhanova, 'Sea buckthorn', 2nd ed., Leningrad, 1991 (in Russian).
5. A. Rousi and H. Aulin, *Ann. Agric. Fenn.,* 1977, **16,** 80.
6. Y. Yao, *Agric. Sci. Finl.,* 1993, **2,** 497.
7. Y. Yao, *Acta Agric. Scand.,* 1992, **42,** 12.

FINNISH VEGETABLES - RAW MATERIAL FOR FUNCTIONAL FOODS

Riitta Puupponen-Pimiä[1], Tuomo Kiutamo[1], Tapani Suortti[1], Pasi Tapanainen[2] and Kirsi-Marja Oksman-Caldentey[1]

[1]VTT Biotechnology and Food Research, P.O. Box 1500, 02044 VTT, Finland
[2]MTT Food Research, 31600 Jokioinen, Finland

I INTRODUCTION

Vegetables, which form an important part of our daily diet, are rich in various health-promoting phytochemicals. Plant food provide good sources of vitamins, minerals and dietary fiber. Furthermore, there are additional groups of compounds in plants, the nutritional functions of which have only recently been realized when their potentially protective effects against various diseases have been demonstrated. These include e.g. many secondary plant products such as large families of phenolic compounds.

In 1997, VTT initiated a four-year research project entitled 'Identification and preservation of health-promoting compounds in food raw-material of plant origin'. The project is financed by TEKES, VTT, MTT and Finnish food industry. The project is a part of the TEKES Uudistuva elintarvike research program as well as VTT's Future Foods Program. In this project VTT is cooperating with MTT Food Research. Our research partners in the Finnish food industry are Lännen Tehtaat Oy, Raision Yhtymä Oyj, Kotimaiset Kasvikset Ry, Saarioisten Säilyke Oy and Atria Oyj.

1.1 Aims of the study

The general aim of the project is to increase knowledge of the health-promoting compounds in Finnish vegetables used by the food industry and to study their preservation during food processing. Special interest is focused on dietary fiber constituents and total antioxidativity of vegetables. Research is directed to the following topics.

• Preservation of health-promoting compounds during food processing and storage
• Distribution of health-promoting compounds in different plant organs
• Utilization of novel plants as raw material for the food industry
• Utilization of novel methods in food processing in order to preserve
  health-promoting compounds
• Novel applications for side streams of the food industry

1.2 Materials and methods

The following vegetables were used in the study: pea, potato, cauliflower, cabbage, broccoli, carrot, spinach, onion, beetroot and albino beetroot. The main process which was studied was blanching-freezing. Other processes studied were preservation in vinegar and grating of fresh vegetables.

Total, insoluble and soluble dietary fiber contents of raw and processed material were determined using the enzymatic gravimetric method (1). Total pentosans and pectin were analysed spectrophotometrically using modified Douglas (2) and Listin *et al.* (3) methods, respectively.

Fiber constituents were determined by size-exclusion chromatography with laser-light scattering and refractive index detection.

Fructo-oligosaccharides were determined using a commercial kit (Fructan Assay Procedure, Megazyme International Ireland Ltd.)

Electrochemically active phenolic compounds were screened and the most common bioflavonoids were determined with reverse chromatography (RS-HPLC) and electrochemical (CEAS) and diode-array (DAD) detection.

Vitamin C and carotene contents were determined according to Hägg *et al.* (4).

1.3 Results

• Preliminary results showed that only slight changes occurred in dietary fiber contents. Chromatographic analysis on the other hand showed that processing did have some effects on the molecular weight distribution of the fiber constituents. This could be explained by mechanical tissue breakage and enzymatic release of fiber constituent.

• Vegetables analysed in the present study contained only minor amounts of fructo-oligosaccharides, with the exception of onion which contained a large amount of inulin.

• Our screening system showed that processing had significant effects on the contents and profile of phenolic compounds acting as antioxidants (Figure 1). This was also supported by the result that the contents of common bioflavonoids found in vegetables decreased during processing.

• Processing also significantly reduced the vitamin C contents of vegetables, although it did not appear to affect carotene contents.

1.4 Conclusions

Food processing had no significant effects on the dietary fiber contents of vegetables. However, the contents of phenolic compounds acting as antioxidants, as well as of vitamin C, decreased in all the processes studied. These results will be utilized in process development of food raw material of plant origin.

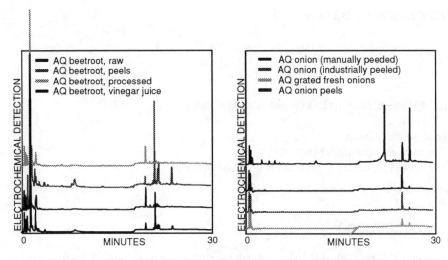

Figure 1. Electrochemically active phenolic compounds in vegetables.

References

1. Asp, N.-G. Johansson, C.-G., Hallmer, H. and Siljeström, M. *J. Agric . Food Chem.*, 1983, **31**, 476.
2. Douglas, S. G. *Food Chem.*, 1981, **7**, 139.
3. List, D., Buddruss, S. and Bodtke, M. *Z. Lebensm. Unter. Forsch.*, 1985, **180**, 48.
4. Hägg, M., Ylikoski, S. and Kumpulainen, J. 1988-1993 and 1992-1993. *J. Food Comp. Anal.*, 1994, **7**, 252.

PROCESSING OF BRASSICA VEGETABLES

Ruud Verkerk, Matthijs Dekker and Wim M.F. Jongen

Food Science Group
Wageningen Agricultural University
P.O. box 8129, 6700 EV Wageningen, The Netherlands

1 INTRODUCTION

Vegetables form a valuable source of many potentially anticarcinogenic constituents, such as fibres, vitamins, minerals and a variety of non-nutritive components. The protective effect of brassica vegetables against cancer has been suggested to be partly due to their relatively high content of glucosinolates which distinguishes them from other vegetables. Brassica vegetables as Brussels sprouts, cabbage, broccoli and cauliflower are frequently consumed by humans from Western and Eastern cultures.

Glucosinolates appear to have little biological impact themselves, but are converted to biologically active products such as isothiocyanates, indoles, and oxazolidinethiones upon enzymatic degradation by the endogenous enzyme myrosinase.[2] The anticarcinogenic mechanisms by which these compounds may act include the induction of detoxification enzymes and the inhibition of activation of promutagens/procarcinogens.[1-4]

During processing and storage of crops post-harvest physiological processes can lead to changes in flavour, palatability, nutritional value and healthiness of the plant products.

In this paper we emphasise the importance of the consequences of processing, preparation and cooking of Brassica vegetables for glucosinolates as quality factors. As an example we describe the changes in levels of glucosinolates, after chopping of white cabbage.

2 QUALITY FACTORS OF GLUCOSINOLATES

2.1 Taste

The typical flavour and odour of Brassicas is largely due to glucosinolate-derived volatiles (isothiocyanates, thiocyanates, nitriles). It has been shown that the glucosinolates sinigrin and progoitrin are involved in the bitterness observed in Brussels sprouts[5]. Van Doorn[6] et al. confirmed the role of sinigrin and progoitrin in taste preference by using taste trials with samples of Brussels sprouts. It appeared that consumers preferred Brussels sprouts with a low sinigrin and progoitrin content. Pungency and bitterness caused by glucosinolate breakdown products are therefore important quality factors for Brassica vegetables.

2.2 Healthiness

The protective effects against cancer can be considered as another notable quality factor of glucosinolates and their derivatives. Brassicaceous vegetables or glucosinolate derivatives have been shown to modify endogenous detoxification processes and, therefore, may interfere in a positive way with the metabolism of chemical carcinogens.[4,7] Isothiocyanates and indoles, two groups of breakdown products of glucosinolates, arise in plants as a result of enzymatic cleavage of glucosinolates by the endogenous enzyme myrosinase. These compounds are attracting increasing attention as chemical and dietary protectors against cancer. Their anticarcinogenic activities have been demonstrated in rodents (mice and rats) with a wide variety of chemical carcinogens.

Indolyl glucosinolates such as glucobrassicin undergo enzymatic hydrolysis to form indol-3-carbinol, 3-indoleacetonitrile and 3,3'-diindolylmethane. Isothiocyanates are usually produced at neutral pH while nitrile production occurs at lower pH. Enhancement of these effects by increasing the levels of specific glucosinolates is of importance for obtaining protective effects at normal consumption levels.

3 EFFECTS OF PROCESSING

All kinds of processes like chopping (cole slaw), fermentation (sauerkraut) and cooking (blanching or pressure cooking) of Brassica vegetables will damage plant cells and brings myrosinase in contact with glucosinolates. This triggers the hydrolysis which influence levels and composition of the glucosinolates. Prior to many processing steps chopping of the vegetable is necessary. Chopping of fresh plant tissues creates optimal conditions for myrosinase and a high degree of glucosinolate hydrolysis can be expected. In contrast to these expectations and reported findings we[8] observed, after chopping of white cabbage and prolonged exposure to air, elevated levels of all indole glucosinolates (Fig.1).

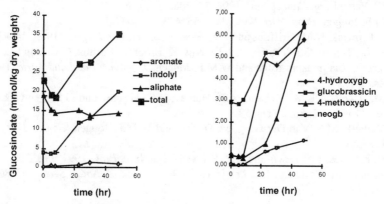

Fig. 1. Levels of glucosinolates in white cabbage after chopping and
prolonged exposure at room temperature.

In white cabbage the largest increase was found for 4-methoxyglucobrassicin which increased 15-fold. Increasing the amount of indole glucosinolates can have large influences on quality factors such as flavour and anticarcinogenicity of Brassica vegetables.

Stress-induced increases in levels of glucosinolates have been demonstrated in different studies like mechanical wounding and infestation[9] for intact plants or UV-irradiation[10] for post-harvest vegetables.

3 DISCUSSION

It is essential to explore and quantify the behaviour of the glucosinolates during processing, preparation and cooking of Brassica vegetables. The influence of processing conditions can have considerable consequences for quality factors such as flavour and healthiness. In this respect the enhancement of indole glucosinolates by pre-chopping cabbage can be of interest for their anti-cancer effects.

Most evidence for anticarcinogenic effects of glucosinolate breakdown products is obtained from studies in animals with high doses of active compounds or extracts of Brassica vegetables. More studies are required in which normal consumption levels are used and the effects of processing on glucosinolates and breakdown products are included. It is clear that factors inducing and directing indole glucosinolate metabolism in plants need to be studied in more detail as evidence increases on the biological activities of these compounds in man. The possibilities of increasing the amount of specific glucosinolates by processing as found in this study opens new ways of improving the healthiness of Brassica vegetables.

References

1.	L.O. Dragsted, M. Strube and J.C. Larsen, *Pharm. and Toxicol.,* 1993, **72** (1), 116.
2.	R. Verkerk, M. Dekker and W.M.F. Jongen, 'Natural Toxicants in Foods', ed. D.H. Watson, Sheffield Academic Press, Sheffield, England, 1998, chapter 3, 29.
3.	L.W. Wattenberg, *Cancer Res.,* 1992, **52**, 2085s.
4.	W.M.F. Jongen, *Proc. Nutr. Soc.,* 1996, **55**, 433.
5.	G.R. Fenwick, N.M. Griffiths and R.K. Heaney, *J. Sci. Food Agric.,* 1983, **34**, 73.
6.	H.E. Van Doorn, G.J. Van Holst, J. De Nijs, K. Broer, E. Postma, G.C. Van Der Kruk, N.C.M.E. Raaijmakers-Ruijs and W.M.F. Jongen, *J. Science Food and Agric.* 1998, in Press.
7.	R. McDanell, A.E.M. McLean, A.B. Hanley, R.K. Heaney and G.R. Fenwick, *Food Chem. Toxicol.,* 1988, **26 (1)**, 59.
8.	R. Verkerk, M.S. Van Der Gaag, M. Dekker and W.M.F. Jongen, *Cancer Lett.,* 1997, **114**, 193.
9.	V.M. Koritsas, J.A. Lewis and G.R. Fenwick, *Ann.Appl. Biol.,* 1991, **118**, 209.
10.	K. Monde, M. Takasugi, J.A. Lewis and G.R. Fenwick, *Z. Natur. Section C Bios.,* **46** (3-4), 189.

Fruit ripening and quality in relation to crop load of apple trees

Gottfried Lafer
Research Station for Fruit Growing Haidegg
Ragnitzstrasse 193
A-8047 Graz, Austria

Additional index words: *Malus domestica*, apple, quality, flesh firmness, refraction, ripening

Abstract

During a period of 4 years trees of *Malus domestica* cv. Jonagold, Elstar and Golden Delicious were loaded with different number of fruits and yield (kg) per cm² trunk circular area, respectively. At harvest and after a storage and shelf-life period fruit quality was ascertained automatically by the laboratory "Pimprenelle". By calculation of the coefficient of regression and the r² the influence of the crop load on the individual quality properties was determined to find out the optimal crop load for the different varieties and types of trees under Styrian growing conditions.

1. Introduction

Internal quality of apples will be an important factor for market acceptance in future. High quality apples with a good flesh firmness and a high content of soluble solids are mostly preferred to others and within the cultivars, firmer fruits with higher sugar content often taste better than underdeveloped fruits (Lafer, 1987).
In many European countries flesh firmness and the content of soluble solids are commonly respected factors to place the fruits better on the national and international market (Perlim in France, Migros in Switzerland, Saphir produce in UK, Greenery in Netherlands, Kesko Ltd. in Finland, Norway, VIP in Italy).

2. Material and methods

Between 1994 – 1997 trees of *Malus domestica* cv. Jonagold and its two mutants Jonica and Jonagored were loaded with 3 levels of specific yield per tree (0,5 kg/cm² trunk circular arena, 1,0 kg/cm² and 1,5 kg/cm²) by hand-thinning at the begin of July. In 1996 the cultivars of cv. Golden Delicious and Elstar were included in the research programme. Fruits of these trees were harvested at the optimal stage of maturity. The optimal harvest date of all cultivars was assessed by the Streif index (Streif, 1990), which is a combination of quality and maturity parameters (firmness/starch index and refraction). After harvest a sample of fruits (Jonagold and Elstar approx. 40 kg) were stored approximately 3 months at 1,5 °C, r.H. 95 % and normal atmosphere (cold storage). In 1996 fruits of cv. Golden Del. were stored in normal atmosphere at 1,5 °C and 1997 under ULO conditions (1 °C, 4 % CO_2 and 1,0 % O_2) for more than 6 month.

After harvest and after the storage including a shelf-life period of 14 days (7 days at 10°c, 7 days at 20 °C) fruit weight (g), flesh firmness (kg/cm²), soluble solids (°Brix) and titratable acids (malic acid, g/l) were evaluated automatically with the laboratory „Pimprenelle" , to estimate the fruit quality. At harvest also starch conversion was estimated visually: Scores from 1 to 10 (1 = no starch conversion, 10 = starch totally converted)

The estimated fruit parameters were used to calculate fruit quality indices as follows:

- THIAULT-index (Th) = Total sugar in gram/litre + malic acid (g/l) * 10 (Thiault, 1970)
- Total sugar = (°Brix * 10,6) - 20,6
- PERLIM-index (PI) = (kg/cm² * 0,5 + °Brix * 0,67 + malic acid g/l * 0,67)
- Ratio between soluble solids and acidity
- STREIF-index: Firmness/(refraction * starch stage)
- **Coloration Index** (a combination of percentage red cover colour on surface and the intensity of coloration)

Fruit coloration (percentage of red colour on surface and the intensity of coloration) was estimated fully automatically by the Aweta machine Rudy II and expressed as **Coloration Index.** After storage the storage disorders **senescent breakdown** and **low temperature breakdown** were estimated.

3. Results and discussion

The experiments demonstrate that there are very high and significant correlation's between the crop load per tree and the external and internal fruit quality. The higher the specific number of fruits per tree (number of fruits per cm² trunk circular area), the lower the content of soluble solids and malic acid expressed by the THIAULT-index (Fig. 1). Firmness of fruit flesh is also influenced by the crop load (Fig. 2). At harvest Golden Delicious showed no correlation between crop load and firmness, but after the storage and shelf-life period the correlation was highly significant. Firmness of fruits and malic acid are correlated with the content of soluble solids strongly (Fig. 3 + 4). The PERLIM-index, a quality index (a combination of firmness, soluble solids and malic acid) is also strongly related to the number of fruits/cm² trunk circular area. Also the STREIF-index is influenced by the crop load, because of earlier and faster starch degeneration in fruits of over-cropped trees (Fig. 5). In practice the variety Jonagold often has problems with the red coloration on surface. Due to reduction of crop load, it was possible to increase fruit coloration. The reduction of number of fruits per tree by thinning is limited by negative side effects, like enhanced storage disorders, such as bitter bite, low temperature and senescent break down (Fig. 6) and oversized fruits (more than 90 mm). The optimal crop load for apple trees under Styrian growing conditions depends on the trunk diameter (Fig. 7). Trees with smaller trunks are able to provide more fruits per cm² trunk circular area than older trees with larger trunk diameters.

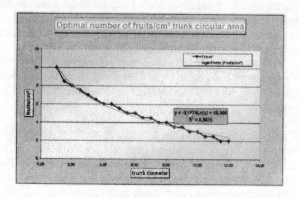

Fig. 7: Optimal crop load per cm² trunk circular area

Fig. 1: Thinning trial - Golden Del. 1997

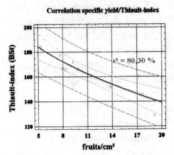

Fig. 2: Correlation specific yield/flesh firmness

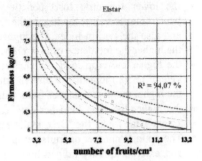

Fig. 3: Thinning trial 1996 - Golden Del.

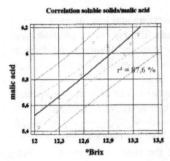

Fig. 4: Thinning trial 1996 - Golden Del.

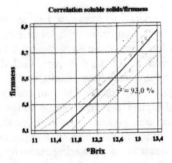

Fig. 5: Correlation specific yield/Streif-index

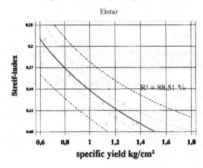

Fig. 6: Storability of Jonagold - different crop load

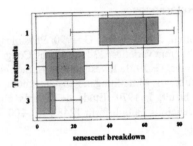

The **results** of this analysis indicated that

- the lower the crop load per tree (less than 3 fruits; 0,5 kg/cm²), the higher the penetrometric value, the higher the °Brix, the higher the malic acid, the higher the Thiault and also the Perlim Index, the higher the coloration Index, the better the grading results, the higher the incidence of low temperature and senescent breakdown.
- the higher the crop load per cm² trunk circular area (specific yield 1,5 kg/cm² and 2,0 kg/cm² resp., more than 9 and 13 fruits/cm² resp.), the lower the fruit flesh firmness and the higher the percentage of fruits with green background colour.
- the higher the crop load per cm² trunk circular area (specific yield 1,5 kg/cm² and 2,0 kg/cm²), the lower the °Brix, malic acid and also the Thiault and Perlim Index
- The optimal number of fruits per cm² trunk circular area to obtain fruits with high internal and external quality ranged between 3 and 9. More fruits per cm² induced inadequate internal quality and reduced also storability (shrinkage of fruits).

4. Conclusion

Fruits from trees with **heavier crop load** had **less coloration** and **less internal quality** than fruits from trees with lower crop load. Any **significant differences in physiological disorders** were observed between 1,0 kg/cm² and 1,5 kg/cm². Storage disorders and diseases increased at 0,5 kg/cm² crop load dramatically (bitter bit, senescent breakdown, low temperature breakdown, spoilage).

It appears, that for the best fruit quality and the best storage behaviour of fruits the **optimal specific yield per tree should range between 1,0 - 1,5 kg/cm²** (6 – 9 fruits/cm²) trunk circular area, depending on the trunk diameter. Trees with lower trunk diameter can provide more fruits/cm² than trees with larger trunk diameters. Crop load per tree also influences the **catabolism of starch during maturity period**. To interpret the starch conversion of fruits correctly, a correction **factor** for different stages of crop load should be considered: **- 1 for over-loaded trees and + 1 for under-cropped trees. Overcrop depress colour and internal fruit quality.** Thiault-, Perlim Index and organoleptical quality are also reduced in fruits from this over-cropped trees.

References

Streif, J. and Bufler, G. 1990. Physiological ripening parameters at optimum picking date of different varieties. 23th Int.Hortic.Congr. Firence, Abstr. No 2395, 636

Lafer, G. 1991. Investigations in the internal quality of four important Austrian table apple cultivars. Mitteillungen Klosterneuburg, Vol 41/1991, 27- 40

Thiault, J. 1970. Etude de criterès objectifs de la qualitè gustative de Pommes Golden Delicious. Bull. Inform. (248), 191-200

Subject Index

RETURN TO ➡

CHEMISTRY LIBRARY
100 Hildebrand Hall • 642-3753

LOAN PERIOD 1	2	3
4	5 1 MONTH	6

ALL BOOKS MAY BE RECALLED AFTER 7 DAYS
Renewable by telephone

DUE AS STAMPED BELOW

NON-CIRCULATING		
UNTIL: 10/22/99		

FORM NO. DD5

UNIVERSITY OF CALIFORNIA, BERKELEY
BERKELEY, CA 94720-6000